D1218621

GENERALIZED INVERSES:
THEORY AND APPLICATIONS

GENERALIZED INVERSES: THEORY AND APPLICATIONS

Adi Ben-Israel
Technion-Israel Institute of Technology and Northwestern University

Thomas N. E. Greville
University of Wisconsin and National Center for Health Statistics

A Wiley-Interscience Publication

JOHN WILEY & SONS
NEW YORK   LONDON   SYDNEY   TORONTO

Copyright © 1974, by John Wiley & Sons, Inc.

Library of Congress Cataloging in Publication Data:
Ben-Israel, Adi.
    Generalized inverses: theory and applications.

    (Pure and applied mathematics)
    "A Wiley-Interscience publication."
    Bibliography: p.
    1. Matrix inversion. I. Greville, Thomas Nall
Eden, 1910-   joint author. II. Title.

QA188.B46   512.9'43   73-23078
ISBN 0-471-06577-3

Printed in the United States of America

10 9 8 7 6 5 4 3 2 1

Dedicated to Abraham Charnes and J. Barkley Rosser

PREFACE

This book is intended to provide a survey of generalized inverses from a unified point of view, illustrating the theory with applications in many areas. It contains more than 450 exercises at different levels of difficulty, many of which are solved in detail. This feature makes it suitable either for reference and self-study or for use as a classroom text. It can be used profitably by graduate students or advanced undergraduates, only an elementary knowledge of linear algebra being assumed.

The book consists of an introduction and eight chapters, seven of which treat generalized inverses of finite matrices, while the eighth introduces generalized inverses of operators between Hilbert spaces. Numerical methods are considered in Chapter 7 and in Section 8.5.

While working in the area of generalized inverses, the authors have had the benefit of conversations and consultations with many colleagues. We would like to thank especially A. Charnes, R. E. Cline, P. J. Erdelsky, I. Erdelyi, J. B. Hawkins, A. S. Householder, A. Lent, C. C. MacDuffee, M. Z. Nashed, P. L. Odell, D. W. Showalter, and S. Zlobec. However, any errors that may have occurred are the sole responsibility of the authors.

The authors were supported during the writing of this book by the National Science Foundation and the Mathematics Research Center, University of Wisconsin. We are especially grateful to B. R. Agins of NSF and J. B. Rosser of MRC.

We wish to thank Julia Berlet and Wilma Wall for their help in typing and preparing the manuscript.

Finally, we are deeply indebted to Beatrice Shube, Editor for Wiley-Interscience Division, for her constant encouragement and patience during the long period of bringing the manuscript to completion.

September 1973
                                                                    A. Ben-Israel
                                                                    T. N. E. Greville

CONTENTS

ix

CONTENTS

GENERALIZED INVERSES:
THEORY AND APPLICATIONS

INTRODUCTION

1. THE INVERSE OF A NONSINGULAR MATRIX

It is well known that every nonsingular matrix A has a unique inverse, usually denoted by A^{-1}, such that

$$AA^{-1} = A^{-1}A = I, \tag{1}$$

where I is the unit matrix. Of the numerous properties of the inverse matrix, we mention a few. Thus,

$$(A^{-1})^{-1} = A,$$

$$(A^T)^{-1} = (A^{-1})^T,$$

$$(A^*)^{-1} = (A^{-1})^*,$$

$$(AB)^{-1} = B^{-1}A^{-1},$$

where A^T and A^*, respectively, denote the transpose and conjugate transpose of A. It will be recalled that a real or complex number λ is called an eigenvalue of a square matrix A, and a nonzero vector x is called an eigenvector of A corresponding to λ, if

$$Ax = \lambda x.$$

Another property of the inverse A^{-1} is that its eigenvalues are the reciprocals of those of A.

2. GENERALIZED INVERSES OF MATRICES

A matrix has an inverse only if it is square, and even then only if it is nonsingular, or, in other words, if its columns (or rows) are linearly

1

independent. In recent years needs have been felt in numerous areas of applied mathematics for some kind of partial inverse of a matrix that is singular or even rectangular. By a generalized inverse of a given matrix A we shall mean a matrix X associated in some way with A that (i) exists for a class of matrices larger than the class of nonsingular matrices, (ii) has some of the properties of the usual inverse, and (iii) reduces to the usual inverse when A is nonsingular. Some writers have used the term "pseudoinverse" rather than "generalized inverse."

As an illustration of part (iii) of our description of a generalized inverse, consider a definition used by a number of writers (e.g., Rohde [1]) to the effect that a generalized inverse of A is any matrix satisfying

$$AXA = A. \tag{2}$$

If A were nonsingular, multiplication by A^{-1} both on the left and on the right would give at once

$$X = A^{-1}.$$

3. ILLUSTRATION: SOLVABILITY OF LINEAR SYSTEMS

Probably the most familiar application of matrices is to the solution of systems of simultaneous linear equations. Let

$$Ax = b \tag{3}$$

be such a system, where b is a given vector and x is an unknown vector. If A is nonsingular, there is a unique solution for x, given by

$$x = A^{-1}b.$$

In the general case, when A may be singular or rectangular, there may sometimes be no solutions or a multiplicity of solutions.

The existence of a vector x satisfying (3) is tantamount to the statement that b is some linear combination of the columns of A. If A is $m \times n$ and of rank less than m, this may not be the case. If it is, there is some vector h such that

$$b = Ah.$$

Now, if X is some matrix satisfying (2), and if we take

$$x = Xb,$$

we have

$$Ax = AXb = AXAh = Ah = b,$$

and so this x satisfies (3).

In the general case, however, when (3) may have many solutions, we may desire not just one solution but a characterization of all solutions. It has been shown (Bjerhammar [3], Penrose [1]) that, if X is any matrix satisfying $AXA = A$, then $Ax = b$ has a solution if and only if

$$AXb = b,$$

in which case the most general solution is

$$x = Xb + (I - XA)y, \tag{4}$$

where y is arbitrary.

We shall see later that for every matrix A there exist one or more matrices X satisfying (2).

EXERCISES

1. If A is nonsingular and has an eigenvalue λ, and x is a corresponding eigenvector, show that λ^{-1} is an eigenvalue of A^{-1} with the same eigenvector x.

2. For any square A, let a "generalized inverse" be defined as any matrix X satisfying $A^{k+1}X = A^k$ for some positive integer k. Show that $X = A^{-1}$ if A is nonsingular.

3. If X satisfies $AXA = A$, show that $Ax = b$ has a solution if and only if $AXb = b$.

4. Show that (4) is the most general solution of $Ax = b$. [Hint: First show that it is a solution; then show that every solution can be expressed in this form. Let x be any solution; then write $x = XAx + (I - XA)x$.]

5. If A is an $m \times n$ matrix of zeros, what is the class of matrices X satisfying $AXA = A$?

6. Let A be an $m \times n$ matrix whose elements are all equal to zero except the (i,j)th element, which is equal to 1. What is the class of matrices satisfying (2)?

7. Let A be given, and let X have the property that $x = Xb$ is a solution of $Ax = b$ for *all* b such that a solution exists. Show that X satisfies $AXA = A$.

4. DIVERSITY OF GENERALIZED INVERSES

From Exercises 3, 4, and 7 the reader will perceive that, for a given matrix A, the matrix equation $AXA = A$ alone characterizes those generalized inverses X that are of use in analyzing the solutions of the linear system $Ax = b$. For other purposes, other relationships play an essential role. Thus, if we are concerned with least-squares properties, (2) is not enough and must be supplemented by further relations. There results a more restricted class of generalized inverses.

If we are interested in spectral properties (i.e., those relating to eigenvalues and eigenvectors), consideration is necessarily limited to square matrices, since only these have eigenvalues and eigenvectors. In this connection, we shall see that (2) plays a role only for a restricted class of matrices A and must be supplanted, in the general case, by other relations.

Thus, unlike the case of the nonsingular matrix, which has a single unique inverse for all purposes, there are different generalized inverses for different purposes. For some purposes, as in the example of solutions of linear systems, there is not a unique inverse, but any matrix of a certain class will suffice.

This book does not pretend to be exhaustive, but seeks to develop and describe in a natural sequence the more interesting and useful kinds of generalized inverses and their properties. For the most part, the discussion is limited to generalized inverses of finite matrices, but extensions to infinite-dimensional spaces and to differential and integral operators are briefly introduced in Chapter 8. Pseudoinverses on general rings and semigroups are not discussed; the interested reader is referred to Drazin [1], Foulis [1], and Munn [1].

The literature on generalized inverses has become so extensive that it would be impossible to do justice to it in a book of moderate size. In particular, applications of generalized inverses in statistics will not be treated here, since they are amply covered in the books by Rao and Mitra [1] and Albert [3]. We have been forced to make a selection of topics to be covered, and it is inevitable that not everyone will agree with the choices we have made. We apologize to those authors whose work has been slighted. A virtually complete bibliography as of 1968 is found in Boullion and Odell [1]. See also the bibliography in Nashed [3].

5. PREPARATION EXPECTED OF THE READER

It is assumed that the reader has a knowledge of linear algebra that would normally result from completion of an introductory course in the subject. In particular, vector spaces will be extensively utilized. Except in Chapter 8, which deals with Hilbert spaces, the vector spaces and linear transformations used are finite-dimensional, real or complex. Familiarity with these topics is assumed, say at the level of Halmos [1] or Noble [2].

6. HISTORICAL NOTE

The concept of a generalized inverse seems to have been first mentioned in print in 1903 by Fredholm [1], where a particular generalized inverse (called by him "pseudoinverse") of an integral operator was given. The class of all pseudoinverses was characterized in 1912 by Hurwitz [1], who used the finite dimensionality of the null spaces of the Fredholm operators to give a simple algebraic construction (see, e.g., Exs. 19–20 in Chapter 8). Generalized inverses of differential operators, already implicit in Hilbert's [1] discussion in 1904 of generalized Green's functions, were consequently studied by numerous authors, in particular Myller (1906), Westfall (1909), Bounitzky (1909), Elliott (1928), and Reid (1931). For a history of this subject see the excellent survey by Reid [4].

Generalized inverses of differential and integral operators thus antedated the generalized inverses of matrices, whose existence was first noted by E. H. Moore, who defined a unique inverse (called by him the "general reciprocal") for every finite matrix (square or rectangular). Although his first publication on the subject (Moore [1]), an abstract of a talk given at a meeting of the American Mathematical Society, appeared in 1920, his results are thought to have been obtained much earlier. One writer (Lanczos [1], p. 676) has assigned the date 1906. Details were published (Moore [2]) only in 1935 after Moore's death. Little notice was taken of Moore's discovery for 30 years after its first publication, during which time generalized inverses were given for matrices by Siegel [1] in 1937, and for operators by Tseng ([1]-1933, [2], [3], [4]-1949), Murray and von Neumann [1] in 1936, Atkinson ([1]-1951, [2]-1954) and others. Revival of interest in the subject in the 1950s centered around the least-squares properties (not mentioned by Moore) of certain generalized inverses. These properties were recognized in 1951 by Bjerhammar, who rediscovered Moore's inverse and also noted the relationship of generalized inverses to solutions of linear systems (Bjerhammar [1], [2], [3]). In 1955 Penrose [1] sharpened and extended Bjerhammar's results on linear systems and showed that Moore's inverse, for a given matrix A is the unique matrix X satisfying the four equations (1)–(4) of the next chapter. The latter discovery has been so important and fruitful that this unique inverse (called by some writers *the* generalized inverse) is now commonly called the Moore–Penrose inverse.

Since 1955 hundreds of papers on various aspects of generalized inverses and their applications have appeared. In view of the vast scope of this literature, we shall not attempt to trace the history of the subject further, but the subsequent chapters will include selected references on particular items.

7. REMARKS ON NOTATION

Equation j of Chapter i is denoted by (j) in Chapter i and by $(i.j)$ in other chapters. Similarly, Theorem j of Chapter i is called Theorem j in Chapter i, and Theorem $i.j$ in other chapters. Similar conventions apply to corollaries, lemmas, and exercises.

References are numbered from 1 on, for each author or ordered set of co-authors. Thus, Doe [i] denotes the ith article or book by Doe cited in the bibliography at the end of the book. By "Doe proved in [i]" is meant "Doe proved in Doe [i]."

SUGGESTED FURTHER READING

Section 2. A ring $\mathcal{R}$ is called *regular* if for every $A \in \mathcal{R}$ there exists an $X \in \mathcal{R}$ satisfying $AXA = A$. See von Neumann ([2] and [5], p. 90), Murray and von Neumann ([1], p. 229), McCoy [1], and Hartwig [4].

Section 4. For generalized inverses in abstract algebraic setting see also Davis and Robinson [1], Gabriel ([1], [2], [3]), Hansen and Robinson [1], Hartwig [4], Munn and Penrose [1], Pearl [8], Rabson [1], and Rado [1].

For applications in statistics see Albert ([3], [4]), Albert and Sittler [1], Banerjee [1], Banerjee and Federer [1], Chernoff [1], Chipman ([1], [2]), Chipman and Rao ([1], [2]) Drygas ([1], [2]), Goldman and Zelen [1], Golub ([2], [3]), Golub and Styan [1], Good [1], Graybill and Marsaglia [1], J. A. John [1], P. W. M. John [1], Meyer and Painter [1], Mitra [1], Mitra and Rao ([1], [2]), Price [1], Rao ([1], [2], [3]), Rao and Mitra [1], Rayner and Pringle [1], Rohde ([1], [4]), Rohde and Harvey [1], Tan [1], Zacks [1], Zyskind [1], and Zyskind and Martin [1].

1
EXISTENCE AND CONSTRUCTION OF GENERALIZED INVERSES

1. THE PENROSE EQUATIONS

In 1955 Penrose [1] showed that, for every finite matrix A (square or rectangular) of real or complex elements, there is a unique matrix X satisfying the four equations

$$AXA = A, \tag{1}$$

$$XAX = X, \tag{2}$$

$$(AX)^* = AX, \tag{3}$$

$$(XA)^* = XA, \tag{4}$$

where A^* denotes the conjugate transpose of A. Because this unique generalized inverse had previously been studied (though defined in a different way) by E. H. Moore [1],[2], it is commonly known as the *Moore–Penrose inverse*, and is often denoted by $A^\dagger$.

If A is nonsingular, it is clear that $X = A^{-1}$ trivially satisfies the four equations. Since the Moore–Penrose inverse is known to be unique (as we shall prove shortly) it follows that the Moore–Penrose inverse of a nonsingular matrix is the same as the ordinary inverse.

Throughout this book we shall be much concerned with generalized inverses that satisfy some, but not all, of the four Penrose equations. As we shall wish to deal with a number of different subsets of the set of four equations, we need a convenient notation for a generalized inverse satisfying certain specified equations. Let $C^{m \times n}$ $[R^{m \times n}]$ denote the class of $m \times n$ complex [real] matrices.

DEFINITION 1.   For any $A \in C^{m \times n}$, let $A\{i,j,\ldots,l\}$ denote the set of matrices X in $C^{n \times m}$ which satisfy equations $(i),(j),\ldots,(l)$ from among the equations $(1),(2),(3),(4)$. A matrix $X \in A\{i,j,\ldots,l\}$ is called an $\{i,j,\ldots,l\}$-*inverse* of A, and also denoted by $A^{(i,j,\ldots,l)}$.

In Chapter 4 we shall extend the scope of this notation* by enlarging the set of four matrix equations to include several further equations, applicable only to square matrices, that will play an essential role in the study of generalized inverses having spectral properties.

EXERCISES

1.   If $A\{1,2,3,4\}$ is nonempty, then it consists of a single element (Penrose [1]).

PROOF.   Let $X, Y \in A\{1,2,3,4\}$. Then

$$X = X(AX)^* = XX^*A^* = X(AX)^*(AY)^*$$

$$= XAY = (XA)^*(YA)^*Y = A^*Y^*Y$$

$$= (YA)^*Y = Y. \quad \blacksquare$$

2.   By means of a (trivial) example, show that $A\{2,3,4\}$ is nonempty.

2.   EXISTENCE AND CONSTRUCTION OF $\{1\}$-INVERSES

Let $C_r^{m \times n}$ $[R_r^{m \times n}]$ denote the class of $m \times n$ complex [real] matrices of rank r.

DEFINITION 2.   A matrix in $C_r^{m \times n}$ is said to be in *Hermite normal form* (also sometimes called row–echelon form) if:

(i) Each of the first r rows contains at least one nonzero element; the remaining rows contain only zeros.

(ii) The first r columns of the unit matrix I_m appear in columns $c_1, c_2, \ldots, c_r$.

---

*Some writers have adopted descriptive names to designate various classes of generalized inverses. However, there is a notable lack of uniformity and consistency in the use of these terms by different writers. Thus, $X \in A\{1\}$ is called a *generalized inverse* (Rao [2]), *pseudo-inverse* (Sheffield [2]), *inverse* (Bjerhammar [3]). $X \in A\{1,2\}$ is called a *semi-inverse* (Frame [1]), *reciprocal inverse* (Bjerhammar), *reflexive generalized inverse* (Rohde [2]). $X \in \{1,2,3\}$ is called a *normalized generalized inverse* (Rohde). $X \in \{1,2,4\}$ is called a *weak generalized inverse* (Goldman and Zelen [1]). $X \in A\{1,2,3,4\}$ is called the *general reciprocal* (Moore [1,2]), *generalized inverse* (Penrose [1]), *pseudoinverse* (Greville [2]), the *Moore–Penrose inverse* (Ben-Israel and Charnes [1]). In view of this diversity of terminology, the unambiguous notation adopted here is considered preferable. This notation also emphasizes the lack of uniqueness of many of the generalized inverses considered.

The preceding definition differs somewhat from that of Marcus and Minc [1].

By a suitable permutation of its columns, a matrix $H \in C_r^{m \times n}$ in Hermite normal form can be brought into the partitioned form

$$R = \begin{bmatrix} I_r & K \\ O & O \end{bmatrix}, \tag{5}$$

where O denotes a null matrix. Such a permutation of the columns of H can be interpreted as postmultiplication of H by a suitable permutation matrix P. If P_j denotes the jth column of P, and e_j the jth column of I_n, we have

$$P_j = e_k \quad \text{where} \quad k = c_j \quad (j = 1, 2, \ldots, r).$$

The remaining columns of P are the remaining unit vectors $\{e_k : k \neq c_j$ $(j = 1, 2, \ldots, r)\}$ in any order.

In particular cases, the partitioned form (5) must be suitably interpreted. If $R \in C_r^{m \times n}$, then the two right-hand submatrices are absent in case $r = n$, and the two lower submatrices are absent if $r = m$.

It is easy to construct a {1}-inverse of the matrix $R \in C_r^{m \times n}$ given by (5). For any $L \in C^{(n-r) \times (m-r)}$, the $n \times m$ matrix

$$S = \begin{bmatrix} I_r & O \\ O & L \end{bmatrix}$$

is a {1}-inverse of (5). If R is of full column [row] rank, the two lower [right-hand] submatrices of S are interpreted as absent.

The construction of {1}-inverses for an arbitrary $A \in C^{m \times n}$ is simplified by transforming A into a Hermite normal form, as we shall now see.

An *elementary row operation* on the matrix A means either (i) multiplication of a given row by a nonzero scalar, or (ii) addition to a given row of a scalar multiple of another row.* Every such operation can be interpreted as premultiplication of A by a suitable nonsingular matrix, called an *elementary row matrix*. An elementary row matrix has only one row that is different from the corresponding row of the unit matrix of the same order. Any matrix can be transformed into its Hermite normal form by a finite sequence of elementary row operations. The Hermite normal form into which a given matrix is transformed is not unique under our Definition 2, but is unique (Noble [2]) under the definition of Marcus and Minc [1].

---

*The interchange of two rows, which can be expressed as a sequence of (four) operations of types (i) and (ii), is sometimes also listed as an elementary row operation (of type (iii)); see, e.g., Noble [2].

Therefore, for any $A \in C_r^{m \times n}$ there is a nonsingular $m \times m$ matrix E (the product of the elementary row matrices mentioned above) and a permutation matrix P of order n, such that

$$EAP = \begin{bmatrix} I_r & K \\ O & O \end{bmatrix}. \tag{6}$$

For a numerical example, see Exs. 6 and 7 below.

The following theorem describes the use of the Hermite normal form to construct a {1}-inverse of an arbitrary finite matrix of rank $r > 0$.

THEOREM 1. Let $A \in C_r^{m \times n}$, and let $E \in C_m^{m \times m}$ and $P \in C_n^{n \times n}$ be such that

$$EAP = \begin{bmatrix} I_r & K \\ O & O \end{bmatrix}. \tag{6}$$

Then, for any $L \in C^{(n-r) \times (m-r)}$, the $n \times m$ matrix

$$X = P \begin{bmatrix} I_r & O \\ O & L \end{bmatrix} E \tag{7}$$

is a {1}-inverse of A. The partitioned matrices in (6) and (7) must be suitably interpreted in case $r = m$ or n.

PROOF. Rewriting (6) as

$$A = E^{-1} \begin{bmatrix} I_r & K \\ O & O \end{bmatrix} P^{-1}, \tag{8}$$

it is easily verified that any X given by (7) satisfies $AXA = A$. ∎

In the trivial case of $r = 0$, when A is therefore the $m \times n$ null matrix, *any* $n \times m$ matrix is a {1}-inverse.

We note that since P and E are both nonsingular, the rank of X as given by (7) is the rank of the partitioned matrix in the right member. In view of the form of the latter matrix,

$$\text{rank } X = r + \text{rank } L. \tag{9}$$

Since L is arbitrary, it follows that a {1}-inverse of A exists having any specified rank between r and $\min\{m, n\}$, inclusive (see also Fisher [1]).

Theorem 1 shows that every finite matrix with elements in the complex field has a {1}-inverse, and suggests how such an inverse can be constructed.

EXERCISES

3. What is the Hermite normal form of a nonsingular matrix A? In this case, what is the matrix E, and what is its relationship to A? What is the permutation matrix P? What is the matrix X given by (7)?

4. An $m \times n$ matrix A has all its elements equal to 0 except the (i,j)th element, which is 1. What is its Hermite normal form? Show that E can be taken as a permutation matrix. What are the simplest choices of E and P? (By "simplest" we mean having the smallest number of elements different from the corresponding elements of the unit matrix of the same order.) Using these choices of E and P, but regarding L as entirely arbitrary, what is the form of the resulting matrix X as given by (7)? Is this X the most general $\{1\}$-inverse of A? (See Ex. 6, Introduction.)

5. Show that every square matrix has a nonsingular $\{1\}$-inverse.

3. PROPERTIES OF {1}-INVERSES

Certain properties of $\{1\}$-inverses are given in Lemma 1. For a given matrix A, we denote any $\{1\}$-inverse by $A^{(1)}$. Note that, in general, $A^{(1)}$ is not a uniquely defined matrix (see Ex. 9 below). For any scalar λ we define $\lambda^\dagger$ by

$$\lambda^\dagger = \begin{cases} \lambda^{-1} & (\lambda \neq 0) \\ 0 & (\lambda = 0) \end{cases}.$$

It will be recalled that a square matrix E is called *idempotent* if $E^2 = E$. Idempotent matrices are intimately related to generalized inverses, and their properties are considered in some detail in Chapter 2.

LEMMA 1. Let $A \in C^{m \times n}$, $\lambda \in C$. Then,

  (a) $(A^{(1)})^* \in A^*\{1\}$.
  (b) If A is nonsingular, $A^{(1)} = A^{-1}$ uniquely (see also Ex. 8 below).
  (c) $\lambda^\dagger A^{(1)} \in (\lambda A)\{1\}$.
  (d) rank $A^{(1)} \geqslant$ rank A.
  (e) If S and T are nonsingular, $T^{-1}A^{(1)}S^{-1} \in SAT\{1\}$.
  (f) $AA^{(1)}$ and $A^{(1)}A$ are idempotent and have the same rank as A.

PROOF. These are immediate consequences of the defining relation (1); (d) and the latter part of (f) depend on the fact that the rank of a product of matrices does not exceed the rank of any factor. ■

If an $m \times n$ matrix A is of full column rank, its $\{1\}$-inverses are its left inverses. If it is of full row rank, its $\{1\}$-inverses are its right inverses.

LEMMA 2. Let $A \in C_r^{m \times n}$. Then,

(a) $A^{(1)}A = I_n$ if and only if $r = n$.

(b) $AA^{(1)} = I_m$ if and only if $r = m$.

PROOF. (a) *If*: Let $A \in C_r^{m \times n}$. Then the $n \times n$ matrix $A^{(1)}A$ is, by Lemma 1 (f), idempotent and nonsingular. Multiplying $(A^{(1)}A)^2 = A^{(1)}A$ by $(A^{(1)}A)^{-1}$ gives $A^{(1)}A = I_n$.

Only if: $A^{(1)}A = I_n \Rightarrow \text{rank } A^{(1)}A = n \Rightarrow \text{rank } A = n$, by Lemma 1 (f).

(b) Similarly proved. ∎

EXERCISES AND EXAMPLES

6. *Transforming a matrix into Hermite normal form.* Let $A \in C^{m \times n}$, and $T_0 = [A \; I_m]$. A matrix E transforming A into a Hermite normal form EA can be found by Gaussian elimination on T_0, where, after the elimination is completed,

$$ET_0 = [EA \quad E],$$

E being recorded as the right-hand $m \times m$ submatrix of ET_0. We illustrate the procedure for the matrix

$$A = \begin{bmatrix} 0 & 2i & i & 0 & 4+2i & 1 \\ 0 & 0 & 0 & -3 & -6 & -3-3i \\ 0 & 2 & 1 & 1 & 4-4i & 1 \end{bmatrix},$$

the pivot being underlined at each step.

$$T_0 = \begin{bmatrix} 0 & \underline{2i} & i & 0 & 4+2i & 1 & \vline & 1 & 0 & 0 \\ 0 & 0 & 0 & -3 & -6 & -3-3i & \vline & 0 & 1 & 0 \\ 0 & 2 & 1 & 1 & 4-4i & 1 & \vline & 0 & 0 & 1 \end{bmatrix},$$

$$T_1 = \begin{bmatrix} 0 & 1 & \frac{1}{2} & 0 & 1-2i & -\frac{1}{2}i & \vline & -\frac{1}{2}i & 0 & 0 \\ 0 & 0 & 0 & \underline{-3} & -6 & -3-3i & \vline & 0 & 1 & 0 \\ 0 & 0 & 0 & 1 & 2 & 1+i & \vline & i & 0 & 1 \end{bmatrix},$$

$$T_2 = 3 \begin{bmatrix} 0 & 1 & \frac{1}{2} & 0 & 1-2i & -\frac{1}{2}i & \vline & -\frac{1}{2}i & 0 & 0 \\ 0 & 0 & 0 & 1 & 2 & 1+i & \vline & 0 & -\frac{1}{3} & 0 \\ 0 & 0 & 0 & 0 & 0 & 0 & \vline & i & \frac{1}{3} & 1 \end{bmatrix}.$$

From $T_2 = [EA \quad E]$ we read

$$E = \begin{bmatrix} -\frac{1}{2}i & 0 & 0 \\ 0 & -\frac{1}{3} & 0 \\ i & \frac{1}{3} & 1 \end{bmatrix}$$

and $r = \text{rank } A = 2$. EA is a Hermite normal form of A.

7. *Computing a {1}-inverse.* This is demonstrated for the matrix A of Ex. 6, using (7) with E as computed in Ex. 6 and an arbitrary $L \in C^{(n-r) \times (m-r)}$. A permutation matrix P such that

$$EAP = \begin{bmatrix} I_r & K \\ O & O \end{bmatrix}$$

is

$$P = [e_2 \quad e_4 \quad e_1 \quad e_3 \quad e_5 \quad e_6],$$

where e_j denotes the jth column of I_6. (A different permutation of the last $n - r = 4$ columns of P results in a different permutation of the columns of K.) Thus,

$$EAP = \begin{bmatrix} 1 & 0 & 0 & \frac{1}{2} & 1-2i & -\frac{1}{2}i \\ 0 & 1 & 0 & 0 & 2 & 1+i \\ 0 & 0 & 0 & 0 & 0 & 0 \end{bmatrix}.$$

We take

$$L = \begin{bmatrix} \alpha \\ \beta \\ \gamma \\ \delta \end{bmatrix},$$

with $\alpha, \beta, \gamma, \delta \in C$, since $m = 3$, $n = 6$, $r = 2$. Equation (7) then gives

$$X = P \begin{bmatrix} I_r & 0 \\ 0 & L \end{bmatrix} E,$$

$$= \begin{bmatrix} 0 & 0 & 1 & 0 & 0 & 0 \\ 1 & 0 & 0 & 0 & 0 & 0 \\ 0 & 0 & 0 & 1 & 0 & 0 \\ 0 & 1 & 0 & 0 & 0 & 0 \\ 0 & 0 & 0 & 0 & 1 & 0 \\ 0 & 0 & 0 & 0 & 0 & 1 \end{bmatrix} \begin{bmatrix} 1 & 0 & 0 \\ 0 & 1 & 0 \\ 0 & 0 & \alpha \\ 0 & 0 & \beta \\ 0 & 0 & \gamma \\ 0 & 0 & \delta \end{bmatrix} \begin{bmatrix} -\frac{1}{2}i & 0 & 0 \\ 0 & -\frac{1}{3} & 0 \\ i & \frac{1}{3} & 1 \end{bmatrix}$$

$$= \begin{bmatrix} i\alpha & \frac{1}{3}\alpha & \alpha \\ -\frac{1}{2}i & 0 & 0 \\ i\beta & \frac{1}{3}\beta & \beta \\ 0 & -\frac{1}{3} & 0 \\ i\gamma & \frac{1}{3}\gamma & \gamma \\ i\delta & \frac{1}{3}\delta & \delta \end{bmatrix}$$

Note that, in general, the quantities $i\alpha$, $i\beta$, $i\gamma$, $i\delta$ are not pure imaginaries since α, β, γ, δ are complex.

8. Let $A = FHG$, where F is of full column rank and G is of full row rank. Then rank $A = $ rank H. (*Hint*: Use Lemma 2.)

4. BASES FOR THE RANGE AND NULL SPACE OF A MATRIX

For any $A \in C^{m \times n}$ we denote by

$$R(A) = \{ y \in C^m : y = Ax \text{ for some } x \in C^n \}, \qquad \text{the } range \text{ of } A,$$

$$N(A) = \{ x \in C^n : Ax = 0 \}, \qquad \text{the } null \ space \text{ of } A.$$

A basis for $R(A)$ is useful in a number of applications, such as, for example, in the numerical computation of the Moore–Penrose inverse, and of the group inverse to be discussed in Chapter 4.

The need for a basis for $N(A)$ is illustrated by the fact that the general solution of the linear inhomogeneous equation

$$Ax = b$$

is the sum of any particular solution x_0 and the general solution of the homogeneous equation

$$Ax = 0.$$

The latter general solution consists of all linear combinations of the elements of any basis for $N(A)$.

A further advantage of the Hermite normal form EA of A (and its column-permuted form EAP) is that from them bases for $R(A)$, $N(A)$, and $R(A^*)$ can be read off directly.

A basis for $R(A)$ consists of the c_1th, c_2th, ..., c_rth columns of A, where c_j is as defined in Definition 2, Section 2 (Willner [1]). To see this, let P_1 denote the submatrix consisting of the first r columns of the permutation matrix P of (6) and (7). Then, because of the way in which these r columns of P were chosen,

$$EAP_1 = \begin{bmatrix} I_r \\ O \end{bmatrix}. \tag{10}$$

Now, AP_1 is an $m \times r$ matrix, and is of rank r, since the right member of (10) is of rank r. But AP_1 is merely the submatrix of A consisting of the c_1th, c_2th, ..., c_rth columns.

It follows from (8) that the columns of the $n \times (n-r)$ matrix

$$P \begin{bmatrix} -K \\ I_{n-r} \end{bmatrix} \tag{11}$$

are a basis for $N(A)$. (The reader should verify this.)

Moreover, it is evident that the first r rows of the Hermite normal form EA are linearly independent, and each is some linear combination of the rows of A. Thus, they are a basis for the space spanned by the rows of A. Consequently, if

$$EA = \begin{bmatrix} G \\ O \end{bmatrix}, \tag{12}$$

then the columns of the $n \times r$ matrix

$$G^* = P \begin{bmatrix} I_r \\ K^* \end{bmatrix}$$

are a basis for $R(A^*)$.

As an example, for the matrix A of Exs. 6 and 7, we note that in its Hermite normal form EA (as exhibited in the left-hand portion of the matrix T_2 of Ex. 6) the two unit vectors of C^2 appear in the *second* and *fourth* columns. Therefore, the second and fourth columns of A form a basis for $R(A)$.

Using (11) with K and P computed as in Ex. 7, we find that the columns of the following matrix form a basis for $N(A)$:

$$P\begin{bmatrix} -K \\ I_{n-r} \end{bmatrix} = \begin{bmatrix} 0 & 0 & 1 & 0 & 0 & 0 \\ 1 & 0 & 0 & 0 & 0 & 0 \\ 0 & 0 & 0 & 1 & 0 & 0 \\ 0 & 1 & 0 & 0 & 0 & 0 \\ 0 & 0 & 0 & 0 & 1 & 0 \\ 0 & 0 & 0 & 0 & 0 & 1 \end{bmatrix} \begin{bmatrix} 0 & -\frac{1}{2} & -1+2i & \frac{1}{2}i \\ 0 & 0 & -2 & -1-i \\ 1 & 0 & 0 & 0 \\ 0 & 1 & 0 & 0 \\ 0 & 0 & 1 & 0 \\ 0 & 0 & 0 & 1 \end{bmatrix}$$

$$= \begin{bmatrix} 1 & 0 & 0 & 0 \\ 0 & -\frac{1}{2} & -1+2i & \frac{1}{2}i \\ 0 & 1 & 0 & 0 \\ 0 & 0 & -2 & -1-i \\ 0 & 0 & 1 & 0 \\ 0 & 0 & 0 & 1 \end{bmatrix}.$$

EXERCISES

9. Show that A is nonsingular if and only if it has a unique $\{1\}$-inverse, which then coincides with A^{-1}.

PROOF. For any $x \in N(A)$ $[y \in N(A^*)]$, adding x $[y^*]$ to any column [row] of an $X \in A\{1\}$ gives another $\{1\}$-inverse of A. The uniqueness of the $\{1\}$-inverse is therefore equivalent to

$$N(A) = \{0\}, \qquad N(A^*) = \{0\},$$

i.e., to the nonsingularity of A. ∎

10. Show that if $A^{(1)} \in A\{1\}$, then $R(AA^{(1)}) = R(A)$, $N(A^{(1)}A) = N(A)$, and $R((A^{(1)}A)^*) = R(A^*)$.

PROOF. We have

$$R(A) \supset R(AA^{(1)}) \supset R(AA^{(1)}A) = R(A).$$

from which the first result follows.

Similarly,

$$N(A) \subset N(A^{(1)}A) \subset N(AA^{(1)}A) = N(A)$$

yields the second equation.

Finally, by Lemma 1(a),

$$R(A^*) \supset R(A^*(A^{(1)})^*) = R((A^{(1)}A)^*) \supset R(A^*(A^{(1)})^*A^*) = R(A^*). \quad \blacksquare$$

11. More generally, show that $R(AB) = R(A)$ if and only if rank AB = rank A, and $N(AB) = N(B)$ if and only if rank AB = rank B.

PROOF. Evidently $R(A) \supset R(AB)$, and these two subspaces are identical if and only if they have the same dimension. But, the rank of any matrix is the dimension of its range.

Similarly, $N(B) \subset N(AB)$. Now, the nullity of any matrix is the dimension of its null space and also the number of columns minus the rank. Thus, $N(B) = N(AB)$ if and only if B and AB have the same nullity, which is equivalent, in this case, to having the same rank, since the two matrices have the same number of columns. $\blacksquare$

12. The answer to the last question in Ex. 4 indicates that, for particular choices of E and P, one does not get all the $\{1\}$-inverses of A merely by varying L in (7). Note, however, that Theorem 1 does not require P to be a permutation matrix. Could one get all the $\{1\}$-inverses by considering all nonsingular P and Q such that

$$QAP = \begin{bmatrix} I_r & O \\ O & O \end{bmatrix}? \tag{13}$$

Given $A \in C_r^{m \times n}$, show that $X \in A\{1\}$ if and only if

$$X = P \begin{bmatrix} I_r & O \\ O & L \end{bmatrix} Q \tag{14}$$

for some L and for some nonsingular P and Q satisfying (13).

PROOF. If (13) and (14) hold, X is a $\{1\}$-inverse of A by Theorem 1.

On the other hand, let $AXA = A$. Then, both AX and XA are indempotent and of rank r, by Lemma 1(f). Since any indempotent matrix E satisfies $E(E - I) = O$, its only eigenvalues are 0 and 1. Thus, the Jordan canonical forms of both AX and XA are of the form

$$\begin{bmatrix} I_r & O \\ O & O \end{bmatrix},$$

being of orders m and n, respectively. Therefore, there exist nonsingular P and R such that

$$R^{-1}AXR = \begin{bmatrix} I_r & O \\ O & O \end{bmatrix}, \quad P^{-1}XAP = \begin{bmatrix} I_r & O \\ O & O \end{bmatrix}.$$

Thus,

$$R^{-1}AP = R^{-1}AXAXAP = (R^{-1}AXR)R^{-1}AP(P^{-1}XAP)$$

$$= \begin{bmatrix} I_r & O \\ O & O \end{bmatrix} R^{-1}AP \begin{bmatrix} I_r & O \\ O & O \end{bmatrix}.$$

It follows that $R^{-1}AP$ is of the form

$$R^{-1}AP = \begin{bmatrix} H & O \\ O & O \end{bmatrix},$$

where $H \in C_r^{r \times r}$, i.e., nonsingular. Let

$$Q = \begin{bmatrix} H^{-1} & O \\ O & I_{m-r} \end{bmatrix} R^{-1}.$$

Then (13) is satisfied. Consider the matrix $P^{-1}XQ^{-1}$. We have

$$\begin{bmatrix} I_r & O \\ O & O \end{bmatrix} (P^{-1}XQ^{-1}) = (QAP)(P^{-1}XQ^{-1}) = QAXQ^{-1}$$

$$= \begin{bmatrix} H^{-1} & O \\ O & I_{m-r} \end{bmatrix} \begin{bmatrix} I_r & O \\ O & O \end{bmatrix} \begin{bmatrix} H & O \\ O & I_{m-r} \end{bmatrix} = \begin{bmatrix} I_r & O \\ O & O \end{bmatrix}$$

and

$$(P^{-1}XQ^{-1}) \begin{bmatrix} I_r & O \\ O & O \end{bmatrix} = (P^{-1}XQ^{-1})(QAP) = P^{-1}XAP$$

$$= \begin{bmatrix} I_r & O \\ O & O \end{bmatrix}.$$

From the latter two equations it follows that

$$P^{-1}XQ^{-1} = \begin{bmatrix} I_r & O \\ O & L \end{bmatrix}$$

for some L. But this is equivalent to (14). ∎

5.　EXISTENCE AND CONSTRUCTION OF $\{1,2\}$-INVERSES

It was first noted by Bjerhammar [3] that the existence of a $\{1\}$-inverse of a matrix A implies the existence of a $\{1,2\}$-inverse. This easily verified observation is stated as a lemma for convenience of reference.

LEMMA 3. Let $Y, Z \in A\{1\}$, and let

$$X = YAZ.$$

Then $X \in A\{1,2\}$.

Since the matrices A and X occur symmetrically in (1) and (2), $X \in A\{1,2\}$ and $A \in X\{1,2\}$ are equivalent statements, and in either case we can say that A and X are $\{1,2\}$ inverses *of each other.*

From (1) and (2) and the fact that the rank of a product of matrices does not exceed the rank of any factor, it follows at once that if A and X are $\{1,2\}$-inverses of each other, they have the same rank. Less obvious is the fact, first noted by Bjerhammar [3], that if X is a $\{1\}$-inverse of A and of the same rank as A, it is a $\{1,2\}$-inverse of A.

THEOREM 2 (Bjerhammar). Given A and $X \in A\{1\}$, $X \in A\{1,2\}$ if and only if rank $X =$ rank A.

PROOF. *If*: Clearly $R(XA) \subset R(X)$. But rank $XA =$ rank A by Lemma 1(f), and so, if rank $X =$ rank A, $R(XA) = R(X)$ by Ex. 11. Thus,

$$XAY = X$$

for some Y. Premultiplication by A gives

$$AX = AXAY = AY,$$

and therefore

$$XAX = X.$$

Only if: This follows at once from (1) and (2). ∎

An equivalent statement is the following:

COROLLARY 1. Any two of the following three statements imply the third:

$$X \in A\{1\},$$

$$X \in A\{2\},$$

$$\text{rank } X = \text{rank } A. \quad ∎$$

In view of Theorem 2, (9) shows that the $\{1\}$-inverse obtained from the Hermite normal form is a $\{1,2\}$-inverse if we take $L = O$. In other words,

$$X = P \begin{bmatrix} I_r & O \\ O & O \end{bmatrix} E$$

is a $\{1,2\}$-inverse of A if E and P are nonsingular and satisfy (6).

6. EXISTENCE AND CONSTRUCTION OF $\{1,2,3\}$-, $\{1,2,4\}$-, AND $\{1,2,3,4\}$-INVERSES

Just as Bjerhammar [3] showed that the existence of a $\{1\}$-inverse implies the existence of a $\{1,2\}$-inverse, Urquhart [1] has shown that the existence of a $\{1\}$-inverse of every finite matrix with elements in C implies the existence of a $\{1,2,3\}$-inverse and a $\{1,2,4\}$-inverse of every such matrix. However, in order to show the nonemptiness of $A\{1,2,3\}$ and $A\{1,2,4\}$ for any given A, we shall utilize the $\{1\}$-inverse not of A itself but of a related matrix. For that purpose we shall need the following lemma.

LEMMA 4. For any finite matrix A,

$$\text{rank } AA^* = \text{rank } A = \text{rank } A^*A.$$

PROOF. If $A \in C^{m \times n}$, both A and AA^* have m rows. Now, the rank of any m-rowed matrix is equal to m minus the number of independent linear relations among its rows. To show that rank $AA^* = $ rank A, it is sufficient, therefore, to show that every linear relation among the rows of A holds for the corresponding rows of AA^*, and vice versa. Any nontrivial linear relation among the rows of a matrix H is equivalent to the existence of a nonzero row vector x^* such that $x^*H = 0$. Now, evidently,

$$x^*A = 0 \Rightarrow x^*AA^* = 0,$$

and, conversely,

$$x^*AA^* = 0 \Rightarrow 0 = x^*AA^*x = (A^*x)^*A^*x$$

$$\Rightarrow A^*x = 0 \Rightarrow 0 = (A^*x)^* = x^*A.$$

Here we have used the fact that, for any column vector y of complex elements, y^*y is the sum of the squares of the absolute values of the elements, and this sum vanishes only if every element is zero.

Finally, applying this result to the matrix A^* gives rank $A^*A = $ rank A^*, and, of course, rank $A^* = $ rank A. ∎

COROLLARY 2. For any finite matrix A, $R(AA^*) = R(A)$ and $N(A^*A) = N(A)$.

PROOF. This follows from Lemma 4 and Ex. 11. ∎

Using the preceding lemma, we can now prove the following theorem.

THEOREM 3 (Urquhart). For every finite matrix A with complex elements,

$$Y = (A^*A)^{(1)}A^* \in A\{1,2,3\} \tag{15}$$

and

$$Z = A^*(AA^*)^{(1)} \in A\{1,2,4\}. \tag{16}$$

PROOF. Applying Corollary 2 to A^* gives

$$R(A^*A) = R(A^*),$$

and so,

$$A^* = A^*AU \tag{17}$$

for some U. Taking conjugate transposes gives

$$A = U^*A^*A. \tag{18}$$

Consequently,

$$AYA = U^*A^*A(A^*A)^{(1)}A^*A = U^*A^*A = A.$$

Thus, $Y \in A\{1\}$. But rank $Y \geqslant$ rank A by Lemma 1 (d), and rank $Y \leqslant$ rank $A^* =$ rank A by the definition of Y. Therefore,

$$\text{rank } Y = \text{rank } A,$$

and, by Theorem 2, $Y \in A\{1,2\}$. Finally, (17) and (18) give

$$AY = U^*A^*A(A^*A)^{(1)}A^*AU = U^*A^*AU,$$

which is clearly Hermitian. Thus, (15) is established.

Relation (16) is similarly proved. ∎

A $\{1,2\}$-inverse of a matrix A is, of course, a $\{2\}$-inverse, and, similarly, a $\{1,2,3\}$-inverse is also a $\{1,3\}$-inverse and a $\{2,3\}$-inverse. Thus, if we can establish the existence of a $\{1,2,3,4\}$-inverse, we will have demonstrated the existence of an $\{i,j,\ldots,l\}$-inverse for all possible choices of one, two or three integers $i,j,\ldots,l$ from the set $\{1,2,3,4\}$. It was shown in Ex. 1 that if a $\{1,2,3,4\}$-inverse exists, it is unique. We know, as a matter of fact, that it does exist, because it is the well-known Moore–Penrose inverse, $A^\dagger$. However, we have not yet proved this. This is done in the next theorem.

THEOREM 4   (Urquhart).   For any finite matrix A of complex elements,

$$A^{(1,4)}AA^{(1,3)} = A^\dagger. \tag{19}$$

PROOF. Let X denote the left member of (19). It follows at once from Lemma 3 that $X \in A\{1,2\}$. Moreover, (19) gives

$$AX = AA^{(1,3)}, \qquad XA = A^{(1,4)}A.$$

But, both $AA^{(1,3)}$ and $A^{(1,4)}A$ are Hermitian, by the definition of $A^{(1,3)}$ and $A^{(1,4)}$. Thus

$$X \in A\{1,2,3,4\}.$$

However, by Ex. 1, $A\{1,2,3,4\}$ contains at most a single element. Therefore, it contains exactly one element, which we denote by $A^\dagger$, and $X = A^\dagger$. ∎

7. FULL-RANK FACTORIZATIONS

A non-null matrix that is not of full (column or row) rank can be expressed as the product of a matrix of full column rank and a matrix of full row rank. Such factorizations turn out to be a powerful tool in the study of generalized inverses.

LEMMA 5. Let $A \in C_r^{m \times n}$, $r > 0$. Then there exist matrices $F \in C_r^{m \times r}$ and $G \in C_r^{r \times n}$, such that

$$A = FG. \tag{20}$$

PROOF. Let F be any matrix whose columns are a basis for $R(A)$. Then $F \in C_r^{m \times r}$. The matrix $G \in C^{r \times n}$ is then uniquely determined by (20), since every column of A is uniquely representable as a linear combination of the columns of F. Finally, rank $G = r$, since

$$\text{rank } G \geqslant \text{rank } FG = r. \quad ∎$$

The columns of F can, in particular, be chosen as any maximal linearly independent set of columns of A. Also, G could be chosen first as any matrix whose rows are a basis for the space spanned by the rows of A, and then F is uniquely determined by (20).

We shall call a factorization (20) with the properties stated in Lemma 5 a *full-rank factorization* of A. When A is of full (column or row) rank, the most obvious factorization is a trivial one, one factor being a unit matrix. Nevertheless, the lemma still holds in this case.

A full-rank factorization of any matrix A is easily read off from its Hermite normal form. Indeed, it was pointed out in Section 4 above that the first r rows of the Hermite normal form EA (i.e., the rows of the matrix G of (12)) form a basis for the space spanned by the rows of A. Thus, this

G can serve also as the matrix G of (20). Consequently, (20) holds for some F. As in Section 4, let P_1 denote the submatrix of P consisting of the first r columns. Because of the way in which these r columns were constructed,

$$GP_1 = I_r.$$

Thus, multiplying (20) on the right by P_1 gives

$$F = AP_1,$$

and so (20) becomes

$$A = (AP_1)G, \tag{21}$$

where P_1 and G are as defined in Section 4. (Indeed it was already noted there that the columns of AP_1 are a basis for $R(A)$.)

For example, for the matrix A of Exs. 6 and 7, (21) gives

$$A = (AP_1)G = \begin{bmatrix} 2i & 0 \\ 0 & -3 \\ 2 & 1 \end{bmatrix} \begin{bmatrix} 0 & 1 & \frac{1}{2} & 0 & 1-2i & -\frac{1}{2}i \\ 0 & 0 & 0 & 1 & 2 & 1+i \end{bmatrix}.$$

8. EXPLICIT FORMULA FOR $A^\dagger$

C.C. MacDuffee apparently was the first to point out, about 1959, that a full-rank factorization of a matrix A leads to an explicit formula for its Moore–Penrose inverse, $A^\dagger$. However, he did so in private communications, and there is no published work that can be cited.

THEOREM 5   (MacDuffee).   If $A \in C_r^{m \times n}$, $r > 0$, has the full-rank factorization

$$A = FG, \tag{22}$$

then

$$A^\dagger = G^*(F^*AG^*)^{-1}F^*. \tag{23}$$

PROOF.   First, we must show that F^*AG^* is nonsingular. By (22),

$$F^*AG^* = (F^*F)(GG^*), \tag{24}$$

and both factors of the right member are $r \times r$ matrices. Also by Lemma 4, both are of rank r. Thus, F^*AG^* is the product of two nonsingular matrices, and therefore nonsingular. Moreover, (24) gives

$$(F^*AG^*)^{-1} = (GG^*)^{-1}(F^*F)^{-1}.$$

Denoting by X the right member of (23), we now have

$$X = G^*(GG^*)^{-1}(F^*F)^{-1}F^*, \tag{25}$$

and it is easily verified that this expression for X satisfies the Penrose equations (1)–(4). As $A^\dagger$ is the sole element of $A\{1,2,3,4\}$, (23) is therefore established. ∎

EXERCISES

13. Theorem 5 provides an alternative proof of the existence of the $\{1,2,3,4\}$-inverse (previously established by Theorem 4). However, Theorem 5 excludes the case $r=0$. Complete the alternative existence proof by showing that if $r=0$, (2) has a unique solution for X, and this X satisfies (1), (3), and (4).

14. Compute $A^\dagger$ for the matrix A of Exs. 6 and 7.

15. What is the most general $\{1,2\}$-inverse of the special matrix A of Ex. 4? What is its Moore–Penrose inverse?

16. Show that if $A = FG$ is a full-rank factorization, then

$$A^\dagger = G^\dagger F^\dagger.$$

17. Show that, for every matrix A,

(a) $(A^\dagger)^\dagger = A$

(b) $(A^*)^\dagger = (A^\dagger)^*$

(c) $(A^T)^\dagger = (A^\dagger)^T$

(d) $A^\dagger = (A^*A)^\dagger A^* = A^*(AA^*)^\dagger.$

18. If a and b are column vectors, then

(a) $a^\dagger = (a^*a)^\dagger a^*,$

(b) $(ab^*)^\dagger = (a^*a)^\dagger (b^*b)^\dagger ba^*.$

19. Show that if H is Hermitian and idempotent, $H^\dagger = H$.

20. Show that $H^\dagger = H$ if and only if H^2 is Hermitian and idempotent and rank $H^2 = $ rank H.

21. If $D = \text{diag}(d_1, d_2, \ldots, d_n)$, show that $D^\dagger = \text{diag}(d_1^\dagger, d_2^\dagger, \ldots, d_n^\dagger)$.

22. If U and V are unitary matrices, show that

$$(UAV)^\dagger = V^*A^\dagger U^*$$

for any matrix A for which the product UAV is defined.

9. CONSTRUCTION OF {2}-INVERSES OF PRESCRIBED RANK

Following the proof of Theorem 1, we described A. G. Fisher's construction of a {1}-inverse of a given $A \in C_r^{m \times n}$ having any prescribed rank between r and $\min(m,n)$, inclusive. From (2) it is easily deduced that

$$\text{rank } A^{(2)} \leqslant r.$$

We note also that the $n \times m$ null matrix is a {2}-inverse of rank 0, and any $A^{(1,2)}$ is a {2}-inverse of rank r, by Theorem 2. For $r > 1$, is there a construction analogous to Fisher's for a {2}-inverse of rank s for arbitrary s between 0 and r? Using the principle of full-rank factorization, we can readily answer the question in the affirmative.

Let $X_0 \in A\{1,2\}$ have a full-rank factorization

$$X_0 = YZ.$$

Then, $Y \in C_r^{m \times r}$ and $Z \in C_r^{r \times n}$, and (2) becomes

$$YZAYZ = YZ.$$

In view of Lemma 2, multiplication on the left by $Y^{(1)}$ and on the right by $Z^{(1)}$ gives (see Stewart [2])

$$ZAY = I_r. \tag{26}$$

Let Y_s denote the submatrix of Y consisting of the first s columns, and Z_s the submatrix of Z consisting of the first s rows. Then, both Y_s and Z_s are of full rank s, and it follows from (26) that

$$Z_s A Y_s = I_s. \tag{27}$$

Now, let

$$X_s = Y_s Z_s.$$

Then, rank $X_s = s$, by Ex. 8 and (27) gives

$$X_s A X_s = X_s.$$

EXERCISES

23. For

$$A = \begin{bmatrix} 1 & 0 & 0 & 1 \\ 1 & 1 & 0 & 0 \\ 0 & 1 & 1 & 0 \\ 0 & 0 & 1 & 1 \end{bmatrix},$$

find elements of $A\{2\}$ of ranks 1, 2, and 3, respectively.

24. With A as in Ex. 23, find a $\{2\}$-inverse of rank 2 having zero elements in the last two rows and the last two columns.

25. Show that there is at most one matrix X satisfying the three equations $AX = B$, $XA = D$, $XAX = X$ (Cline; see Cline and Greville [1]).

26. Let $A = FG$ be a full-rank factorization of $A \in C_r^{m \times n}$, i.e., $F \in C_r^{m \times r}$, $G \in C_r^{r \times n}$. Then

(a) $G^{(i)} F^{(1)} \in A\{i\}$ $(i = 1, 2, 4)$,

(b) $G^{(1)} F^{(j)} \in A\{j\}$ $(j = 1, 2, 3)$.

PROOF.   (a) $i = 1$:

$$FGG^{(1)} F^{(1)} FG = FG,$$

since

$$F^{(1)} F = GG^{(1)} = I_r$$

by Lemma 2.

  $i = 2$:

$$G^{(2)} F^{(1)} FGG^{(2)} F^{(1)} = G^{(2)} F^{(1)},$$

since

$$F^{(1)} F = I_r, \qquad G^{(2)} GG^{(2)} = G^{(2)}.$$

  $i = 4$:

$$G^{(4)} F^{(1)} FG = G^{(4)} G = (G^{(4)} G)^*.$$

(b) Similarly proved, with the roles of F and G interchanged.   ∎

27. Let A, F, G be as in Ex. 26. Then,

$$A^\dagger = G^\dagger F^{(1,3)} = G^{(1,4)} F^\dagger.$$

10. AN APPLICATION OF {2}-INVERSES IN ITERATIVE METHODS FOR SOLVING NONLINEAR EQUATIONS

One of the best-known methods for solving a single (nonlinear) equation in a single variable, say

$$f(x) = 0, \tag{28}$$

is *Newton's* (also *Newton–Raphson*) *method*

$$x_{k+1} = x_k - \frac{f(x_k)}{f'(x_k)} \qquad (k = 0, 1, \dots). \tag{29}$$

Under suitable conditions on the function f and the initial approximation x_0, the sequence (29) converges to a solution of (28); see, e.g., Ortega and Rheinboldt [1]. The *modified Newton method* uses the iteration

$$x_{k+1} = x_k - \frac{f(x_k)}{f'(x_0)} \qquad (k = 0, 1, \dots), \tag{30}$$

instead of (29).

Newton's method for solving a system of m equations in n variables

$$
\begin{aligned}
f_1(x_1, \dots, x_n) &= 0 \\
&\vdots \qquad\qquad \text{or} \quad f(x) = 0 \\
f_m(x_1, \dots, x_n) &= 0
\end{aligned}
\tag{31}
$$

is similarly given, for the case $m = n$, by

$$x_{k+1} = x_k - f'(x_k)^{-1} f(x_k) \qquad (k = 0, 1, \dots), \tag{32}$$

where $f'(x_k)$ is the *derivative of f at x_k*, represented by the matrix of partial derivatives (the *Jacobian* matrix)

$$f'(x_k) = \left[\frac{\partial f_i}{\partial x_j}(x_k) \right]. \tag{33}$$

The reader is referred to the excellent recent texts by Ortega and Rheinboldt [1] and Rall [1], for iterative methods in nonlinear analysis, and in particular, for the many variations and extensions of Newton's method (32).

If the nonsingularity of $f'(x_k)$ cannot be assumed for every x_k, and in particular, if the number of equations (31) is different from the number of unknowns, then it is natural to inquire whether a generalized inverse of $f'(x_k)$ can be used in (32), still resulting in a sequence converging to a solution of (31).

In this section we illustrate the use of $\{2\}$-inverses in a modified Newton method (Theorem 6 below) and in a Newton method (Ex. 28 below) for solving the nonlinear equations (31). Other applications of generalized inverses in the iterative methods of nonlinear analysis are Leach [1], Altman [1] and [2], Ben-Israel [2], [3], and [9], Rheinboldt ([1] especially Theorem 3.5), and Fletcher [1].

Readers not familiar with vector and matrix norms used in this section may consult Exs. 29–39 below for a brief introduction to norms.

Throughout this section we denote by $\| \ \|$ both a given (but arbitrary) vector norm in C^n, and a matrix norm in $C^{m \times n}$ consistent with it; see, e.g., Ex. 34. For a given point $x_0 \in C^n$ and a positive scalar r we denote by

$$B(x_0, r) = \left\{ x \in C^n : \|x - x_0\| < r \right\}$$

the *open ball with center* x_0 *and radius* r. The *closed ball* with the same center and radius is

$$\overline{B(x_0, r)} = \left\{ x \in C^n : \|x - x_0\| \leqslant r \right\}.$$

THEOREM 6.   Let the following be given:

$$x_0 \in C^n, \qquad r > 0,$$

$$f : B(x_0, r) \to C^m \text{ a function,}$$

$$A \in C^{m \times n}, \qquad T \in C^{n \times m} \text{ matrices,}$$

$$\epsilon > 0, \qquad \delta > 0 \text{ positive scalars,}$$

such that:

$$\|f(u) - f(v) - A(u - v)\| \leqslant \epsilon \|u - v\|$$

$$\text{for all } u, v \in B(x_0, r), \tag{34}$$

$$TAT = T, \tag{35}$$

$$\epsilon \|T\| = \delta < 1, \tag{36}$$

$$\|T\| \, \|f(x_0)\| < (1 - \delta) r. \tag{37}$$

Then the sequence

$$x_{k+1} = x_k - Tf(x_k) \qquad (38)$$

converges to a point

$$x_* \in \overline{B(x_0, r)} \qquad (39)$$

satisfying

$$Tf(x) = 0. \qquad (40)$$

PROOF.   Using induction on k we prove that the sequence (38) satisfies for $k = 0, 1, \ldots$

$$x_k \in B(x_0, r), \qquad (41.k)$$

$$\|x_{k+1} - x_k\| \leqslant \delta^k (1 - \delta) r. \qquad (42.k)$$

Now (42.0) and (41.1) follow from (37). Assuming (42.j) for $0 \leqslant j \leqslant k - 1$ we get

$$\|x_k - x_0\| \leqslant \sum_{j=0}^{k-1} \|x_{j+1} - x_j\| \leqslant (1 - \delta) r \sum_{j=0}^{k-1} \delta^j = (1 - \delta^k) r,$$

which proves (41.k). To prove (42.k) we write

$$x_{k+1} - x_k = -Tf(x_k)$$
$$= -Tf(x_{k-1}) - T[f(x_k) - f(x_{k-1})]$$
$$= T[A(x_k - x_{k-1}) - f(x_k) + f(x_{k-1})],$$

$$\text{by (35) and (38)}.$$

From (34) and (36) it therefore follows that

$$\|x_{k+1} - x_k\| \leqslant \|T\| \|f(x_k) - f(x_{k-1}) - A(x_k - x_{k-1})\|$$

$$\leqslant \delta \|x_k - x_{k-1}\|,$$

proving (42.k).   ∎

Remarks.   (a) f is *differentiable at* x_0 and the linear transformation A is its *derivative at* x_0, if

$$\lim_{x \to x_0} \frac{\|f(x) - f(x_0) - A(x - x_0)\|}{\|x - x_0\|} = 0.$$

Comparing this with (34) we conclude that the linear transformation A in Theorem 6 is an "approximate derivative," and can be chosen as the derivative of f at x_0 if f is continuously differentiable in $B(x_0,r)$.

(b) The limit x_* of the sequence (38) is a solution of (40), but in general not of (31), unless T is of full column rank in which case (40) and (31) are equivalent. Thus, the choice of the $\{2\}$-inverse T in Theorem 6, which by Section 9 can have any rank between 0 and rank A, will determine the extent to which x_* can be called a solution of (31). The "worst" choice of T is the trivial choice $T = O$, in which case any x is a solution of (40) and the iteration (38) stops at x_0.

(c) For any nontrivial T, the inequality (37) shows that x_0 is an approximate solution of (31),

$$\|f(x_0)\| < \frac{(1-\delta)r}{\|T\|}.$$

(d) Note that (34) needs to hold only for $u,v \in B(x_0,r)$ such that $u - v \in R(T)$, and that the limit x_* of (38) lies in

$$\overline{B(x_0,r)} \cap \{x_0 + R(T)\}.$$

EXERCISES

28. *A Newton's method using $\{2\}$-inverses.* Let the following be given:

$$x_0 \in C^n, \qquad r > 0,$$

$$f: B(x_0,r) \to C^m \qquad \text{a function,}$$

$$\epsilon > 0, \qquad \delta > 0, \qquad \eta > 0 \qquad \text{positive scalars,}$$

and for any $x \in \overline{B(x_0,r)}$ let

$$A_x \in C^{m \times n}, \qquad T_x \in C^{n \times m}$$

be matrices satisfying for all $u,v \in B(x_0,r)$:

$$\|f(u) - f(v) - A_v(u-v)\| \leqslant \epsilon\|u-v\|, \qquad (43)$$

$$T_u A_u T_u = T_u, \qquad (44)$$

$$\|(T_u - T_v)f(v)\| \leqslant \eta\|u-v\|, \qquad (45)$$

$$\epsilon\|T_u\| + \eta \leqslant \delta < 1, \qquad (46)$$

and

$$\| T_{x_0} \| \, \| f(x_0) \| < (1-\delta)r. \tag{47}$$

Then the sequence

$$x_{k+1} = x_k - T_{x_k} f(x_k) \qquad (k=0,1,\dots) \tag{48}$$

converges to a point

$$x_* \in \overline{B(x_0,r)} \tag{39}$$

which is a solution of

$$T_{x_*} f(x) = 0. \tag{49}$$

PROOF. As in the proof of Theorem 6 we use induction on k to prove that the sequence (48) satisfies

$$x_k \in B(x_0,r), \tag{41.k}$$

$$\| x_{k+1} - x_k \| \leqslant \delta^k (1-\delta)r. \tag{42.k}$$

Again (42.0) and (41.1) follow from (47), and assuming (42.j) for $0 \leqslant j \leqslant k-1$ we get (41.k). To prove (42.k) we write

$$x_{k+1} - x_k = - T_{x_k} f(x_k)$$

$$= x_k - x_{k-1} - T_{x_k} f(x_k) + T_{x_{k-1}} f(x_{k-1}), \qquad \text{by (48)},$$

$$= T_{x_{k-1}} A_{x_{k-1}} (x_k - x_{k-1}) - T_{x_k} f(x_k) + T_{x_{k-1}} f(x_{k-1}),$$

since $TAT = T$ implies $TAx = x$ for every $x \in R(T)$

$$= T_{x_{k-1}} [A_{x_{k-1}} (x_k - x_{k-1}) - f(x_k) + f(x_{k-1})]$$

$$+ (T_{x_{k-1}} - T_{x_k}) f(x_k).$$

Therefore

$$\| x_{k+1} - x_k \| \leqslant (\epsilon \| T_{x_{k-1}} \| + \eta) \| x_k - x_{k-1} \|, \qquad \text{by (43) and (45)},$$

$$\leqslant \delta \| x_k - x_{k-1} \|, \qquad \text{by (46)},$$

which proves (42.k). ∎

29. *Vector norms.* A real valued function, $\| \ \|$, on C^n is called a

(*vector*) *norm* on C^n if it satisfies

$$\|x\| \geqslant 0, \qquad \|x\| = 0 \text{ only if } x = 0, \tag{N1}$$

$$\|\alpha x\| = |\alpha| \, \|x\|, \tag{N2}$$

$$\|x + y\| \leqslant \|x\| + \|y\|, \tag{N3}$$

for all $x, y \in C^n$, $\alpha \in C$.

Let $\| \ \|_{(1)}, \| \ \|_{(2)}$ be two norms on C^n and let α_1, α_2 be positive scalars. Show that the following functions

$$\text{(a)} \qquad \max\{\|x\|_1, \|x\|_2\},$$

$$\text{(b)} \qquad \alpha_1\|x\|_{(1)} + \alpha_2\|x\|_{(2)}$$

are norms on C^n.

30. *The l_p-norms.* For any $p \geqslant 1$ the function

$$\|x\|_p = \left(\sum_{j=1}^{n} |x_j|^p \right)^{1/p} \tag{50.p}$$

is a norm on C^n, called the l_p-*norm*.

Hint. The statement that (50.p) satisfies ($N3$) for $p \geqslant 1$ is the classical Minkowski's inequality; see, e.g., Beckenbach and Bellman [1].

31. The most popular l_p-norms are the choices $p = 1, 2$, and ∞,

$$\|x\|_1 = \sum_{j=1}^{n} |x_j|, \qquad \text{the } l_1\text{-}norm, \tag{50.1}$$

$$\|x\|_2 = \left(\sum_{j=1}^{n} |x_j|^2 \right)^{1/2} = (x^*x)^{1/2}, \quad \text{the } l_2\text{-}norm \text{ or the } Euclidean \ norm, \tag{50.2}$$

$$\|x\|_\infty = \max\{|x_j| : j = 1, \ldots, n\},$$

$$\text{the } l_\infty\text{-}norm \text{ or the } Tchebycheff \ norm. \tag{50.∞}$$

Is $\|x\|_\infty = \lim_{p \to \infty} \|x\|_p$?

32. Let $\| \ \|_{(1)}, \| \ \|_{(2)}$ be any two norms on C^n. Show that there exist positive scalars α, β such that

$$\alpha\|x\|_{(1)} \leqslant \|x\|_{(2)} \leqslant \beta\|x\|_{(1)} \tag{51}$$

for all $x \in C^n$.

Hint.

$$\alpha = \inf \{ \|x\|_{(2)} : \|x\|_{(1)} = 1 \}$$

and

$$\beta = \sup \{ \|x\|_{(2)} : \|x\|_{(1)} = 1 \}.$$

Remark. Two norms, $\| \ \|_{(1)}$ and $\| \ \|_{(2)}$ are called *equivalent* if there exist positive scalars α, β such that (51) holds for all $x \in C^n$. From Ex. 32, *any* two norms on C^n are equivalent. Therefore, if a sequence $\{x_k\} \subset C^n$ satisfies

$$\lim_{k \to \infty} \|x_k\| = 0 \tag{52}$$

for some norm, then (52) holds for any norm. Topological concepts like convergence and continuity, defined by limiting expressions like (52), are therefore independent of the norm used in their definition. Thus we say that a sequence $\{x_k\} \subset C^n$ *converges to a point x_** if

$$\lim_{k \to \infty} \|x_k - x_*\| = 0,$$

for some norm.

33. *Matrix norms.* A real valued function $\| \ \|$ on $C^{m \times n}$ is called a (*matrix*) *norm* if

$$\|A\| \geqslant 0, \qquad \|A\| = 0 \quad \text{only if} \quad A = O, \tag{M1}$$

$$\|\alpha A\| = |\alpha| \|A\|, \tag{M2}$$

$$\|A + B\| \leqslant \|A\| + \|B\|, \qquad \text{for all} \quad A, B \in C^{m \times n}, \qquad \alpha \in C. \tag{M3}$$

If, in addition,

$$\|AB\| \leqslant \|A\| \|B\| \tag{M4}$$

whenever the matrix product AB is defined, then $\| \ \|$ is called a *multiplicative norm*. Some authors (see, e.g., Householder [1] Section 2.2) define a matrix norm as a function having all four properties $(M1)$–$(M4)$.

Show that the functions

$$\left(\sum_{i=1}^{m} \sum_{j=1}^{n} |a_{ij}|^2 \right)^{1/2} = (\text{trace} \, A^* A)^{1/2} \tag{53}$$

and

$$\max \{ |a_{ij}| : i = 1, \ldots, m; j = 1, \ldots, n \} \tag{54}$$

are matrix norms. Which of these functions is a multiplicative norm?

34. *Consistent norms.* A vector norm $\| \ \|$ and a matrix norm $\| \ \|$ are called *consistent* if for any vector x and matrix A such that Ax is defined,

$$\|Ax\| \leqslant \|A\| \, \|x\|. \tag{55}$$

Given a vector norm $\| \ \|_*$, show that

$$\|A\|_* = \sup_{x \neq 0} \frac{\|Ax\|_*}{\|x\|_*} \tag{56}$$

is a multiplicative matrix norm consistent with $\|x\|_*$, and that any other matrix norm $\| \ \|$ consistent with $\|x\|_*$, satisfies

$$\|A\| \geqslant \|A\|_*, \qquad \text{for all } A. \tag{57}$$

The norm $\|A\|_*$, defined by (56), is called the *matrix norm corresponding to the vector norm* $\| \ \|_*$, or the *bound* of A with respect to $K = \{x : \|x\|_* \leqslant 1\}$; see, e.g., Householder [1] Section 2.2.

35. Show that (56) is the same as

$$\|A\|_* = \sup_{\|x\|_* \leqslant 1} \frac{\|Ax\|_*}{\|x\|_*} = \sup_{\|x\|_* = 1} \|Ax\|_*. \tag{58}$$

36. Given a multiplicative matrix norm $\| \ \|_*$, find a vector norm consistent with it.

37. Show that the matrix norm on $C^{m \times n}$, corresponding to the vector norm

$$\|x\|_1 = \sum_{j=1}^{n} |x_j| \tag{50.1}$$

is

$$\|A\|_1 = \max_{1 \leqslant j \leqslant n} \sum_{i=1}^{m} |a_{ij}|. \tag{59}$$

PROOF. Follows from (58) since for any $x \in C^n$

$$\|Ax\|_1 = \sum_{i=1}^{m} \left| \sum_{j=1}^{n} a_{ij} x_j \right| \leqslant \sum_{i=1}^{m} \sum_{j=1}^{n} |a_{ij}| \, |x_j|$$

$$\leqslant \sum_{j=1}^{n} |x_j| \sum_{i=1}^{m} |a_{ij}|$$

$$\leqslant \left(\max_{1 \leqslant j \leqslant n} \sum_{i=1}^{m} |a_{ij}| \right) (\|x\|_1)$$

with equality if x is the kth unit vector, where k is any j for which the maximum in (59) is attained

$$\sum_{i=1}^{m} |a_{ik}| = \max_{1 \leqslant j \leqslant n} \sum_{i=1}^{m} |a_{ij}|.$$

38. Show that the matrix norm on $C^{m \times n}$, corresponding to the vector norm

$$\|x\|_\infty = \max_{1 \leqslant j \leqslant n} |x_j|, \tag{50.∞}$$

is

$$\|A\|_\infty = \max_{1 \leqslant i \leqslant n} \sum_{j=1}^{n} |a_{ij}|. \tag{60}$$

39. Show that the matrix norm on $C^{m \times n}$, corresponding to the Euclidean norm

$$\|x\|_2 = \left(\sum_{j=1}^{n} |x_j|^2 \right)^{1/2}, \tag{50.2}$$

is

$$\|A\|_2 = \max\left\{ \sqrt{\lambda} : \lambda \text{ an eigenvalue of } A^*A \right\}. \tag{61}$$

Note that (61) is different from (53), which is the Euclidean norm of the mn-dimensional vector obtained by listing all components of A in some order. The norm $\| \quad \|_2$ given by (61) is called the *spectral norm*.

40. For any matrix norm $\| \quad \|$ on $C^{n \times n}$, consistent with some vector norm, the norm of the unit matrix satisfies

$$\|I_n\| \geqslant 1.$$

In particular, if $\| \quad \|_*$ is a matrix norm, computed by (56) from a corresponding vector norm, then

$$\|I\|_* = 1. \tag{62}$$

41. A matrix norm $\| \quad \|$ on $C^{m \times n}$ is called *unitarily invariant* if for any two unitary matrices $U \in C^{m \times m}$ and $V \in C^{n \times n}$

$$\|UAV\| = \|A\| \qquad \text{for all } A \in C^{m \times n}.$$

Show that the matrix norms (53) and (61) are unitarily invariant.

42. *Spectral radius.* The *spectral radius* $\rho(A)$ of a square matrix $A \in C^{n \times n}$ is the maximal value among the n moduli of the eigenvalues of A,

$$\rho(A) = \max\{\, |\lambda| : \lambda \text{ an eigenvalue of } A \,\}. \tag{63}$$

Let $\|\ \ \|$ be any multiplicative matrix norm on $C^{n \times n}$. Then for any $A \in C^{n \times n}$

$$\rho(A) \leqslant \|A\|. \tag{64}$$

PROOF. Let $\|\ \ \|$ denote both a given multiplicative matrix norm, and a vector norm consistent with it; see, e.g., Ex. 36. Then

$$Ax = \lambda x$$

implies

$$|\lambda|\, \|x\| = \|Ax\| \leqslant \|A\|\, \|x\|.$$

43. For any $A \in C^{n \times n}$ and any $\epsilon > 0$, there exists a multiplicative matrix norm $\|\ \ \|$ such that

$$\|A\| \leqslant \rho(A) + \epsilon \qquad \text{(Householder [1], p. 46)}.$$

44. If A is a square matrix,

$$\rho(A^k) = \rho^k(A), \qquad k = 0, 1, \ldots. \tag{65}$$

45. For any $A \in C^{m \times n}$, the spectral norm $\|\ \ \|_2$ of (61) equals

$$\|A\|_2 = \rho^{1/2}(A^*A) = \rho^{1/2}(AA^*). \tag{66}$$

In particular, if A is Hermitian then

$$\|A\|_2 = \rho(A). \tag{67}$$

In general the spectral norm $\|A\|_2$ and the spectral radius $\rho(A)$ may be quite apart; see, e.g., Noble [2], p. 430.

46. *Convergent matrices.* A square matrix A is called *convergent* if

$$A^k \to O \qquad \text{as} \qquad k \to \infty. \tag{68}$$

Show that $A \in C^{n \times n}$ is convergent if and only if

$$\rho(A) < 1. \tag{69}$$

PROOF. *If.* From (69) and Ex. 43 it follows that there exists a multiplicative matrix norm $\| \quad \|$ such that $\|A\| < 1$. Then

$$\|A^k\| \leqslant \|A\|^k \to 0 \qquad \text{as} \qquad k \to \infty,$$

proving (68).

Only if. If $\rho(A) \geqslant 1$, then, by (65), so is $\rho(A^k)$ for $k = 0, 1, \ldots$, contradicting (68). ■

47. A square matrix A is convergent if and only if the sequence of partial sums

$$S_k = I + A + A^2 + \cdots + A^k = \sum_{j=0}^{k} A^j$$

converges, in which case it converges to $(I - A)^{-1}$, i.e.,

$$(I - A)^{-1} = I + A + A^2 + \cdots = \sum_{j=0}^{\infty} A^j \qquad (\text{Householder } [1], \text{ p. 54}).$$

(70)

48. Let A be convergent. Then

$$(I + A)^{-1} = I - A + A^2 - \cdots = \sum_{j=0}^{\infty} (-1)^j A^j. \tag{71}$$

49. Let $A \in C^{n \times n}$ be nonsingular, and let $\| \quad \|$ be any multiplicative matrix norm. Then $A + B$ is nonsingular for any matrix B satisfying

$$\|B\| < \frac{1}{\|A^{-1}\|}. \tag{72}$$

PROOF.   From

$$A + B = A(I + A^{-1}B)$$

and Ex. 48, it follows that $A + B$ is nonsingular if $A^{-1}B$ is convergent which, by Ex. 42, is implied by

$$\|A^{-1}B\| < 1$$

and hence by

$$\|A^{-1}\| \|B\| < 1.$$

See also Exs. 4.56 and 6.23. ■

50. *Stein's Theorem.* A square matrix A is convergent if and only if there exists a positive definite matrix H such that $H - A^*HA$ is also positive definite (Stein [1], Taussky [2]).

SUGGESTED FURTHER READING

Section 2.   Rao [2], Sheffield [2].
Section 3.   Rao ([1], [4]).
Section 5.   Deutsch [1], Frame [1], Greville [8], Hartwig [1], Przeworska-Rolewicz and Rolewicz ([1], p. 96).
Section 6.   Hearon and Evans [1], Rao [4], Sibuya [1].
Section 7.   Hartwig [3].
Section 10.   See also Burmeister [1], Fletcher [3], Golub and Pereyra [1].

2
LINEAR SYSTEMS AND CHARACTERIZATION OF GENERALIZED INVERSES

1. SOLUTION OF LINEAR SYSTEMS

As already indicated in Section 3, Introduction, the principal application of {1}-inverses is to the solution of linear systems, where they are used in much the same way as ordinary inverses in the nonsingular case. The main result of this section is the following theorem of Penrose [1], to whom the proof is also due.

THEOREM 1. Let $A \in C^{m \times n}$, $B \in C^{p \times q}$, $D \in C^{m \times q}$. Then the matrix equation

$$AXB = D \tag{1}$$

is consistent if and only if for some $A^{(1)}, B^{(1)}$,

$$AA^{(1)}DB^{(1)}B = D, \tag{2}$$

in which case the general solution is

$$X = A^{(1)}DB^{(1)} + Y - A^{(1)}AYBB^{(1)} \tag{3}$$

for arbitrary $Y \in C^{n \times p}$.

PROOF. If (2) holds, then $X = A^{(1)}DB^{(1)}$ is a solution of (1). Conversely, if X is any solution of (1), then

$$D = AXB = AA^{(1)}AXBB^{(1)}B = AA^{(1)}DB^{(1)}B.$$

Moreover, it follows from (2) and the definition of $A^{(1)}$ and $B^{(1)}$ that every matrix X of the form (3) satisfies (1). On the other hand, let X be any solution of (1). Then, clearly

$$X = A^{(1)}DB^{(1)} + X - A^{(1)}AXBB^{(1)},$$

which is of the form (3). ■

The following characterization of the set $A\{1\}$ in terms of an arbitrary element $A^{(1)}$ of the set is due essentially to Bjerhammar [3].

COROLLARY 1.   Let $A \in C^{m \times n}$, $A^{(1)} \in A\{1\}$. Then

$$A\{1\} = \{A^{(1)} + Z - A^{(1)}AZAA^{(1)} : Z \in C^{n \times m}\}. \tag{4}$$

PROOF.   The set described in the right member of (4) is obtained by writing $Y = A^{(1)} + Z$ in the set of solutions of $AXA = A$ as given by Theorem 1. ■

Specializing Theorem 1 to ordinary systems of linear equations gives:

COROLLARY 2.   Let $A \in C^{m \times n}$, $b \in C^m$. Then the equation

$$Ax = b \tag{5}$$

is consistent if and only if for some $A^{(1)}$

$$AA^{(1)}b = b, \tag{6}$$

in which case the general solution of (5) is

$$x = A^{(1)}b + (I - A^{(1)}A)y \tag{7}$$

for arbitrary $y \in C^n$.

The following theorem appears in the doctoral dissertation of C. A. Rohde [1], who attributes it to R. C. Bose. It is an alternative characterization of $A\{1\}$.

THEOREM 2.   Let $A \in C^{m \times n}$, $X \in C^{n \times m}$. Then $X \in A\{1\}$ if and only if for all b such that $Ax = b$ is consistent, $x = Xb$ is a solution.

PROOF.   *If*: Let a_j denote the jth column of A. Then

$$Ax = a_j$$

is consistent, and Xa_j is a solution, i.e.,

$$AXa_j = a_j \qquad (j = 1, 2, \ldots, n).$$

Therefore

$$AXA = A.$$

Only if: This follows from (6). ∎

EXERCISES AND EXAMPLES

1. Show that the general solution of the system $Ax = b$, where A is the matrix A of Ex. 1.6 and

$$b = \begin{bmatrix} 14 + 5i \\ -15 + 3i \\ 10 - 15i \end{bmatrix}$$

can be written in the form

$$x = \begin{bmatrix} 0 \\ \frac{5}{2} - 7i \\ 0 \\ 5 - i \\ 0 \\ 0 \end{bmatrix} + \begin{bmatrix} 1 & 0 & 0 & 0 & 0 & 0 \\ 0 & 0 & -\frac{1}{2} & 0 & -1 + 2i & \frac{1}{2}i \\ 0 & 0 & 1 & 0 & 0 & 0 \\ 0 & 0 & 0 & 0 & -2 & -1 - i \\ 0 & 0 & 0 & 0 & 1 & 0 \\ 0 & 0 & 0 & 0 & 0 & 1 \end{bmatrix} \begin{bmatrix} y_1 \\ y_2 \\ y_3 \\ y_4 \\ y_5 \\ y_6 \end{bmatrix},$$

where $y_1, y_2, \ldots, y_6$ are arbitrary.

2. *Kronecker products.* The *Kronecker product* $A \otimes B$ of the two matrices $A = [a_{ij}] \in C^{m \times n}$, $B \in C^{p \times q}$ is the $mp \times nq$ matrix expressible in partitioned form as

$$A \otimes B = \begin{bmatrix} a_{11}B & a_{12}B & \cdots & a_{1n}B \\ a_{21}B & a_{22}B & \cdots & a_{2n}B \\ \cdot & \cdot & \cdots & \cdot \\ a_{m1}B & a_{m2}B & \cdots & a_{mn}B \end{bmatrix}.$$

The properties of this product (e.g., Marcus and Minc [1]) include

$$(A \otimes B)(P \otimes Q) = AP \otimes BQ \tag{8}$$

for every $A \in C^{m \times n}$, $B \in C^{p \times q}$, $P \in C^{n \times r}$, and $Q \in C^{q \times s}$,

$$(A \otimes B)^* = A^* \otimes B^*, \qquad (A \otimes B)^T = A^T \otimes B^T. \tag{9}$$

An important application of the Kronecker product is rewriting a matrix equation

$$AXB = D \tag{1}$$

as a vector equation. For any $X = [x_{ij}] \in C^{m \times n}$, let the vector $v(X) = [v_k] \in C^{mn}$ be the transpose of the row vector obtained by placing the rows of X end to end with the first row on the left and the last row on the right. In other words,

$$v_{n(i-1)+j} = x_{ij} \qquad (i = 1, 2, \ldots, m; j = 1, 2, \ldots, n).$$

The energetic reader should now verify that

$$v(AXB) = (A \otimes B^T) v(X). \tag{10}$$

By using (10), the matrix equation (1) can be rewritten as the vector equation

$$(A \otimes B^T) v(X) = v(D). \tag{11}$$

Theorem 1 must therefore be equivalent to Corollary 2 applied to the vector equation (11). To demonstrate this we need the following two results:

$$A^{(1)} \otimes B^{(1)} \in (A \otimes B)\{1\} \qquad \text{(follows from (8))}, \tag{12}$$

$$(A^{(1)})^T \in A^T\{1\}. \tag{13}$$

Now, (1) is consistent if and only if (11) is consistent, and the latter statement

$$\Leftrightarrow (A \otimes B^T)(A \otimes B^T)^{(1)} v(D) = v(D) \qquad \text{(by Corollary 2)}$$

$$\Leftrightarrow (A \otimes B^T)\left(A^{(1)} \otimes (B^{(1)})^T\right) v(D) = v(D) \qquad \text{(by (12), (13))}$$

$$\Leftrightarrow \left(AA^{(1)} \otimes (B^{(1)}B)^T\right) v(D) = v(D) \qquad \text{(by (8))}$$

$$\Leftrightarrow AA^{(1)} DB^{(1)}B = D \qquad \text{(by (10))}.$$

The other statements of Theorem 1 can be shown similarly to follow from their counterparts in Corollary 2. The two results are thus equivalent.

3. $(A \otimes B)^{\dagger} = A^{\dagger} \otimes B^{\dagger}$ (Greville [4]).

PROOF.   Upon replacing A by $A \otimes B$ and X by $A^{\dagger} \otimes B^{\dagger}$ in (1.1)–(1.4) and making use of (8) and (9), it is easily verified that (1.1)–(1.4) are satisfied. ∎

4.   The matrix equations

$$AX = B, \quad XD = E \tag{14}$$

have a common solution if and only if each equation separately has a solution and

$$AE = BD.$$

PROOF   (Penrose [1]).   *If*: For any $A^{(1)}, D^{(1)}$,

$$X = A^{(1)}B + ED^{(1)} - A^{(1)}AED^{(1)}$$

is a common solution of both equations (14) provided $AE = BD$ and

$$AA^{(1)}B = B, \quad ED^{(1)}D = E.$$

By Theorem 1, the latter two equations are equivalent to the consistency of equations (14) considered separately.
    Only if: Obvious.   ∎

5.   Let equations (14) have a common solution $X_0 \in C^{m \times n}$. Then, show that the general common solution is

$$X = X_0 + (I - A^{(1)}A)Y(I - DD^{(1)}) \tag{15}$$

for arbitrary $A^{(1)} \in A\{1\}$, $D^{(1)} \in D\{1\}$, $Y \in C^{m \times n}$. [Hint: First, show that the right member of (15) is a common solution. Then, if X is any common solution, evaluate the right member of (15) for $Y = X - X_0$.]

2.   CHARACTERIZATION OF $A\{1,3\}$ AND $A\{1,4\}$

The set $A\{1\}$ is completely characterized in Corollary 1. Let us now turn our attention to $A\{1,3\}$. The key to its characterization is the following theorem.

THEOREM 3.   The set $A\{1,3\}$ consists of all solutions for X of

$$AX = AA^{(1,3)}, \tag{16}$$

where $A^{(1,3)}$ is an arbitrary element of $A\{1,3\}$.

PROOF.   If X satisfies (16), then clearly

$$AXA = AA^{(1,3)}A = A,$$

and, moreover, AX is Hermitian since $AA^{(1,3)}$ is Hermitian by definition. Thus, $X \in A\{1,3\}$.

On the other hand, if $X \in A\{1,3\}$, then

$$AA^{(1,3)} = AXAA^{(1,3)} = (AX)^*AA^{(1,3)} = X^*A^*(A^{(1,3)})^*A^*$$

$$= X^*A^* = AX,$$

where we have used Lemma 1.1(a). ∎

COROLLARY 3.   Let $A \in C^{m \times n}$, $A^{(1,3)} \in A\{1,3\}$. Then

$$A\{1,3\} = \{ A^{(1,3)} + (I - A^{(1,3)}A)Z : Z \in C^{n \times m} \}. \tag{17}$$

PROOF.   Applying Theorem 1 to (16) and substituting $Z + A^{(1,3)}$ for Y gives (17). ∎

The following theorem and its corollary are obtained in a manner analogous to the proofs of Theorem 3 and Corollary 3.

THEOREM 4.   The set $A\{1,4\}$ consists of all solutions for X of

$$XA = A^{(1,4)}A.$$

COROLLARY 4.   Let $A \in C^{m \times n}$, $A^{(1,4)} \in A\{1,4\}$. Then

$$A\{1,4\} = \{ A^{(1,4)} + Y(I - AA^{(1,4)}) : Y \in C^{n \times m} \}.$$

Other characterizations of $A\{1,3\}$ and $A\{1,4\}$ based on their least-squares properties will be given in Chapter 3.

EXERCISES

6. Prove Theorem 4 and Corollary 4.
7. If A is the matrix A of Ex. 1.6, show that $A\{1,3\}$ is the set of matrices

X of the form

$$X = \frac{1}{38}\begin{bmatrix} 0 & 0 & 0 \\ -10i & 3 & 9 \\ 0 & 0 & 0 \\ 2i & -12 & 2 \\ 0 & 0 & 0 \\ 0 & 0 & 0 \end{bmatrix}$$

$$+ \begin{bmatrix} 1 & 0 & 0 & 0 & 0 & 0 \\ 0 & 0 & -\frac{1}{2} & 0 & -1+2i & \frac{1}{2}i \\ 0 & 0 & 1 & 0 & 0 & 0 \\ 0 & 0 & 0 & 0 & -2 & -1-i \\ 0 & 0 & 0 & 0 & 1 & 0 \\ 0 & 0 & 0 & 0 & 0 & 1 \end{bmatrix} Z,$$

where Z is an arbitrary element of $C^{6\times3}$.

8. For the matrix A of Ex. 7 show that $A\{1,4\}$ is the set of matrices Y of the form

$$Y = \frac{1}{276}\begin{bmatrix} 0 & 0 & 0 \\ 0 & 20-18i & 42 \\ 0 & 10-9i & 21 \\ 0 & -29-9i & -9-27i \\ 0 & -2+4i & 24+30i \\ 0 & -29+30i & -36+3i \end{bmatrix} + Z\begin{bmatrix} 1 & -\frac{1}{3}i & -i \\ 0 & 0 & 0 \\ 0 & 0 & 0 \end{bmatrix},$$

where Z is an arbitrary element of $C^{6\times3}$.

9. Using Theorem 1.4 and the results of Exs. 7 and 8, calculate $A^{\dagger}$. (Since *any* $A^{(1,4)}$ and $A^{(1,3)}$ will do, choose the simplest.)

10. Give an alternative proof of Theorem 1.4, using Theorems 3 and 4. (Hint: Take $X = A^\dagger$).

11. By applying Ex. 5 show that if $A \in C^{m \times n}$ and $A^{(1,3,4)} \in A\{1,3,4\}$, then

$$A\{1,3,4\} = \left\{ A^{(1,3,4)} + (I - A^{(1,3,4)}A) Y (I - AA^{(1,3,4)}) : Y \in C^{n \times m} \right\}.$$

12. Show that if $A \in C^{m \times n}$, $A^{(1,2,3)} \in A\{1,2,3\}$, then

$$A\{1,2,3\} = \left\{ A^{(1,2,3)} + (I - A^{(1,2,3)}A) Z A^{(1,2,3)} : Z \in C^{n \times n} \right\}.$$

13. Similarly, show that if $A \in C^{m \times n}$, $A^{(1,2,4)} \in A\{1,2,4\}$, then

$$A\{1,2,4\} = \left\{ A^{(1,2,4)} + A^{(1,2,4)} Z (I - AA^{(1,2,4)}) : Z \in C^{m \times m} \right\}.$$

3. CHARACTERIZATION OF $A\{2\}$, $A\{1,2\}$ AND OTHER SUBSETS OF $A\{2\}$

Since

$$XAX = X \tag{1.2}$$

involves X nonlinearly, a characterization of $A\{2\}$ is not obtained by merely applying Theorem 1. However, such a characterization can be reached by utilizing a full-rank factorization of X. The rank of X will play an important role, and it will be convenient to let $A\{i,j,\dots,l\}_s$ denote the subset of $A\{i,j,\dots,l\}$ consisting of matrices of rank s.

We remark that the sets $A\{2\}_0$, $A\{2,3\}_0$, $A\{2,4\}_0$, and $A\{2,3,4\}_0$ are identical and contain a single element. For $A \in C^{m \times n}$ this sole element is the $n \times m$ matrix of zeros. Having thus disposed of the case of $s = 0$, we shall consider only positive s in the remainder of this section.

The following theorem has been stated by G. W. Stewart [2], who attributes it to R. E. Funderlic.

THEOREM 5. Let $A \in C_r^{m \times n}$ and $0 < s \leqslant r$. Then

$$A\{2\}_s = \left\{ YZ : Y \in C^{n \times s}, Z \in C^{s \times m}, ZAY = I_s \right\}. \tag{18}$$

PROOF. Let

$$X = YZ, \tag{19}$$

where the conditions on Y and Z in the right member of (18) are satisfied.

Then Y and Z are of rank s, and X is of rank s by Ex. 1.8. Moreover,

$$XAX = YZAYZ = YZ = X.$$

On the other hand, let $X \in A\{2\}_s$, and let (19) be a full-rank factorization. Then $Y \in C_s^{n \times s}$, $Z \in C_s^{s \times m}$ and

$$YZAYZ = YZ. \tag{20}$$

Moreover, if $Y^{(1)}$ and $Z^{(1)}$ are any $\{1\}$-inverses, then by Lemma 1.2

$$Y^{(1)}Y = ZZ^{(1)} = I_s.$$

Thus, multiplying (20) on the left by $Y^{(1)}$ and on the right by $Z^{(1)}$ gives

$$ZAY = I_s. \quad \blacksquare$$

COROLLARY 5. Let $A \in C_r^{m \times n}$. Then

$$A\{1,2\} = \left\{ YZ : Y \in C^{n \times r}, Z \in^{r \times m}, ZAY = I_r \right\}.$$

PROOF. By Theorem 1.2,

$$A\{1,2\} = A\{2\}_r. \quad \blacksquare$$

The relation $ZAY = I_s$ of (18) implies that $Z \in (AY)\{1,2,4\}$. This remark suggests the approach to the characterization of $A\{2,3\}$ on which the following theorem is based.

THEOREM 6. Let $A \in C_r^{m \times n}$ and $0 < s \leqslant r$. Then

$$A\{2,3\}_s = \left\{ Y(AY)^\dagger : AY \in C_s^{m \times s} \right\}.$$

PROOF. Let $X = Y(AY)^\dagger$, where $AY \in C_s^{m \times s}$. Then we have

$$AX = AY(AY)^\dagger. \tag{21}$$

The right member is Hermitian by (1.3), and

$$XAX = Y(AY)^\dagger AY(AY)^\dagger = Y(AY)^\dagger = X.$$

Thus, $X \in A\{2,3\}$. Finally, since $X \in A\{2\}$, $A \in X\{1\}$, and (21) and Lemma 1.1(f) give

$$s = \operatorname{rank} AY = \operatorname{rank} AX = \operatorname{rank} X.$$

On the other hand, let $X \in A\{2,3\}_s$. Then AX is Hermitian and idempotent, and is of rank s by Lemma 1.1(f), since $A \in X\{1\}$. By Ex. 1.19

$$(AX)^\dagger = AX,$$

and so

$$X(AX)^\dagger = XAX = X.$$

Thus X is of the form described in the theorem. ∎

The following theorem is proved in an analogous fashion.

THEOREM 7.   Let $A \in C_r^{m \times n}$ and $0 < s \leqslant r$. Then

$$A\{2,4\}_s = \left\{ (YA)^\dagger Y : YA \in C_s^{s \times n} \right\}.$$

EXERCISES AND EXAMPLES

14.   Could Theorem 6 be sharpened by replacing $(AY)^\dagger$ by $(AY)^{(i,j,k)}$ for some i, j, k? (Which properties are actually used in the proof?) Note that AY is of full column rank; what bearing, if any, does this have on the answer to the question?

15.   Show that if $A \in C_r^{m \times n}$,

$$A\{1,2,3\} = \left\{ Y(AY)^\dagger : AY \in C_r^{m \times r} \right\}$$

$$A\{1,2,4\} = \left\{ (YA)^\dagger Y : YA \in C_r^{r \times n} \right\}.$$

(Compare these results with Exs. 12 and 13.)

16.   The characterization of $A\{2,3,4\}$ is more difficult, and will be postponed until later in this chapter. Show, however, that if rank $A = 1$, $A\{2,3,4\}$ contains exactly two elements, $A^\dagger$ and O.

4.   IDEMPOTENT MATRICES AND PROJECTORS

A comparison of Eq. (1) of the Introduction with Lemma 1.1(f) suggests that the role played by the unit matrix in connection with the ordinary inverse of a nonsingular matrix is, in a sense, assumed by idempotent matrices in relation to generalized inverses. As the properties of idempotent matrices are likely to be treated in a cursory fashion in an introductory course in linear algebra, some of them are listed in the following lemma.

LEMMA 1.   Let $E \in C^{n \times n}$ be idempotent. Then:
(a) E^* and $I - E$ are idempotent.
(b)   The eigenvalues of E are 0 and 1. The multiplicity of the eigenvalue 1 is rank E.
(c)   rank $E =$ trace E.
(d) $E(I - E) = (I - E)E = O$.
(e) $Ex = x$ if and only if $x \in R(E)$.
(f) $E \in E\{1,2\}$.
(g) $N(E) = R(I - E)$.

PROOF.   Parts (a) to (f) are immediate consequences of the definition of idempotency: (c) follows from (b) and the fact that the trace of any square matrix is the sum of its eigenvalues counting multiplicities; (g) is obtained by applying Corollary 2 to the equation $Ex = 0$.   ∎

LEMMA 2   (Langenhop [1]).   Let a square matrix E have the full-rank factorization .

$$E = FG.$$

Then E is idempotent if and only if $GF = I$.

PROOF.   If $GF = I$, then clearly

$$(FG)^2 = FGFG = FG. \qquad (22)$$

On the other hand, since F is of full column rank and G is of full row rank,

$$F^{(1)}F = GG^{(1)} = I$$

by Lemma 1.2. Thus if (22) holds, multiplication on the left by $F^{(1)}$ and on the right by $G^{(1)}$ gives $GF = I$.   ∎

   For any two sets L, M in C^n, we shall define the *sum of L and M*, denoted by $L + M$, as

$$L + M = \{ y + z : y \in L, z \in M \}.$$

If L and M are subspaces of C^n, then $L + M$ is also a subspace of C^n.
   If, in addition, $L \cap M = \{0\}$, i.e., the only vector common to L and M is the zero vector, the reader will recall that the sum $L + M$ is called the *direct sum of L and M*, denoted by $L \oplus M$. It will be recalled also that two subspaces L and M of C^n are called *complementary* if

$$C^n = L \oplus M. \qquad (23)$$

   When this is the case (see Ex. 20 below), every $x \in C^n$ can be expressed uniquely as a sum

$$x = y + z \qquad (y \in L, z \in M). \tag{24}$$

We shall then call y the *projection of x on L along M.*

Let $P_{L,M}$ denote the transformation that carries any $x \in C^n$ into its projection on L along M. It is easily verified that this transformation is linear (see Ex. 21) and also idempotent (see Ex. 22). We shall call the transformation $P_{L,M}$ the *projector on L along M.*

It is well known (see, e.g., Halmos [1]) that every linear transformation from one finite-dimensional vector space to another can be represented by a matrix, which is uniquely determined by the linear transformation and by the choice of bases for the spaces involved. Except where otherwise specified, the basis for any finite-dimensional vector space, used in this book, is the standard basis of unit vectors. Having thus fixed the bases, there is a one-to-one correspondence between $C^{m \times n}$, the $m \times n$ *complex matrices*, and $\mathcal{L}(C^n, C^m)$, the *space of linear transformations mapping C^n into C^m.* This correspondence permits using the same symbol, say A, to denote both the linear transformation $A \in \mathcal{L}(C^n, C^m)$ and its matrix representation $A \in C^{m \times n}$. Thus the matrix–vector equation

$$Ax = y \qquad (A \in C^{m \times n}, x \in C^n, y \in C^m)$$

can equally be regarded as a statement that the linear transformation A maps x into y. The notion of a matrix as representing a linear transformation will be utilized in an important way in Chapter 6; see, e.g., Exs. 6.1–8.

In particular, linear transformations mapping C^n into itself are represented by the square matrices of order n. Specializing further, the next theorem establishes a one-to-one correspondence between the idempotent matrices of order n and the projectors $P_{L,M}$ where $L \oplus M = C^n$. Moreover, for any two complementary subspaces L and M, a method for computing $P_{L,M}$ is given by (29) below.

THEOREM 8. For every idempotent matrix $E \in C^{n \times n}$, $R(E)$ and $N(E)$ are complementary subspaces with

$$E = P_{R(E), N(E)}. \tag{25}$$

Conversely, if L and M are complementary subspaces, there is a unique idempotent $P_{L,M}$ such that $R(P_{L,M}) = L$, $N(P_{L,M}) = M$.

PROOF. Let E be idempotent of order n. Then it follows from Lemma 1(e) and 1(g) and from the equation

$$x = Ex + (I - E)x \tag{26}$$

that C^n is the sum of $R(E)$ and $N(E)$. Moreover, $R(E) \cap N(E) = \{0\}$, since

$$Ex = (I - E)y \Rightarrow Ex = E^2 x = E(I - E)y = 0,$$

by Lemma 1(d). Thus, $R(E)$ and $N(E)$ are complementary, and (26) shows that, for every x, Ex is the projection of x on $R(E)$ along $N(E)$. This establishes (25).

On the other hand let $\{x_1, x_2, \ldots, x_l\}$ and $\{y_1, y_2, \ldots, y_m\}$ be any bases for L and M, respectively. Then, $P_{L,M}$ if it exists, is uniquely determined by

$$\begin{cases} P_{L,M} x_i = x_i & (i = 1, 2, \ldots, l) \\ P_{L,M} y_i = 0 & (i = 1, 2, \ldots, m) \end{cases}. \tag{27}$$

Let $X = [x_1 \; x_2 \ldots x_l]$ denote the matrix whose columns are the vectors x_i. Similarly, let $Y = [y_1 \; y_2 \ldots y_m]$. Then (27) is equivalent to

$$P_{L,M}[X \; Y] = [X \; O]. \tag{28}$$

Since $[X \; Y]$ is nonsingular, the unique solution of (28), and therefore of (27), is

$$P_{L,M} = [X \; O][X \; Y]^{-1}. \tag{29}$$

Since (27) implies

$$P_{L,M}[X \; O] = [X \; O],$$

$P_{L,M}$ as given by (29) is clearly idempotent. ∎

The relation between the direct sum (23) and the projector* $P_{L,M}$ is given in the following.

COROLLARY 6.   Let L and M be complementary subspaces of C^n.

---

*Our use of the term "projector" to denote either the linear transformation $P_{L,M}$ or its idempotent matrix representation is not standard in the literature. Many writers have used "projection" in the same sense. The latter usage, however, seems to us to lead to undesirable ambiguity, since "projection" also describes the image $P_{L,M}x$ of the vector x under the transformation $P_{L,M}$. The use of "projection" in the sense of "image" is clearly much older (e.g., in elementary geometry) than its use in the sense of "transformation." "Projector" describes more accurately than "projection" what is meant here, and has been used in this sense by Afriat [1], de Boor [1], Bourbaki ([1], Ch. I. Def. 6, p. 16; [2], Ch. VIII, Section 1), Greville [1], Przeworska-Rolewicz and Rolewicz [1], Schwerdtfeger [1] and Ward, Boullion and Lewis [1]. Still other writers use "projector" to designate the orthogonal projector to be discussed in Section 6. This is true of Householder [1], Yosida [1], Kantorovich and Akilov [1], and numerous other Russian writers. We are indebted to de Boor for several of the preceding references.

Then, for every $x \in C^n$, the unique decomposition (24) is given by

$$P_{L,M}x = y, \qquad (I - P_{L,M})x = z.$$

If $A^{(1)} \in A\{1\}$, we know from Lemma 1.1(f) that both $AA^{(1)}$ and $A^{(1)}A$ are idempotent, and therefore are projectors. It is of interest to find out what we can say about the subspaces associated with these projectors. In fact, we already know from Ex. 1.10 that

$$R(AA^{(1)}) = R(A), \qquad N(A^{(1)}A) = N(A), \qquad R((A^{(1)}A)^*) = R(A^*).$$

$$(30)$$

The following is an immediate consequence of these results.

COROLLARY 7. If A and X are $\{1,2\}$-inverses of each other, AX is the projector on $R(A)$ along $N(X)$, and XA is projector on $R(X)$ along $N(A)$.

An important application of projectors is to the class of diagonable matrices. (The reader will recall that a square matrix is called *diagonable* if it is similar to a diagonal matrix.) It is easily verified (see Ex. 25) that a matrix $A \in C^{n \times n}$ is diagonable if and only if it has n linearly independent eigenvectors. The latter fact will be used in the proof of the following theorem, which expresses an arbitrary diagonable matrix as a linear combination of projectors.

THEOREM 9   (Spectral Theorem for Diagonable Matrices). Let $A \in C^{n \times n}$ with k distinct eigenvalues $\lambda_1, \lambda_2, \ldots, \lambda_k$. Then A is diagonable if and only if there exist projectors $E_1, E_2, \ldots, E_k$ such that

$$E_i E_j = O \qquad \text{if } i \neq j, \tag{31}$$

$$I_n = \sum_{i=1}^{k} E_i, \tag{32}$$

$$A = \sum_{i=1}^{k} \lambda_i E_i. \tag{33}$$

PROOF. *If:* For $i = 1, 2, \ldots, k$, let $r_i = \text{rank } E_i$ and let $X_i \in C^{n \times r_i}$ be a matrix whose columns are a basis for $R(E_i)$. Let

$$X = [X_1 X_2 \cdots X_k].$$

Then, by Lemma 1(c), the number of columns of X is

$$\sum_{i=1}^{k} r_i = \sum_{i=1}^{k} \text{trace } E_i = \text{trace} \sum_{i=1}^{k} E_i = \text{trace } I_n = n,$$

by (32). Thus X is square of order n. By the definition of X_i, there exists for each i a Y_i such that

$$E_i = X_i Y_i.$$

Let

$$Y = \begin{bmatrix} Y_1 \\ Y_2 \\ \vdots \\ Y_k \end{bmatrix}.$$

Then

$$XY = \sum_{i=1}^{k} X_i Y_i = \sum_{i=1}^{k} E_i = I_n,$$

by (32). Therefore X is nonsingular. By Lemma 1(e),

$$E_i X_i = X_i,$$

and therefore by (31) and (33),

$$AX = \sum_{i=1}^{k} \lambda_i E_i X = [\lambda_1 X_1 \quad \lambda_2 X_2 \cdots \lambda_k X_k]$$

$$= XD, \tag{34}$$

where

$$D = \operatorname{diag}(\lambda_1 I_1, \lambda_2 I_2, \ldots, \lambda_k I_k), \tag{35}$$

I_i being used to denote the unit matrix of order r_i. Since X is nonsingular, it follows from (34) that A and D are similar.

Only if: If A is diagonable,

$$AX = XD, \tag{36}$$

where X is nonsingular, and D can be expressed in the form (35). Let X then be partitioned by columns into $X_1, X_2, \ldots, X_k$ in conformity with the diagonal blocks of D, and, for $i = 1, 2, \ldots, k$, let

$$E_i = [O \ldots O \ X_i \ O \ldots O] X^{-1}.$$

In other words, $E_i = \tilde{X}_i X^{-1}$, where $\tilde{X}_i$ denotes the matrix obtained from X

by replacing all its columns except the columns of X_i by columns of zeros. It is then easily verified that E_i is idempotent, and that (31) and (32) hold. Finally,

$$\sum_{i=1}^{k} \lambda_i E_i = [\lambda_1 X_1 \ \ \lambda_2 X_2 \cdots \lambda_k X_k] X^{-1} = XDX^{-1} = A,$$

by (36). ∎

The idempotent matrices $\{E_i : i = 1, 2, \ldots, k\}$ (shown in Ex. 27 below to be uniquely determined by the diagonable matrix A) are called its *principal idempotents*. Relation (33) is called the *spectral decomposition* of A. Further properties of this decomposition are studied in Exs. 27–29.

Note that $R(E_i)$ is the *eigenspace* of A (space spanned by the eigenvectors) associated with the eigenvalue λ_i, while, because of (31), $N(E_i)$ is the direct sum of the eigenspaces associated with all eigenvalues of A other than λ_i.

EXERCISES AND EXAMPLES

17. Show that $I_n\{2\}$ consists of all idempotent matrices of order n.

18. If E is idempotent, $X \in E\{2\}$ and $R(X) \subset R(E)$, show that X is idempotent.

19. Let $E \in C_r^{n \times n}$. Then E is idempotent if and only if its Jordan canonical form can be written as

$$\begin{bmatrix} I_r & O \\ O & O \end{bmatrix},$$

where the null matrices are interpreted as absent if $r = n$.

20. *Direct sums.* Let L and M be subspaces of a vector space V. Then the following statements are equivalent:

(a) $V = L \oplus M$.

(b) Every vector $x \in V$ is uniquely represented as

$$x = y + z \qquad (y \in L, z \in M).$$

(c) $\dim V = \dim L + \dim M$, $L \cap M = \{0\}$.

(Here, dim means *dimension*.)

(d) If $\{x_1, x_2, \ldots, x_l\}$ and $\{y_1, y_2, \ldots, y_m\}$ are bases for L and M, respectively, then $\{x_1, x_2, \ldots, x_l, y_1, y_2, \ldots, y_m\}$ is a basis for V.

21. A transformation T from a vector space V to a vector space W is called *linear* if

$$T(\alpha x_1 + x_2) = \alpha T x_1 + T x_2$$

for every scalar α and all $x_1, x_2 \in V$.

Let L and M be complementary subspaces of C^n. Show that the projector $P_{L,M}$, which carries $x \in C^n$ into its projection on L along M, is a linear transformation (from C^n to L).

22. Let L and M be complementary subspaces of C^n, let $x \in C^n$, and let y be the projection of x on L along M. What is the unique expression for y as the sum of a vector in L and a vector in M? What, therefore, is $P_{L,M} y = P_{L,M}^2 x$, the projection of y on L along M? Show, therefore, that the transformation $P_{L,M}$ is idempotent.

23. Show that $P_{L,M} A = A$ if and only if $R(A) \subset L$ and $A P_{L,M} = A$ if and only if $N(A) \supset M$.

PROOF. The first part of the statement follows from Lemma 1(e). By Lemma 1(g), $R(I - P_{L,M}) = N(P_{L,M}) = M$. Thus, if $N(A) \supset M$,

$$O = A(I - P_{L,M}) = A - A P_{L,M}. \quad \blacksquare$$

24. $AB(AB)^{(1)}A = A$ if and only if rank $AB = $ rank A, and $B(AB)^{(1)}AB = B$ if and only if rank $AB = $ rank B. (*Hint:* Use Exs. 23, 1.10 and 1.11.)

25. A matrix $A \in C^{n \times n}$ is diagonable if and only if it has n linearly independent eigenvectors.

PROOF. Diagonability of A is equivalent to the existence of a nonsingular X such that $X^{-1}AX = D$, which in turn is equivalent to $AX = XD$. But the latter equation expresses the fact that each column of X is an eigenvector of A, and X is nonsingular if and only if its columns are linearly independent. $\quad \blacksquare$

26. Show that $I - P_{L,M} = P_{M,L}$.

27. *Principal idempotents.* Let $A \in C^{n \times n}$ be a diagonable matrix with k distinct eigenvalues $\lambda_1, \lambda_2, \dots, \lambda_k$. Then the idempotents $E_1, E_2, \dots, E_k$ satisfying (31)–(33) are uniquely determined by A.

PROOF. Let $\{F_i : i = 1, 2, \dots, k\}$ be any idempotent matrices satisfying

$$F_i F_j = O \quad \text{if } i \neq j, \tag{$31^\triangledown$}$$

$$I_n = \sum_{i=1}^k F_i, \tag{$32^\triangledown$}$$

$$A = \sum_{i=1}^k \lambda_i F_i. \tag{$33^\triangledown$}$$

From (31) and (33) it follows that

$$E_i A = A E_i = \lambda_i E_i \qquad (i = 1, 2, \ldots, k). \tag{37}$$

Similarly from $(31^{\triangledown})$ and $(33^{\triangledown})$

$$F_j A = A F_j = \lambda_j F_j \qquad (j = 1, 2, \ldots, k) \tag{$37^{\triangledown}$}$$

so that

$$E_i (A F_j) = \lambda_j E_i F_j$$

and

$$(E_i A) F_j = \lambda_i E_i F_j,$$

proving that

$$E_i F_j = O \qquad \text{if } i \neq j. \tag{38}$$

The uniqueness of $\{E_i : i = 1, 2, \ldots, k\}$ now follows:

$$E_i = E_i \sum_{j=1}^{k} F_j, \qquad \text{by } (32^{\triangledown})$$

$$= E_i F_i, \qquad \text{by } (38)$$

$$= \left(\sum_{j=1}^{k} E_j \right) F_i, \qquad \text{by } (38)$$

$$= F_i, \qquad \text{by } (32). \qquad \blacksquare$$

28. Let $A \in C^{n \times n}$ be a diagonable matrix with k distinct eigenvalues $\lambda_1, \lambda_2, \ldots, \lambda_k$. Then the principal idempotents of A are given by

$$E_i = \frac{p_i(A)}{p_i(\lambda_i)} \qquad (i = 1, 2, \ldots, k), \tag{39}$$

where

$$p_i(\lambda) = \prod_{\substack{j=1 \\ j \neq i}}^{k} (\lambda - \lambda_j). \tag{40}$$

PROOF. Let G_i $(i = 1, 2, \ldots, k)$ denote the right member of (39) and let

$E_1, E_2, \ldots, E_k$ be the principal idempotents of A. For any $i, j = 1, 2, \ldots, k$

$$G_i E_j = \frac{1}{p_i(\lambda_i)} \prod_{\substack{h=1 \\ h \neq i}}^{k} (A - \lambda_h I) E_j$$

$$= \frac{1}{p_i(\lambda_i)} \prod_{\substack{h=1 \\ h \neq i}}^{k} (\lambda_j - \lambda_h) E_j, \qquad \text{by (37)}$$

$$= \begin{cases} O & \text{if } i \neq j \\ E_i & \text{if } i = j \end{cases}.$$

Therefore, $G_i = G_i \sum_{j=1}^{k} E_j = E_i \; (i = 1, 2, \ldots, k)$. ∎

29. Let A be a diagonable matrix with k distinct eigenvalues $\lambda_1, \lambda_2, \ldots, \lambda_k$ and principal idempotents $E_1, E_2, \ldots, E_k$. Then:

(a) If $f(\lambda)$ is any polynomial,

$$f(A) = \sum_{i=1}^{k} f(\lambda_i) E_i.$$

(b) Any matrix commutes with A if and only if it commutes with every $E_i \; (i = 1, 2, \ldots, k)$.

PROOF. (a) Follows from (31), (32), and (33).

(b) Follows from (33) and (39) which express A as a linear combination of the $\{E_i : i = 1, 2, \ldots, k\}$ and each E_i as a polynomial in A. ∎

30. Prove the following analog of Theorem 5 for $\{1\}$-inverses: Let $A \in C_r^{m \times n}$ with $r < s \leqslant \min(m, n)$. Then

$$A\{1\}_s = \left\{ YZ : Y \in C_s^{n \times s}, \; Z \in C_s^{s \times m}, \; ZAY = \begin{bmatrix} I_r & O \\ O & O \end{bmatrix} \right\}. \quad (41)$$

PROOF. Let $X = YZ$, where the conditions on Y and Z in the right member of (41) are satisfied. Then rank $X = s$ by Ex. 1.8. Let

$$Y = [Y_1 \;\; Y_2], \qquad Z = \begin{bmatrix} Z_1 \\ Z_2 \end{bmatrix},$$

where Y_1 denotes the first r columns of Y and Z_1 the first r rows of Z.

Then (41) gives

$$Z_1 A Y_1 = I_r, \qquad Z_1 A Y_2 = O. \tag{42}$$

Let $X_1 = Y_1 Z_1$. Then it follows from the first equation (42) that $X_1 \in A\{2\}$. Since, by Ex. 1.8, rank $X_1 = r = $ rank A, $X_1 \in A\{1\}$ by Theorem 1.2. Thus

$$AXA = AX_1 AXA = AY_1 (Z_1 AY) ZA = AY_1 [I_r \quad O] \begin{bmatrix} Z_1 \\ Z_2 \end{bmatrix} A$$

$$= AY_1 Z_1 A = AX_1 A = A.$$

On the other hand, let $X \in A\{1\}_s$, and let $X = UV$ be a full-rank factorization. Then $U \in C_s^{n \times s}$, $V \in C_s^{s \times m}$, and

$$VAUVAU = VAU$$

and so VAU is idempotent, and is of rank r by Ex. 1.8. Thus, by Ex. 19, there is a nonsingular T such that

$$TVAUT^{-1} = \begin{bmatrix} I_r & O \\ O & O \end{bmatrix}.$$

If we now take

$$Y = UT^{-1}, \qquad Z = TV,$$

then

$$Y \in C_s^{n \times s}, \qquad Z \in C_s^{s \times m},$$

$$ZAY = \begin{bmatrix} I_r & O \\ O & O \end{bmatrix}, \qquad \text{and} \qquad YZ = UV = X. \qquad \blacksquare$$

5. GENERALIZED INVERSES WITH PRESCRIBED RANGE AND NULL SPACE

Let $A \in C^{m \times n}$ and let $A^{(1)}$ be an arbitrary element of $A\{1\}$. Let $R(A) = L$ and $N(A) = M$. By Lemma 1.1(f), $AA^{(1)}$ and $A^{(1)}A$ are idempotent. By (30) and Theorem 8,

$$AA^{(1)} = P_{L,S}, \qquad A^{(1)}A = P_{T,M},$$

where S is some subspace of C^m complementary to L, and T is some subspace of C^n complementary to M.

If we choose arbitrary subspaces S and T complementary to L and M, respectively, does there exist a $\{1\}$-inverse $A^{(1)}$ such that $N(AA^{(1)}) = S$ and $R(A^{(1)}A) = T$? The following theorem (parts of which have appeared previously in the work of Robinson [1], Langenhop [1], and Milne [1]) answers the question in the affirmative.

THEOREM 10. Let $A \in C_r^{m \times n}$, $R(A) = L$, $N(A) = M$, $L \oplus S = C^m$, and $M \oplus T = C^n$. Then:

(a) X is a $\{1\}$-inverse of A such that $N(AX) = S$ and $R(XA) = T$ if and only if

$$AX = P_{L,S}, \qquad XA = P_{T,M}. \tag{43}$$

(b) The general solution of (43) is

$$X = P_{T,M}A^{(1)}P_{L,S} + (I_n - A^{(1)}A)Y(I_m - AA^{(1)}), \tag{44}$$

where $A^{(1)}$ is a fixed (but arbitrary) element of $A\{1\}$ and Y is an arbitrary element of $C^{n \times m}$.

(c) $A_{T,S}^{(1,2)} = P_{T,M}A^{(1)}P_{L,S}$ is the unique $\{1,2\}$-inverse of A having range T and null space S.

PROOF. (a). The "if" part of the statement follows at once from Theorem 8 and Lemma 1(e), the "only if" part from Lemma 1.1(f), (30), and Theorem 8.

(b) By repeated use of Ex. 23, along with (30), we can easily verify that (43) is satisfied by $X = P_{T,M}A^{(1)}P_{L,S}$. The result then follows from Ex. 5.

(c) Since $P_{T,M}A^{(1)}P_{L,S}$ is a $\{1\}$-inverse of A, its rank is at least r by Lemma 1.1(d), while its rank does not exceed r, since rank $P_{L,S} = r$ by (43) and Lemma 1.1(f). Thus it has the same rank as A, and is therefore a $\{1,2\}$-inverse, by Theorem 1.2. It follows from parts (a) and (b) that it has the required range and null space.

On the other hand, a $\{1,2\}$-inverse of A having range T and null space S satisfies (43) and also

$$XAX = X. \tag{1.2}$$

By Ex. 1.25, these three equations have at most one common solution. ∎

COROLLARY 8.   Under the hypotheses of Theorem 10, let $A_{T,S}^{(1)}$ be some $\{1\}$-inverse of A such that $R(A_{T,S}^{(1)}A) = T$, $N(AA_{T,S}^{(1)}) = S$, and let $A\{1\}_{T,S}$ denote the class of such $\{1\}$-inverses of A. Then

$$A\{1\}_{T,S} = \left\{ A_{T,S}^{(1)} + \left(I_n - A_{T,S}^{(1)}A \right) Y \left(I_m - AA_{T,S}^{(1)} \right) : Y \in C^{n \times m} \right\}. \quad (45)$$

For a subspace L of C^m, a complementary space of particular interest is the orthogonal complement, denoted by $L^\perp$, which consists of all vectors in C^m orthogonal to L. If in Theorem 10 we take $S = L^\perp$ and $T = M^\perp$, the class of $\{1\}$-inverses given by (45) is the class of $\{1,3,4\}$-inverses, and $A_{T,S}^{(1,2)} = A^\dagger$.

The formulas in Theorem 10 generally are not convenient for computational purposes. When this is the case, the following theorem (which extends results due to Urquhart [1]) may be resorted to.

THEOREM 11.   Let $A \in C_r^{m \times n}$, $U \in C^{n \times p}$, $V \in C^{q \times m}$, and

$$X = U(VAU)^{(1)}V,$$

where $(VAU)^{(1)}$ is a fixed, but arbitrary element of $(VAU)\{1\}$. Then:
(a) $X \in A\{1\}$ if and only if rank $VAU = r$.
(b) $X \in A\{2\}$ and $R(X) = R(U)$ if and only if rank $VAU = $ rank U.
(c) $X \in A\{2\}$ and $N(X) = N(V)$ if and only if rank $VAU = $ rank V.
(d) $X = A_{R(U),N(V)}^{(1,2)}$ if and only if rank $U = $ rank $V = $ rank $VAU = r$.

PROOF OF (a).   If: We have rank $AU = r$, since

$$r = \text{rank } VAU \leqslant \text{rank } AU \leqslant \text{rank } A = r.$$

Therefore, by Ex. 1.11, $R(AU) = R(A)$, and so $A = AUY$ for some Y. Thus by Ex. 24,

$$AXA = AU(VAU)^{(1)}VAUY = AUY = A.$$

Only if: Since $X \in A\{1\}$,

$$A = AXAXA = AU(VAU)^{(1)}VAU(VAU)^{(1)}VA,$$

and therefore rank $VAU = $ rank $A = r$.

PROOF OF (b).   If: By Ex. 24,

$$XAU = U(VAU)^{(1)}VAU = U,$$

from which it follows that $XAX = X$, and also rank $X = $ rank U. By Ex. 1.11, $R(X) = R(U)$.

Only if: Since $X \in A\{2\}$,

$$X = XAX = U(VAU)^{(1)}VAU(VAU)^{(1)}V.$$

Therefore

$$\text{rank } X \leqslant \text{rank } VAU \leqslant \text{rank } U = \text{rank } X.$$

PROOF OF (*c*).   Similar to (b).

PROOF OF (*d*).   Follows from (a), (b), and (c). ■

Note that if we require only a $\{1\}$-inverse X such that $R(X) \subset R(U)$ and $N(X) \supset N(V)$, part (a) of the theorem is sufficient.

Theorem 11 can be used to prove the following modified analog of Theorem 10(c) for all $\{2\}$-inverses, and not merely $\{1,2\}$-inverses.

THEOREM 12.   Let $A \in C_r^{m \times n}$, let T be a subspace of C^n of dimension $s \leqslant r$, and let S be a subspace of C^m of dimension $m - s$. Then, A has a $\{2\}$-inverse X such that $R(X) = T$ and $N(X) = S$ if and only if

$$AT \oplus S = C^m, \tag{46}$$

in which case X is unique.

PROOF.   *If*: Let the columns of $U \in C_s^{n \times s}$ be a basis for T, and let the columns of $V^* \in C_s^{m \times s}$ be a basis for $S^{\perp}$. Then the columns of AU span AT. Since it follows from (46) that $\dim AT = s$,

$$\text{rank } AU = s. \tag{47}$$

A further consequence of (46) is

$$AT \cap S = \{0\}. \tag{48}$$

Moreover, the $s \times s$ matrix VAU is nonsingular (i.e., of rank s) because

$$VAUy = 0 \Rightarrow AUy \perp S^{\perp} \Rightarrow AUy \in S$$

$$\Rightarrow AUy = 0 \quad (\text{by } (48))$$

$$\Rightarrow y = 0 \quad (\text{by } (47)).$$

Therefore, by Theorem 11,

$$X = U(VAU)^{-1}V$$

is a $\{2\}$-inverse of A having range T and null space S (see also Stewart [2]).

Only if: Since $A \in X\{1\}$, AX is idempotent by Lemma 1.1(f). Moreover,

$AT = R(AX)$ and $S = N(X) = N(AX)$ by (30). Thus (46) follows from Theorem 8.

Proof of uniqueness: Let X_1, X_2 be $\{2\}$-inverses of A having range T and null space S. By Lemma 1.1(f) and (30), $X_1 A$ is a projector with range T and AX_2 is a projector with null space S. Thus, by Ex. 23,

$$X_2 = (X_1 A) X_2 = X_1 (AX_2) = X_1. \quad \blacksquare$$

COROLLARY 9. Let $A \in C_r^{m \times n}$, let T be a subspace of C^n of dimension r, and let S be a subspace of C^m of dimension $m - r$. Then, the following three statements are equivalent:

(a) $AT \oplus S = C^m$.

(b) $R(A) \oplus S = C^m$ and $N(A) \oplus T = C^n$.

(c)   There exists an $X \in A\{1,2\}$ such that $R(X) = T$ and $N(X) = S$.

EXERCISES

31. Show that $A_{T,S}^{(1,2)}$ is the unique matrix X satisfying the three equations

$$AX = P_{L,S}, \qquad XA = P_{T,M}, \qquad XP_{L,S} = X.$$

(For the Moore–Penrose inverse this was shown by Petryshyn [2]. Compare Ex. 1.25.)

32. For any given matrix A, $A^\dagger$ is the unique matrix $X \in A\{1,2\}$ such that $R(X) = R(A^*)$ and $N(X) = N(A^*)$.

33. Derive the formula of Zlobec [2],

$$A^\dagger = A^* Y A^*,$$

where Y is an arbitrary element of $(A^* A A^*)\{1\}$.

34. Penrose [1] showed that the Moore–Penrose inverse of a product of two Hermitian idempotent matrices is idempotent. Prove this, using Zlobec's formula (Ex. 33).

35. Let A be the matrix of Ex. 1.6, and let S be the subspace spanned by the vector

$$\begin{bmatrix} 0 \\ 0 \\ 1 \end{bmatrix}$$

and T the subspace spanned by the columns of

$$
\begin{bmatrix}
0 & 0 \\
0 & 0 \\
1 & 0 \\
0 & 1 \\
0 & 0 \\
0 & 0
\end{bmatrix}.
$$

Calculate $A_{T,S}^{(1,2)}$.

36. If E is idempotent and the columns of F and G^* are bases for $R(E)$ and $R(E^*)$, respectively, show that $E = F(GF)^{-1}G$.

37. If A is square and $A = FG$ is a full-rank factorization, show that A has a $\{1,2\}$-inverse X with $R(X) = R(A)$ and $N(X) = N(A)$ if and only if GF is nonsingular, in which case $X = F(GF)^{-2}G$ (Cline [4]).

6. ORTHOGONAL PROJECTIONS AND ORTHOGONAL PROJECTORS

Given a vector $x \in C^n$ and a subspace L of C^n, there is in L a unique vector u_x that is "closest" to x in the sense that the "distance" $\|x - u\|$ is smaller for $u = u_x$ than for any other $u \in L$. Here, $\|v\|$ denotes the *Euclidean norm* of the vector v, i.e., for $v, w \in C^n$,

$$
\|v\| = +\sqrt{(v,v)} = +\sqrt{v^*v} = +\sqrt{\sum_{j=1}^{n} |v_j|^2} \, ,
$$

and (v, w) denotes the *standard inner product* defined as

$$
(v, w) = w^*v = \sum_{j=1}^{n} \overline{w}_j v_j.
$$

Not surprisingly, the vector u_x that is "closest" to x of all vectors in L is uniquely characterized (see Ex. 39) by the fact that $x - u_x$ is orthogonal to u_x, which we shall denote by

$$
x - u_x \perp u_x.
$$

We shall therefore call the "closest" vector u_x the *orthogonal projection of x on L*.

The transformation that carries each $x \in C^n$ into its orthogonal projection on L we shall denote by P_L and shall call the *orthogonal projector on L* (sometimes abbreviated to "o.p. on L"). Comparison with the earlier definition of the projector on L along M (see Section 4) shows that the orthogonal projector on L is the same as the projector on L along $L^\perp$. (As previously noted, some writers call the orthogonal projector on L simply the projector on L.)

Being a particular case of the more general projector, the orthogonal projector is representable by a square matrix, which, in this case, is not only idempotent but also Hermitian.

In order to prove this, we shall need the relation

$$N(A) = R(A^*)^\perp, \tag{49}$$

which, in fact, arises frequently in the study of generalized inverses. Two proofs of (49) are given in Ex. 40. The first, using inner products, is immediately generalizable to transformations on Hilbert spaces, which will be discussed in Chapter 8. The second proof shows that, in the more restricted context of finite matrices, (49) is a consequence of the matrix equation $Ax = 0$, which defines $N(A)$.

Let L and M be complementary subspaces of C^n, and consider the matrix $P^*_{L,M}$. By Lemma 1(a), it is idempotent and therefore a projector, by Theorem 8. By the use of (49) and its dual

$$N(A^*) = R(A)^\perp \tag{50}$$

(obtained by replacing A by A^* in (49)), it is readily found that

$$R(P^*_{L,M}) = M^\perp, \qquad N(P^*_{L,M}) = L^\perp.$$

Thus, by Theorem 8,

$$P^*_{L,M} = P_{M^\perp, L^\perp}, \tag{51}$$

from which the next lemma follows easily.

LEMMA 3. Let $C^n = L \oplus M$. Then $M = L^\perp$ if and only if $P_{L,M}$ is Hermitian.

Just as there is a one-to-one correspondence between projectors and idempotent matrices, Lemma 3 shows that there is a one-to-one correspondence between orthogonal projectors and Hermitian idempotents. Matrices of the latter class have many striking properties, some of which are noted in the remainder of this section (including the exercises).

For any subspace L for which a basis is available, it is easy to construct the matrix P_L. The basis must first be orthonormalized (e.g., by Gram–Schmidt orthogonalization). Let $\{x_1, x_2, \ldots, x_l\}$ be an orthonormal basis for L. Then,

$$P_L = \sum_{j=1}^{l} x_j x_j^* \qquad (52)$$

The reader should verify that the right member of (52) is the o.p. on L, and that (29) reduces to (52) if $M = L^{\perp}$ and the basis is orthonormal.

In the preceding section diagonable matrices were studied in relation to projectors. The same relations will now be shown to hold between normal matrices (a subclass of diagonable matrices) and orthogonal projectors. This constitutes the *spectral theory for normal matrices*. We recall that a square matrix A is called *normal* if it commutes with its conjugate transpose

$$AA^* = A^*A .$$

It is well known that every normal matrix is diagonable. A normal matrix A also has the property (see Ex. 43) that the eigenvalues of A^* are the conjugates of those of A, and every eigenvector of A associated with the eigenvalue λ is also an eigenvector of A^* associated with the eigenvalue $\bar{\lambda}$.

The following spectral theorem relates normal matrices to orthogonal projectors, in the same way that diagonable matrices and projectors are related in Theorem 9.

THEOREM 13 (Spectral Theorem for Normal Matrices). Let $A \in C^{n \times n}$ with k distinct eigenvalues $\lambda_1, \lambda_2, \ldots, \lambda_k$. Then A is normal if and only if there exist orthogonal projectors $E_1, E_2, \ldots, E_k$ such that

$$E_i E_j = O \qquad \text{if } i \neq j, \qquad (53)$$

$$I_n = \sum_{i=1}^{k} E_i \qquad (54)$$

$$A = \sum_{i=1}^{k} \lambda_i E_i. \qquad (55)$$

PROOF. *If*: Let A be given by (55), where the principal idempotents E_i are Hermitian. Then,

$$AA^* = \left(\sum_{i=1}^{k} \lambda_i E_i \right) \left(\sum_{j=1}^{k} \bar{\lambda}_j E_j \right)$$

$$= \sum_{i=1}^{k} |\lambda_i|^2 E_i = A^*A.$$

Only if: Since A is normal, it is diagonable; let $E_1, E_2, \ldots, E_k$ be its principal idempotents. We must show that they are Hermitian. By Ex. 43, $R(E_i)$, the eigenspace of A associated with the eigenvalue λ_i is the same as the eigenspace of A^* associated with $\bar{\lambda}_i$. Because of (53), the null spaces of corresponding principal idempotents of A and A^* are also the same (for a given $i = h$, $N(E_h)$ is the direct sum of the eigenspaces $R(E_i)$ for all $i \neq h$, i.e.,

$$N(E_h) = \sum_{\substack{i=1 \\ i \neq h}}^{k} \oplus R(E_i) \qquad (h = 1, 2, \ldots, k).$$

Therefore, A and A^* have the same principal idempotents, by Theorem 8. Consequently,

$$A^* = \sum_{i=1}^{k} \bar{\lambda}_i E_i,$$

by Theorem 9. But taking conjugate transposes in (55) gives

$$A^* = \sum_{i=1}^{k} \bar{\lambda}_i E_i^*,$$

and it is easily seen that the idempotents E_i^* satisfy (53) and (54). Since the spectral decomposition is unique by Ex. 27, we must have

$$E_i = E_i^* \qquad (i = 1, 2, \ldots, k). \quad \blacksquare$$

EXERCISES AND EXAMPLES

38. *Orthogonal subspaces* or the *Pythagorean theorem*. Let Y and Z be subspaces of C^n. Then $Y \perp Z$ if and only if

$$\|y + z\|^2 = \|y\|^2 + \|z\|^2 \qquad \text{(for all } y \in Y, z \in Z). \tag{56}$$

PROOF. *If*: Let $y \in Y$, $z \in Z$. Then (56) implies that

$$(y,y) + (z,z) = \|y\|^2 + \|z\|^2 = \|y + z\|^2$$

$$= (y + z, y + z) = (y,y) + (z,z) + (y,z) + (z,y),$$

and therefore

$$(y,z) + (z,y) = 0. \tag{57}$$

Now, since Z is a subspace, $iz \in Z$, and replacing z by iz in (57) gives

$$0 = (y, iz) + (iz, y) = i(z, y) - i(y, z). \tag{58}$$

(Here we have used the fact that $(\alpha v, w) = \alpha(v, w)$ and $(v, \beta w) = \bar{\beta}(v, w)$.) It follows from (58) that

$$(y, z) - (z, y) = 0,$$

which, in conjunction with (57) gives

$$(y, z) = (z, y) = 0,$$

i.e., $y \perp z$.

Only if: Let $Y \perp Z$. Then, for arbitrary $y \in Y$, $z \in Z$

$$\|y + z\|^2 = (y + z, y + z)$$

$$= (y, y) + (z, z) \qquad (\text{since } (y, z) = (z, y) = 0)$$

$$= \|y\|^2 + \|z\|^2.$$

39. *Orthogonal projections.* Let L be a subspace of C^n. Then, for every $x \in C^n$ there is a unique vector u_x in L such that for all $u \in L$ different from u_x

$$\|x - u_x\| < \|x - u\|.$$

Among the vectors $u \in L$, u_x is uniquely characterized by the fact that

$$x - u_x \perp u_x.$$

PROOF. Let $x \in C^n$. Since L and $L^\perp$ are complementary subspaces, there exist uniquely determined vectors $x_1 \in L$, $x_2 \in L^\perp$ such that

$$x = x_1 + x_2. \tag{59}$$

Therefore for arbitrary $u \in L$,

$$\|x - u\|^2 = \|x_1 + x_2 - u\|^2$$

$$= \|x_1 - u\|^2 + \|x_2\|^2, \tag{60}$$

by Ex. 38, since $x_1 - u \in L$, $x_2 \in L^\perp$. Consequently, there is a unique $u \in L$, namely $u_x = x_1$, for which (60) is smallest.

By the uniqueness of the decomposition (59), $u_x = x_1$ is the only vector

$u \in L$ satisfying

$$x - u \perp u. \quad \blacksquare$$

40. $N(A) = R(A^*)^{\perp}$.

Because of the importance of this relation, we give two proofs, one in terms of inner products, and the other based on matrix multiplication.

FIRST PROOF.   Let $A \in C^{n \times n}$, and recall that for all $x \in C^n$, $y \in C^n$

$$(Ax,y) = (x, A^*y). \tag{61}$$

Let $x \in N(A)$. Then, the left member of (61) vanishes for all $y \in C^m$. From (61) it follows then that $x \perp A^*y$ for all $y \in C^m$, or, in other words, $x \perp R(A^*)$. This proves that $N(A) \subset R(A^*)^{\perp}$.

Conversely, let $x \in R(A^*)^{\perp}$, so that the right member of (61) vanishes for all $y \in C^m$. Then (61) implies that $Ax \perp y$ for all $y \in C^m$. Therefore $Ax = 0$. This proves that $R(A^*)^{\perp} \subset N(A)$, and completes the proof of the original relation.

SECOND PROOF.   By definition of $N(A)$, a vector $x \in C^n$ is in $N(A)$ if and only if $Ax = 0$. But this is equivalent to the statement that $x \in N(A)$ if and only if each row of A postmultiplied by x gives the product 0. Now, the rows of A are the conjugate transposes of the columns of A^*, and therefore $x \in N(A)$ if and only if it is orthogonal to every column of A^*. But this is equivalent to the statement that $x \in N(A)$ if and only if x is orthogonal to the space spanned by the columns of A^*, i.e., $x \perp R(A^*)$. $\blacksquare$

41.   Let $x \in C^n$ and let L be an arbitrary subspace of C^n. Then

$$\|P_L x\| \leqslant \|x\|, \tag{62}$$

with equality if and only if $x \in L$. See also Ex. 55.

PROOF.   We have

$$x = P_L x + (I - P_L)x = P_L x + P_{L^{\perp}} x,$$

by Ex. 26. Then by Ex. 38,

$$\|x\|^2 = \|P_L x\|^2 + \|P_{L^{\perp}} x\|^2,$$

from which (62) follows.

Equality holds in (62) if and only if $P_{L^{\perp}} x = 0$, which is equivalent to $x \in L$. $\blacksquare$

42.   Let A be a square singular matrix, let $\{u_1, \dots, u_n\}$ and $\{x_1, \dots, x_n\}$ be orthonormal bases of $N(A^*)$ and $N(A)$, respectively, and let $\{\alpha_1, \dots, \alpha_n\}$ be nonzero scalars. Then the matrix

$$A_0 = A + \sum_{i=1}^{n} \alpha_i u_i x_i^*$$

is nonsingular, and its inverse is

$$A_0^{-1} = A^\dagger + \sum_{i=1}^{n} \frac{1}{\alpha_i} x_i u_i^*.$$

PROOF. Let X denote the expression given for A_0^{-1}. Then, from $x_i^* x_j = \delta_{ij}$ $(i,j=1,\ldots,n)$, it follows that

$$A_0 X = AA^\dagger + \sum_{i=1}^{n} u_i u_i^*$$

$$= AA^\dagger + P_{N(A^*)} \quad \text{(by (52))}$$

$$= AA^\dagger + I_n - AA^\dagger \quad \text{(by Lemma 1(g))}$$

$$= I_n.$$

Therefore A_0 is nonsingular and $X = A_0^{-1}$. ■

43. If A is normal, $Ax = \lambda x$ if and only if $A^* x = \bar{\lambda} x$.

44. If L is a subspace of C^n and the columns of F are a basis for L, show that

$$P_L = FF^\dagger = F(F^*F)^{-1}F^*.$$

(This may be simpler computationally than orthonormalizing the basis and using (52).)

45. Let L be a subspace of C^n. Then

$$P_{L^\perp} = I_n - P_L.$$

(See Ex. 26.)

46. Let $A \in C^{m \times n}$, $X \in C^{n \times m}$. Then $X \in A\{2\}$ if and only if it is of the form

$$X = (EAF)^\dagger$$

where E and F are suitable Hermitian idempotents (Greville [8]).

PROOF.   *If*: By Ex. 32,

$$R\big((EAF)^\dagger\big) \subset R(F), \qquad N\big((EAF)^\dagger\big) \supset N(E).$$

Therefore, by Ex. 23,

$$X = (EAF)^\dagger = F(EAF)^\dagger = (EAF)^\dagger E.$$

Consequently,

$$XAX = (EAF)^\dagger EAF(EAF)^\dagger = (EAF)^\dagger = X.$$

Only if: By Theorem 10(c) and Ex. 32,

$$X^\dagger = P_{R(X^*)} A P_{R(X)},$$

and, therefore, by Ex. 1.17,

$$X = \big(P_{R(X^*)} A P_{R(X)}\big)^\dagger. \tag{63}$$

Remark.   Equation (63) can be described as stating that if $X \in A\{2\}$, then X is the Moore–Penrose inverse of a modification of A obtained by projecting its columns on $R(X^*)$ and its rows on $R(X)$.

47.   It follows from Exs. 31 and 1.25 that, for arbitrary A, $A^\dagger$ is the unique matrix X satisfying

$$AX = P_{R(A)}, \qquad XA = P_{R(A^*)}, \qquad XAX = X.$$

48.   By means of Exs. 47 and 23, derive (63) directly from $XAX = X$ without using Theorem 10(c).

49.   Prove the following amplification of Penrose's result stated in Ex. 34: A square matrix E is idempotent if and only if it can be expressed in the form

$$E = (FG)^\dagger$$

where F and G are Hermitian idempotents. (*Hint*: Use Ex. 17.)

In particular, derive the formula (Greville [8])

$$P_{L,M} = (P_{M^\perp} P_L)^\dagger = ((I - P_M) P_L)^\dagger. \tag{64}$$

50.   Let S and T be subspaces of C^m and C^n, respectively, such that

$$AT \oplus S = C^m,$$

and let $A_{T,S}^{(2)}$ denote the unique {2}-inverse of A having range T and null space S (see Theorem 12). Then

$$A_{T,S}^{(2)} = (P_{S^\perp} A P_T)^\dagger.$$

51. Show that $P_L + P_M$ is an o.p. if and only if $L \perp M$, in which case

$$P_L + P_M = P_{L+M}.$$

52. Show that $P_L P_M$ is an o.p. if and only if P_L and P_M commute, in which case $P_L P_M = P_{L \cap M}$.

53. Show that $L = L \cap M \oplus L \cap M^\perp$ if and only if P_L and P_M commute.

54. For a Hermitian matrix H we denote $H \geqslant 0$ the fact that H is *non-negative definite*, i.e., $(Hx, x) \geqslant 0$ for all x. For any two Hermitian matrices G, H, $G \geqslant H$ denotes $G - H \geqslant 0$. The relation $\geqslant$ is a partial order on the set of Hermitian matrices.

Let P_L and P_M be orthogonal projectors on the subspaces L and M of C^n, respectively. Then the following statements are equivalent:

(a) $P_L - P_M$ is an o.p.

(b) $P_L \geqslant P_M$.

(c) $\|P_L x\| \geqslant \|P_M x\|$ for all $x \in C^n$.

(d) $M \subset L$.

(e) $P_L P_M = P_M$.

(f) $P_M P_L = P_M$.

55. Let $P \in C^{n \times n}$ be a projector. Then P is an orthogonal projector if and only if

$$\|Px\| \leqslant \|x\| \text{ for all } x \in C^n. \tag{65}$$

PROOF. P is an o.p. if and only if $I - P$ is an o.p. By the equivalence of statements (a) and (c) in Ex. 54, $I - P$ is an o.p. if and only if (65) holds. ∎

Note that for any non-Hermitian idempotent P (i.e., for any projector P which is not an orthogonal projector) there is by this exercise a vector x whose length is increased when multiplied by P, i.e., $\|Px\| > \|x\|$. For

$$P = \begin{bmatrix} 1 & 1 \\ 0 & 0 \end{bmatrix}, \text{ such a vector is } x = \begin{bmatrix} 1 \\ 1 \end{bmatrix}.$$

56. Let $P \in C^{n \times n}$. Then P is an o.p. if and only if

$$P = P^* P.$$

57. It may be asked to what extent the results of Exs. 51–53 carry over to general projectors. This question is explored in this and the two following exercises. Let

$$C^n = L \oplus M = Q \oplus S.$$

Then show that $P_{L,M} + P_{Q,S}$ is a projector if and only if $M \supset Q$ and $S \supset L$, in which case

$$P_{L,M} + P_{Q,S} = P_{L+Q, M \cap S}.$$

Solution. Let $P_1 = P_{L,M}$, $P_2 = P_{Q,S}$. Then,

$$(P_1 + P_2)^2 = P_1 + P_2 + P_1 P_2 + P_2 P_1.$$

Therefore, $P_1 + P_2$ is a projector if and only if

$$P_1 P_2 + P_2 P_1 = O. \tag{66}$$

Now if $M \supset Q$ and $S \supset L$, each term of the left member of (66) is O.

On the other hand, if (66) holds, multiplication by P_1 on the left and on the right, respectively, gives

$$P_1 P_2 + P_1 P_2 P_1 = O = P_1 P_2 P_1 + P_2 P_1.$$

Subtraction then yields

$$P_1 P_2 - P_2 P_1 = O, \tag{67}$$

and (66) and (67) together imply

$$P_1 P_2 = P_2 P_1 = O,$$

from which it follows by Lemma 1(e) that $M \supset Q$ and $S \supset L$. It is then fairly easy to show that

$$P_1 + P_2 = P_{L+Q, M \cap S}.$$

58. With L, M, Q, S as in Ex. 57 show that if $P_{L,M}$ and $P_{Q,S}$ commute, then

$$P_{L,M} P_{Q,S} = P_{Q,S} P_{L,M} = P_{L \cap Q, M+S}. \tag{68}$$

59. If only one of the products in (68) is equal to the projector on the right, it does not necessarily follow that the other product is the same. Instead we have the following result: With L, M, Q, S as in Ex. 57, $P_{L,M}P_{Q,S} = P_{L \cap Q, M+S}$ if and only if $Q = L \cap Q \oplus M \cap Q$. Similarly, $P_{Q,S}P_{L,M} = P_{L \cap Q, M+S}$ if and only if $L = L \cap Q \oplus L \cap S$.

PROOF. Since $L \cap M = \{0\}$, $(L \cap Q) \cap (M \cap Q) = \{0\}$. Therefore $L \cap Q + M \cap Q = L \cap Q \oplus M \cap Q$. Since $M + S \supset M + Q$ and $L + S \supset L \cap Q$, Ex. 57 gives

$$P_{L \cap Q, M+S} + P_{M \cap Q, L+S} = P_{T,U},$$

where $T = L \cap Q \oplus M \cap Q$, $U = (L+S) \cap (M+S)$. Clearly $Q \supset T$ and $U \supset S$. Multiplying on the left by $P_{L,M}$ gives

$$P_{L,M}P_{T,U} = P_{L \cap Q, M+S}. \tag{69}$$

Thus, if $T = Q$, we have $U = S$, and

$$P_{L,M}P_{Q,S} = P_{L \cap Q, M+S}. \tag{70}$$

On the other hand, if (70) holds, (69) and (70) give

$$P_{Q,S} = P_{T,U} + H, \tag{71}$$

where $P_{L,M}H = O$. This implies $R(H) \subset M$. Also, since $T \subset Q$, (71) implies $R(H) \subset Q$, and therefore $R(H) \subset M \cap Q$. Consequently, $R(H) \subset T$ and therefore (71) gives $P_{T,U}P_{Q,S} = P_{Q,S}$. This implies rank $P_{Q,S} \leqslant$ rank $P_{T,U}$. Since $Q \supset T$ it follows that $T = Q$. This proves the first statement, and the proof of the second is similar. ∎

60. The characterization of $A\{2,3,4\}$ was postponed until o.p.'s had been studied. This will now be dealt with in three stages in this exercise and Exs. 61 and 63. If E is Hermitian idempotent, show that $X \in E\{2,3,4\}$ if and only if X is Hermitian idempotent and $R(X) \subset R(E)$.

PROOF. *If*: Since $R(X) \subset R(E)$, $EX = X$ by Lemma 1(e), and taking conjugate transposes gives $XE = X$. Since X is Hermitian, EX and XE are Hermitian. Finally, $XEX = X^2 = X$, since X is idempotent. Thus, $X \in E\{2,3,4\}$.

Only if: Let $X \in E\{2,3,4\}$. Then $X = XEX = EX^*X$. Therefore $R(X) \subset R(E)$. Then $EX = X$ by Lemma 1(e). But EX is Hermitian idempotent, since $X \in E\{2,3\}$. Therefore X is Hermitian idempotent. ∎

61. Let H be Hermitian non-negative definite, with spectral decomposition as in (33) with o.p.'s as its principal idempotents. Thus,

$$H = \sum_{i=1}^{k} \lambda_i E_i. \tag{72}$$

Then $X \in H\{2,3,4\}$ if and only if

$$X = \sum_{i=1}^{k} \lambda_i^\dagger F_i, \tag{73}$$

where, for each i, $F_i \in E_i\{2,3,4\}$.

PROOF. *If*: Since E_i is Hermitian idempotent, $R(F_i) \subset R(E_i)$ by Ex. 60. Therefore (31) gives

$$E_i F_j = F_j E_i = O \qquad (i \neq j), \tag{74}$$

and by Lemma 1(e)

$$E_i F_i = F_i E_i = F_i \qquad (i = 1, 2, \ldots, k).$$

Consequently,

$$HX = \sum_{i=1}^{k} {}' F_i = XH,$$

where $\sum'$ denotes the omission of that index i, if any, such that $\lambda_i = 0$. Since each F_i is Hermitian by Ex. 60, $HX = XH$ is Hermitian. Now,

$$F_i F_j = F_i E_j F_j = O \qquad (i \neq j)$$

by (74), and therefore

$$XHX = \sum_{i=1}^{k} \lambda_i^\dagger F_i^2 = X$$

by (73), since each F_i is idempotent.
  Only if: Let $X \in H\{2,3,4\}$. Then, by (32)

$$X = IXI = \sum_{i=1}^{k} \sum_{j=1}^{k} E_i X E_j. \tag{75}$$

Now, (72) gives

$$HX = \sum_{i=1}^{k} \lambda_i E_i X = \sum_{i=1}^{k} \lambda_i X^* E_i, \tag{76}$$

since $HX = X^*H$. Similarly,

$$XH = \sum_{i=1}^{k} \lambda_i XE_i = \sum_{i=1}^{k} \lambda_i E_i X^*. \tag{77}$$

Multiplying by E_s on the left and by E_t on the right in both (76) and (77) and making use of (31) and the idempotency of E_s and E_t gives

$$\lambda_s E_s XE_t = \lambda_t E_s X^* E_t \tag{78}$$

$$\lambda_t E_s XE_t = \lambda_s E_s X^* E_t, \qquad s,t = 1,2,\ldots,k. \tag{79}$$

Adding and subtracting (78) and (79) gives

$$(\lambda_s + \lambda_t) E_s XE_t = (\lambda_s + \lambda_t) E_s X^* E_t, \tag{80}$$

$$(\lambda_s - \lambda_t) E_s XE_t = -(\lambda_s - \lambda_t) E_s X^* E_t. \tag{81}$$

The λ_i are distinct, and are also non-negative because H is Hermitian non-negative definite. Thus, if $s \neq t$, neither of the quantities $\lambda_s + \lambda_t$ and $\lambda_s - \lambda_t$ vanishes. Therefore, (80) and (81) give

$$E_s XE_t = E_s X^* E_t = -E_s XE_t = O \qquad (s \neq t). \tag{82}$$

Consequently, (75) reduces to

$$X = \sum_{i=1}^{k} E_i XE_i. \tag{83}$$

Now, (72) gives

$$X = XHX = \sum_{i=1}^{k} \lambda_i XE_i X,$$

and therefore by (82)

$$E_s XE_s = \lambda_s E_s XE_s XE_s = \lambda_s (E_s XE_s)^2, \tag{84}$$

from which it follows that $E_s XE_s = O$ if $\lambda_s = 0$. Now, take

$$F_i = \lambda_i E_i XE_i \qquad (i = 1,2,\ldots,k). \tag{85}$$

Then (83) becomes (73), and we have only to show that $F_i \in E_i\{2,3,4\}$. This is trivially true for that i, if any, such that $\lambda_i = 0$. For other i, we deduce from (78) that F_i is Hermitian and from (84) that it is idempotent.

Finally, (85) gives $R(F_i) \subset R(E_i)$, and the desired conclusion follows from Ex. 60.

62. Prove the following corollary of Ex. 61. If H is Hermitian non-negative definite and $X \in H\{2,3,4\}$, then X is Hermitian non-negative definite, and every nonzero eigenvalue of X is the reciprocal of an eigenvalue of H.

63. For arbitrary $A \in C^{m \times n}$

$$A\{2,3,4\} = \{ YA^* : Y \in (A^*A)\{2,3,4\} \}.$$

64. $A\{2.3,4\}$ is a finite set if and only if the nonzero eigenvalues of A^*A are distinct (i.e., each eigenspace associated with a nonzero eigenvalue of A^*A is of dimension one). If this is the case and if there are k such eigenvalues, $A\{2,3,4\}$ contains exactly 2^k elements.

65. Show that the matrix

$$A = \frac{1}{10} \begin{bmatrix} 9-3i & 12-4i & 10-10i \\ 3-3i & 4-4i & 0 \\ 6+6i & 8+8i & 0 \\ 6 & 8 & 0 \end{bmatrix}$$

has exactly four $\{2,3,4\}$-inverses, namely,

$$X_1 = A^\dagger = \frac{1}{70} \begin{bmatrix} 0 & 6+6i & 12-12i & 12 \\ 0 & 8+8i & 16-16i & 16 \\ 35+35i & -5-15i & -30+10i & -20-10i \end{bmatrix},$$

$$X_2 = \frac{1}{60} \begin{bmatrix} -9-3i & 3+3i & 6-6i & 6 \\ -12-4i & 4+4i & 8-8i & 8 \\ 25+25i & -5-15i & -30+10i & -20-10i \end{bmatrix},$$

$$X_3 = \frac{1}{420} \begin{bmatrix} 63+21i & 15+15i & 30-30i & 30 \\ 84+28i & 20+20i & 40-40i & 40 \\ 35+35i & 5+15i & 30-10i & 20+10i \end{bmatrix},$$

$$X_4 = O.$$

7. EFFICIENT CHARACTERIZATION OF CLASSES OF GENERALIZED INVERSES

In the preceding sections of this chapter, characterizations of certain classes of generalized inverses of a given matrix have been given. Most of these characterizations involve one or more matrices with arbitrary elements. In general, the number of such arbitrary elements far exceeds the actual number of degrees of freedom available.

For example, in Section 1 we obtained the characterization

$$A\{1\} = \{A^{(1)} + Z - A^{(1)}AZAA^{(1)} : Z \in C^{n \times m}\}. \tag{4}$$

Now, as Z ranges over the entire class $C^{n \times m}$, every $\{1\}$-inverse of A will be obtained repeatedly an infinite number of times unless A is a matrix of zeros. In fact, the expression in the right member of (4) is unchanged if Z is replaced by $Z + A^{(1)}AWAA^{(1)}$, where W is an arbitrary element of $C^{n \times m}$. We shall now see how in some cases this redundancy in the number of arbitrary parameters can be eliminated. The cases of particular interest are $A\{1\}$ because of its role in the solution of linear systems, $A\{1,2\}$ because of the symmetry inherent in the relation

$$X \in A\{1,2\} \Leftrightarrow A \in X\{1,2\},$$

and $A\{1,3\}$ and $A\{1,4\}$ because of their minimization properties, which will be studied in the next chapter.

As in (4), let $A^{(1)}$ be a fixed, but arbitrary element of $A\{1\}$, where $A \in C_r^{m \times n}$. Also, let $F \in C_{n-r}^{n \times (n-r)}$, $K^* \in C_{m-r}^{m \times (m-r)}$, $B \in C_r^{n \times r}$ be given matrices whose columns are bases for $N(A)$, $N(A^*)$, and $R(A^{(1)}A)$, respectively. We shall show that the general solution of

$$AXA = A \tag{1.1}$$

is

$$X = A^{(1)} + FY + BZK, \tag{86}$$

where $Y \in C^{(n-r) \times m}$ and $Z \in C^{r \times (m-r)}$ are arbitrary.

Clearly $AF = O$ and $KA = O$. Therefore the right member of (86) satisfies (1.1). Since $R(I_n - A^{(1)}A) = N(A)$ and $R((I_m - AA^{(1)})^*) = N(A^*)$ by (30) and Lemma 1(g), there exist uniquely defined matrices G, H, D such that

$$FG = I_n - A^{(1)}A, \qquad HK = I_m - AA^{(1)}, \qquad BD = A^{(1)}A. \tag{87}$$

Since these products are idempotent, we have, by Lemma 2,

$$GF = DB = I_n, \qquad KH = I_m. \tag{88}$$

Moreover, it is easily verified that

$$GB = O, \qquad DF = O. \tag{89}$$

Using (88) and (89), we obtain easily from (86)

$$Y = G(X - A^{(1)}), \qquad Z = D(X - A^{(1)})H. \tag{90}$$

Now, let X be an arbitrary element of $A\{1\}$. Upon substituting in (86) the expressions (90) for Y and Z, it is found that (86) is satisfied. We have shown, therefore, that (86) does indeed give the general solution of (1.1).

We recall that $A^{(1)}, F, G, H, K, B, D$ are fixed matrices. Therefore, not only does (86) give X uniquely in terms of Y and Z, but also (90) gives Y and Z uniquely in terms of X. Therefore, different choices of Y and Z in (86) must yield different $\{1\}$-inverses X. Thus, the characterization (86) is completely efficient, and contains the smallest possible number of arbitrary parameters.

It is interesting to compare the number of arbitrary elements in the characterizations (4) and (86). In (4) this is mn, the number of elements of Z. In (86) it is $mn - r^2$, the total number of elements in Y and Z. Clearly (86) contains fewer arbitrary elements, except in the trivial case $r = 0$, as previously noted.

The case of $A\{1,3\}$ is easier. If, as before, the columns of F are a basis for $N(A)$, it is readily seen that (17) can be written in the alternative form

$$A\{1,3\} = \{A^{(1,3)} + FY : Y \in C^{(n-r) \times m}\}. \tag{91}$$

This is easily shown to be an efficient characterization. Here the number of arbitrary parameters is $m(n-r)$. Evidently this is less than the number in the efficient characterization (86) of $A\{1\}$, unless $r = m$, in which case every $\{1\}$-inverse is a $\{1,3\}$-inverse, since $AA^{(1)} = I_m$ by Lemma 1.2.

Similarly, if the columns of K^* are a basis for $N(A^*)$

$$A\{1,4\} = \{A^{(1,4)} + YK : Y \in C^{n \times (m-r)}\}, \tag{92}$$

where $A^{(1,4)}$ is a fixed, but arbitrary element of $A\{1,4\}$.

Efficient characterization of $A\{1,2\}$ is somewhat more difficult. Let $A^{(1,2)}$ be a fixed, but arbitrary element of $A\{1,2\}$, and let

$$A^{(1,2)} = Y_0 Z_0$$

be a full-rank factorization. As before, let the columns of F and K^* form bases for the null spaces of A and A^*, respectively. Then we shall show that

$$A\{1,2\} = \{ (Y_0 + FU)(Z_0 + VK) : U \in C^{(n-r) \times r}, V \in C^{r \times (m-r)} \}. \quad (93)$$

Indeed, it is easily seen that (1.1) and (1.2) are satisfied if X is taken as the product expression in the right member of (93). Moreover, if

$$FG = I_n - A^{(1,2)}A, \qquad HK = I_m - AA^{(1,2)},$$

it can be shown that

$$U = GXAY_0, \qquad V = Z_0AXH. \quad (94)$$

It is found that the product in the right member of (93) reduces to X if the expressions in (94) are substituted for U and V.

Relation (93) contains $r(m + n - 2r)$ arbitrary parameters. This is less than the number in the efficient characterization (86) of $A\{1\}$ by $(m-r)(n-r)$, which vanishes only if A is of full (row or column) rank, in which case every $\{1\}$-inverse is a $\{1,2\}$-inverse.

EXERCISES

66. In (87) obtain explicit formulas for G, H, and D in terms of A, $A^{(1)}$, F, K, B, and $\{1\}$-inverses of the latter three matrices.

67. Consider the problem of obtaining all $\{1\}$-inverses of the matrix A of Ex. 1.6. Note that the parametric representation of Ex. 1.7 does not give all $\{1\}$-inverses. (In this connection see Ex. 1.12.) Obtain in two ways parametric representations that do in fact give all $\{1\}$-inverses: first by (4) and then by (86). Note that a very simple $\{1\}$-inverse (in fact, a $\{1,2\}$-inverse) is obtained by taking all the arbitrary parameters equal to zero in the representation of Ex. 1.7. Verify that possible choices of F and K are

$$F = \begin{bmatrix} 1 & 0 & 0 & 0 \\ 0 & 1 & -1+2i & 0 \\ 0 & -2 & 0 & i \\ 0 & 0 & -2 & -1-i \\ 0 & 0 & 1 & 0 \\ 0 & 0 & 0 & 1 \end{bmatrix}, \qquad K = [3i \ \ 1 \ \ 3].$$

Compare the number of arbitrary parameters in the two representations.

68. Under the hypotheses of Theorem 10, let F and K^* be matrices whose columns are bases for $N(A)$ and $N(A^*)$, respectively. Then, (44) can be written in the alternative form

$$X = A_{T,S}^{(1,2)} + FZK, \tag{95}$$

where Z is an arbitrary element of $C^{(n-r)\times(m-r)}$. Moreover,

$$\text{rank } X = r + \text{rank } Z. \tag{96}$$

PROOF. Clearly the right member of (95) satisfies (43). On the other hand, substituting in (44) the first two equations (87) gives (95) with $Z = GYH$.

Moreover, (95) and Theorem 10(c) give

$$XP_{L,S} = A_{T,S}^{(1,2)},$$

and therefore

$$X(I_m - P_{L,S}) = FZK.$$

Consequently, $R(X)$ contains the range of each of the two terms of the right member of (95). Furthermore, the intersection of the latter two ranges is $\{0\}$, since $R(F) = N(A) = M$, which is a subspace complementary to $T = R(A_{T,S}^{(1,2)})$. Therefore, $R(X)$ is the direct sum of the two ranges mentioned, and, by statement (c) of Ex. 20, rank X is the sum of the ranks of the two terms of the right member of (95).

Now, the first term is a $\{1,2\}$-inverse of A, and its rank is therefore r by Theorem 1.2, while the rank of the second term is rank Z by Ex. 1.8. This establishes (96).

69. Exercise 68 gives

$$A\{1\}_{T,S} = \left\{ A_{T,S}^{(1,2)} + FZK : Z \in C^{(n-r)\times(m-r)} \right\},$$

where $A\{1\}_{T,S}$ is defined in Corollary 8. Show that this characterization is efficient.

70. Show that if $A \in C^{m\times n}$, $A\{1\}_{T,S}$ contains matrices of all ranks from r to $\min(m,n)$.

71. Let $A = ST$ be a full-rank factorization of $A \in C_r^{m\times n}$, let Y_0 and Z_0 be particular $\{1\}$-inverses of T and S, respectively, and let F and K be defined as in Ex. 68. Then, show that:

$$S\{1\} = \left\{ Z_0 + VK : V \in C^{r\times(m-r)} \right\},$$

$$T\{1\} = \left\{ Y_0 + FU : U \in C^{(n-r)\times r} \right\},$$

$$AA\{1\} = SS\{1\} = \{ S(Z_0 + VK) : V \in C^{r \times (m-r)} \},$$

$$A\{1\}A = T\{1\}T = \{ (Y_0 + FU)T : U \in C^{(n-r) \times r} \},$$

$$A\{1\} = \{ Y_0 Z_0 + Y_0 VK + FUZ_0 + FWK : U \in C^{(n-r) \times r},$$

$$V \in C^{r \times (m-r)} W \in C^{(n-r) \times (m-r)} \},$$

$$= A\{1,2\} + \{ FXK : X \in C^{(n-r) \times (m-r)} \}.$$

Show that all the preceding characterizations are efficient.

72. For the matrix A of Exs. 67 and 1.6, obtain all the characterizations of Ex. 71. *Hint*: Use the full-rank factorization of A given at the end of Section 1.7 and take

$$Z_0 = \begin{bmatrix} -\tfrac{1}{2}i & 0 & 0 \\ 0 & -\tfrac{1}{3} & 0 \end{bmatrix}.$$

8. RESTRICTED GENERALIZED INVERSES.

In a linear equation

$$Ax = b,$$

with given $A \in C^{m \times n}$ and $b \in C^m$, the points x are sometimes constrained to lie in a given subspace S of C^n, resulting in a "constrained" linear equation

$$Ax = b \qquad \text{and} \qquad x \in S. \tag{97}$$

In principle, this situation presents no difficulty since (97) is equivalent to the following, "unconstrained" but larger, linear system

$$\begin{bmatrix} A \\ P_{S^\perp} \end{bmatrix} x = \begin{bmatrix} b \\ 0 \end{bmatrix}, \qquad \text{where } P_{S^\perp} = I - P_S.$$

Another approach to the solution of (97) that does not increase the size of the problem is to interpret A as representing an element of $\mathcal{L}(S, C^m)$, the *space of linear transformations from S to C^m*, instead of an element of $\mathcal{L}(C^n, C^m)$; see, e.g., Sections 4 and 6.1. This interpretation calls for the following definitions.

Let $A \in \mathcal{L}(C^n, C^m)$, and let S be a subspace of C^n. The *restriction of A to S*, denoted by $A_{[S]}$, is the linear transformation from S to C^m defined by

$$A_{[S]}x = Ax, \qquad x \in S. \tag{98}$$

Conversely, let $B \in \mathcal{L}(S, C^m)$. The *extension of B to C^n*, denoted by ext B, is the linear transformation from C^n to C^m defined by

$$(\text{ext}\, B)x = \begin{cases} Bx & \text{if} \quad x \in S \\ 0 & \text{if} \quad x \in S^\perp \end{cases}. \tag{99}$$

Restricting an $A \in \mathcal{L}(C^n, C^m)$ to S and then extending to C^n results in $\text{ext}(A_{[S]}) \in L(C^n, C^m)$ given by

$$\text{ext}(A_{[S]})x = \begin{cases} Ax & \text{if} \quad x \in S \\ 0 & \text{if} \quad x \in S^\perp \end{cases}. \tag{100}$$

From (100) it should be clear that if $A \in \mathcal{L}(C^n, C^m)$ is represented by the matrix $A \in C^{m \times n}$, then $\text{ext}(A_{[S]})$ is represented by AP_S. The following lemma is then obvious.

LEMMA 4.   Let $A \in C^{m \times n}$, $b \in C^m$, and let S be a subspace of C^n. The system

$$Ax = b, \qquad x \in S \tag{97}$$

is consistent if and only if the system

$$AP_S z = b \tag{101}$$

is consistent, in which case x is a solution of (97) if and only if

$$x = P_S z,$$

where z is a solution of (101). ∎

From Lemma 4 and Corollary 2 it follows that the general solution of (97) is

$$x = P_S(AP_S)^{(1)}b + P_S\big(I - (AP_S)^{(1)}AP_S\big)y,$$

for arbitrary $(AP_S)^{(1)} \in (AP_S)\{1\}$ and $y \in C^n$. (102)

We are thus led to study generalized inverses of $\text{ext}(A_{[S]}) = AP_S$, and from (102) it appears that $P_S(AP_S)^{(1)}$, rather than $A^{(1)}$, plays the role of a $\{1\}$-inverse in solving the linear system (97); hence the following definition.

DEFINITION 1. Let $A \in C^{m \times n}$ and let S be a subspace of C^n. A matrix $X \in C^{n \times m}$ is an *S-restricted $\{i,j,\dots,l\}$-inverse of A* if

$$X = P_S(AP_S)^{(i,j,\dots,l)} \qquad (103)$$

for any $(AP_s)^{(i,j,\dots,l)} \in (AP_S)\{i,j,\dots,l\}$.

The role that S-restricted generalized inverses play in constrained problems is completely analogous to the role played by the corresponding generalized inverses in the unconstrained situation. Thus, for example, the following result is the constrained analog of Corollary 2.

COROLLARY 10.   Let $A \in C^{m \times n}$, $b \in C^m$ and let S be a subspace of C^n. Then the equation

$$Ax = b, \qquad x \in S \qquad (97)$$

is consistent if and only if

$$AXb = b,$$

where X is any S-restricted $\{1\}$-inverse of A. If consistent, the general solution of (97) is

$$x = Xb + (I - XA)y$$

with X as above, and arbitrary $y \in S$. ∎

EXERCISES

73.   Let I be the identity transformation in $\mathcal{L}(C^n, C^n)$ and let S be a subspace of C^n. Show that

$$\text{ext}(I_{[S]}) = P_S.$$

74.   Let $A \in \mathcal{L}(C^n, C^m)$. Show that $A_{[R(A^*)]}$, the restriction of A to $R(A^*)$, is a one-to-one mapping of $R(A^*)$ onto $R(A)$.

Solution.   We show first that $A_{[R(A^*)]}$ is one-to-one on $R(A^*)$. Clearly it suffices to show that A is one-to-one on $R(A^*)$. Let $u,v \in R(A^*)$ and suppose that $Au = Av$, i.e., u and v are mapped to the same point. Then

$A(u-v)=0$, i.e.,

$$u-v\in N(A).$$

But we also have

$$u-v\in R(A^*),$$

since u and v are in $R(A^*)$. Therefore

$$u-v\in N(A)\cap R(A^*)$$

and by (49), $u=v$, proving that A is one-to-one on $R(A^*)$.

We show next that $A_{[R(A^*)]}$ is a mapping onto $R(A)$, i.e., that

$$R(A_{[R(A^*)]})=R(A).$$

This follows since for any $x\in C^n$

$$Ax=AA^\dagger Ax=AP_{R(A^*)}x=A_{[R(A^*)]}x.$$

75.   Let $A\in C^{m\times n}$. Show that

$$\text{ext}(A_{[R(A^*)]})=A. \tag{104}$$

76.   From Ex. 74 it follows that the linear transformation

$$A_{[R(A^*)]}\in \mathcal{L}(R(A^*),R(A))$$

has an inverse

$$(A_{[R(A^*)]})^{-1}\in \mathcal{L}(R(A),R(A^*)).$$

Show that this inverse is the restriction of $A^\dagger$ to $R(A)$, namely

$$(A^\dagger)_{[R(A)]}=(A_{[R(A^*)]})^{-1}. \tag{105}$$

Solution.   From Exs. 74, 32, and 47 it follows that, for any $y\in R(A)$, $A^\dagger y$ is the unique element of $R(A^*)$ satisfying

$$Ax=y.$$

Therefore

$$A^\dagger y=(A_{[R(A^*)]})^{-1}y \qquad \text{for all } y\in R(A).$$

77.   Show that the extension of $(A_{[R(A^*)]})^{-1}$ to C^m is the Moore–Penrose inverse of A,

$$\text{ext}((A_{[R(A^*)]})^{-1})=A^\dagger. \tag{106}$$

Compare with (104).

78. Let each of the following two linear equations be consistent:

$$A_1x = b_1, \tag{107}$$

$$A_2x = b_2. \tag{108}$$

Show that (107) and (108) have a common solution if and only if the linear equation

$$A_2 P_{N(A_1)} y = b_2 - A_2 A_1^{(1)} b_1$$

is consistent, in which case the general common solution of (107) and (108) is

$$x = A_1^{(1)} b_1 + P_{N(A_1)} \left(A_2 P_{N(A_1)} \right)^{(1)} \left(b_2 - A_2 A_1^{(1)} b_1 \right) + N(A_1) \cap N(A_2)$$

or equivalently

$$x = A_2^{(1)} b_2 + P_{N(A_2)} \left(A_1 P_{N(A_2)} \right)^{(1)} \left(b_1 - A_1 A_2^{(1)} b_2 \right) + N(A_1) \cap N(A_2).$$

Hint. Substitute the general solution of (107)

$$x = A_1^{(1)} b_1 + P_{N(A_1)} y, \qquad y \text{ arbitrary,}$$

in (108).

79. Exercise 78 illustrates the need for $P_{N(A_1)}(A_2 P_{N(A_1)})^{(1)}$, an $N(A_1)$-restricted $\{1\}$-inverse of A_2. Other applications call for other, similarly restricted, generalized inverses. The $N(A_1)$-restricted $\{1,2,3,4\}$-inverse of A_2 was studied for certain Hilbert space operators, by Minamide and Nakamura ([1] and [2]), who characterized it as the unique solution X of the five equations

$$A_1 X = O,$$

$$A_2 X A_2 = A_2 \text{ on } N(A_1),$$

$$X A_2 X = X,$$

$$(A_2 X)^* = A_2 X,$$

and

$$P_{N(A_1)}(X A_2)^* = X A_2 \text{ on } N(A_1).$$

Show that $P_{N(A_1)}(A_2 P_{N(A_1)})^\dagger$ is the unique solution of these five equations.

9. THE BOTT–DUFFIN INVERSE

Consider the constrained system

$$Ax + y = b, \qquad x \in L, \qquad y \in L^{\perp}, \tag{109}$$

with given $A \in C^{n \times n}$, $b \in C^n$, and a subspace L of C^n. Such systems arise in electrical network theory; see, e.g., Bott and Duffin [1] and Section 12 below. As in Section 8 we conclude that the consistency of (109) is equivalent to the consistency of the following system:

$$(AP_L + P_{L^{\perp}})z = b \tag{110}$$

and that $\begin{bmatrix} x \\ y \end{bmatrix}$ is a solution of (109) if and only if

$$x = P_L z, \qquad y = P_{L^{\perp}} z = b - AP_L z, \tag{111}$$

where z is a solution of (110).

If the matrix $(AP_L + P_{L^{\perp}})$ is nonsingular, then (109) is consistent for all $b \in C^n$ and the solution

$$x = P_L(AP_L + P_{L^{\perp}})^{-1}b, \qquad y = b - Ax$$

is unique. The transformation

$$P_L(AP_L + P_{L^{\perp}})^{-1}$$

was introduced and studied by Bott and Duffin [1], who called it the *constrained inverse* of A. Since it exists only when $(AP_L + P_{L^{\perp}})$ is nonsingular, one may be tempted to introduce generalized inverses of this form, namely

$$P_L(AP_L + P_{L^{\perp}})^{(i,j,\dots,l)} \qquad (1 \leqslant i,j,\dots,l \leqslant 4),$$

which do exist for all A and L. This section, however, is restricted to the Bott–Duffin inverse.

DEFINITION 2. Let $A \in C^{n \times n}$ and let L be a subspace of C^n. If $(AP_L + P_{L^{\perp}})$ is nonsingular, the *Bott–Duffin inverse of A with respect to L*, denoted by $A_{(L)}^{(-1)}$, is defined by

$$A_{(L)}^{(-1)} = P_L(AP_L + P_{L^{\perp}})^{-1}. \tag{112}$$

Some properties of $A_{(L)}^{(-1)}$ are collected in

THEOREM 14 (Bott and Duffin [1]).   Let $(AP_L + P_{L^\perp})$ be nonsingular. Then:

(a) The equation

$$Ax + y = b, \qquad x \in L, \qquad y \in L^\perp \tag{109}$$

has for every b, the unique solution

$$x = A_{(L)}^{(-1)} b, \tag{113}$$

$$y = \left(I - AA_{(L)}^{(-1)}\right) b. \tag{114}$$

(b) A, P_L, and $A_{(L)}^{(-1)}$ satisfy

$$P_L = A_{(L)}^{(-1)} AP_L = P_L AA_{(L)}^{(-1)}, \tag{115}$$

$$A_{(L)}^{(-1)} = P_L A_{(L)}^{(-1)} = A_{(L)}^{(-1)} P_L. \tag{116}$$

PROOF. (a) This follows from the equivalence of (109) and (110)–(111).

(b) From (112), $P_L A_{(L)}^{(-1)} = A_{(L)}^{(-1)}$. Postmultiplying $A_{(L)}^{(-1)}(AP_L + P_{L^\perp}) = P_L$ by P_L gives $A_{(L)}^{(-1)} AP_L = P_L$. Therefore $A_{(L)}^{(-1)} P_{L^\perp} = O$ and $A_{(L)}^{(-1)} P_L = A_{(L)}^{(-1)}$. Multiplying (114) by P_L gives $(P_L - P_L AA_{(L)}^{(-1)})b = 0$ for all b, thus $P_L = P_L AA_{(L)}^{(-1)}$. ∎

From these results it follows that the Bott–Duffin inverse $A_{(L)}^{(-1)}$, whenever it exists, is the $\{1,2\}$-inverse of $(P_L AP_L)$ having range L and null space $L^\perp$.

COROLLARY 11. If $(AP_L + P_{L^\perp})$ is nonsingular, then
(a) $A_{(L)}^{(-1)} = (AP_L)_{L,L^\perp}^{(1,2)} = (P_L A)_{L,L^\perp}^{(1,2)} = (P_L AP_L)_{L,L^\perp}^{(1,2)}$,
(b) $(A_{(L)}^{(-1)})_{(L)}^{(-1)} = P_L AP_L$.

PROOF. (a) From (115), $\dim L = \operatorname{rank} P_L \leqslant \operatorname{rank} A_{(L)}^{(-1)}$. Similarly from (116), $\operatorname{rank} A_{(L)}^{(-1)} \leqslant \dim L$, $R(A_{(L)}^{(-1)}) \subset R(P_L) = L$ and $N(A_{(L)}^{(-1)}) \supset N(P_L) = L^\perp$. Therefore

$$\operatorname{rank} A_{(L)}^{(-1)} = \dim L \tag{117}$$

and

$$R\left(A_{(L)}^{(-1)}\right) = L, \qquad N\left(A_{(L)}^{(-1)}\right) = L^\perp. \tag{118}$$

Now, $A_{(L)}^{(-1)}$ is a $\{1,2\}$-inverse of AP_L:

$$AP_L A_{(L)}^{(-1)} AP_L = AP_L \qquad \text{by (115),}$$

and

$$A_{(L)}^{(-1)} AP_L A_{(L)}^{(-1)} = A_{(L)}^{(-1)} \qquad \text{by (115) and (116).}$$

That $A_{(L)}^{(-1)}$ is a $\{1,2\}$-inverse of $P_L A$ and of $P_L AP_L$ is similarly proved.

(b) We show first that $(A_{(L)}^{(-1)})_{(L)}^{(-1)}$ is defined, i.e, that $(A_{(L)}^{(-1)} P_L + P_{L^\perp})$ is nonsingular. From (116), $A_{(L)}^{(-1)} P_L + P_{L^\perp} = A_{(L)}^{(-1)} + P_{L^\perp}$, which is a nonsingular matrix since its columns span $L + L^\perp = C^n$, by (118). Now $P_L AP_L$ is a $\{1,2\}$-inverse of $A_{(L)}^{(-1)}$, by (a), and therefore by Theorem 1.2 and (117),

$$\operatorname{rank} P_L AP_L = \operatorname{rank} A_{(L)}^{(-1)} = \dim L.$$

This result, together with

$$R(P_L AP_L) \subset R(P_L) = L, \qquad N(P_L AP_L) \supset N(P_L) = L^\perp,$$

shows that

$$R(P_L AP_L) = L, \qquad N(P_L AP_L) = L^\perp,$$

proving that

$$P_L AP_L = \left(A_{(L)}^{(-1)}\right)_{L,L^\perp}^{(1,2)}$$

$$= \left(A_{(L)}^{(-1)}\right)_{(L)}^{(-1)}. \qquad \blacksquare$$

EXCERCISES

80. Show that the following statements are equivalent, for any $A \in C^{n \times n}$ and a subspace $L \subset C^n$.

(a) $AP_L + P_{L^\perp}$ is nonsingular.

(b) $C^n = AL \oplus L^\perp$, i.e., $AL = \{Ax : x \in L\}$ and $L^\perp$ are complementary subspaces of C^n.

(c) $C^n = P_L R(A) \oplus L^\perp$.

(d) $C^n = P_L AL \oplus L^\perp$.

(e) $\operatorname{rank} P_L AP_L = \dim L$.

Thus, each of the above conditions is necessary and sufficient for the existence of $A_{(L)}^{(-1)}$, the Bott–Duffin inverse of A with respect to L.

81. *A converse to Corollary 11.* If any one of the following three $\{1,2\}$-inverses exists,

$$(AP_L)^{(1,2)}_{L,L^\perp}, \qquad (P_LA)^{(1,2)}_{L,L^\perp}, \qquad (P_LAP_L)^{(1,2)}_{L,L^\perp},$$

then all three exist, $AP_L + P_{L^\perp}$ is nonsingular, and

$$(AP_L)^{(1,2)}_{L,L^\perp} = (P_LA)^{(1,2)}_{L,L^\perp} = (P_LAP_L)^{(1,2)}_{L,L^\perp} = A^{(-1)}_{(L)}.$$

Hint. Condition (b) in Ex. 80 is equivalent to the existence of $(AP_L)^{(1,2)}_{L,L^\perp}$.

82. Let K be a matrix whose columns form a basis for L. Then $A^{(-1)}_{(L)}$ exists if and only if K^*AK is nonsingular, in which case

$$A^{(-1)}_{(L)} = K(K^*AK)^{-1}K^* \qquad \text{(Bott and Duffin [1]).}$$

PROOF. Follows from Corollary 11 and Theorem 11(d).

83. If A is Hermitian and $A^{(-1)}_{(L)}$ exists, then $A^{(-1)}_{(L)}$ is Hermitian.

84. Using the notation

$$A = [a_{ij}] \qquad (i,j = 1,\dots,n),$$

$$A^{(-1)}_{(L)} = [t_{ij}] \qquad (i,j = 1,\dots,n),$$

$$d_{A,L} = \det(AP_L + P_{L^\perp}), \tag{119}$$

$$\psi_{A,L} = \log d_{A,L} \tag{120}$$

show that

(a) $\qquad \dfrac{\partial \psi_{A,L}}{\partial a_{ij}} = t_{ji} \qquad (i,j = 1,\dots,n)$

(b) $\qquad \dfrac{\partial t_{kl}}{\partial a_{ij}} = t_{ki}t_{jl} \qquad (i,j,k,l = 1,\dots,n).$

(Bott and Duffin [1], Theorem 3.)

Bott and Duffin called $d_{A,L}$ the *discriminant* of A, and $\psi_{A,L}$ the *potential* of $A^{(-1)}_{(L)}$.

85. Let $A \in C^{n \times n}$ be nonsingular, and let L be a subspace of C^n. Then $A^{(-1)}_{(L)}$ exists if and only if $(A^{-1})^{(-1)}_{(L^\perp)}$ exists.

Hint. Use $A^{-1}P_{L^\perp} + P_L = A^{-1}(AP_L + P_{L^\perp})$ to show that $(A^{-1}P_{L^\perp} + P_L)^{-1} = (AP_L + P_{L^\perp})^{-1}A.$

86. Let $A \in C^{n \times n}$ be nonsingular, let L be a subspace of C^n, let $d_{A,L}$ and $\psi_{A,L}$ be given by (119) and (120), respectively, and similarly define

$$d_{A^{-1}, L^\perp} = \det(A^{-1}P_{L^\perp} + P_L)$$

$$\psi_{A^{-1}, L^\perp} = \log d_{A^{-1}, L^\perp}.$$

Then

(a) $d_{A^{-1}, L^\perp} = \dfrac{d_{A,L}}{\det A}$

(b) $(A^{-1})^{(-1)}_{\langle L^\perp \rangle} = A - AA^{(-1)}_{\langle L \rangle}A.$

(Bott and Duffin [1], Theorem 4.)

87. If $\operatorname{Re}(Au, u) > 0$ for every nonzero vector u, then $d_{A,L} \neq 0$, $\operatorname{Re}(A^{(-1)}_{\langle L \rangle}u, u) \geqslant 0$ for every vector u and $\operatorname{Re}(t_{ii}) \geqslant 0$, where $A^{(-1)}_{\langle L \rangle} = [t_{ij}]$ (Bott and Duffin [1], Theorem 6).

88. Let $A, B \in C^{n \times n}$ and let L be a subspace of C^n such that both $A^{(-1)}_{\langle L \rangle}$ and $B^{(-1)}_{\langle L \rangle}$ exist. Then

$$B^{(-1)}_{\langle L \rangle}A^{(-1)}_{\langle L \rangle} = (AP_L B)^{(-1)}_{\langle L \rangle}.$$

10. AN APPLICATION OF {1}-INVERSES IN INTERVAL LINEAR PROGRAMMING

For two vectors $u, v \in R^m$ let

$$u \leqslant v$$

denote the fact that $u_i \leqslant v_i$ for $i = 1, \ldots, m$. A linear programming problem of the form

$$\text{maximize } \{ c^T x : a \leqslant Ax \leqslant b \}, \tag{121}$$

with given $a, b \in R^m$; $c \in R^n$; $A \in R^{m \times n}$, is called an *interval linear program* (also a *linear program with two-sided constraints*) and denoted by IP-(a, b, c, A) or simply by IP. Any linear programming problem with bounded constraint set can be written as an IP; see, e.g., Robers and Ben-Israel [2].

In this section, which is based on the work of Ben-Israel and Charnes [3], the optimal solutions of (121) are obtained by using {1}-inverses of A, in the special case where A is of full row rank. More general cases were studied by Zlobec and Ben-Israel [1], [2] (see also Exs. 89 and 90), and an iterative method for solving the general IP appears in Robers and Ben-

Israel [2]. Applications of interval programming are given in Ben-Israel, Charnes, and Robers [1], and Robers and Ben-Israel [1]. References for other applications of generalized inverses in linear programming are Pyle [4] and Cline and Pyle [1].

The IP (121) is called *consistent* (also *feasible*) if the set

$$F = \{ x \in R^n : a \leqslant Ax \leqslant b \} \tag{122}$$

is nonempty, in which case the elements of F are called the *feasible solutions* of IP(a,b,c,A). A consistent IP(a,b,c,A) is called *bounded* if

$$\max \{ c^T x : x \in F \}$$

is finite, in which case the *optimal solutions* of IP(a,b,c,A) are its feasible solutions x_0 which satisfy

$$c^T x_0 = \max \{ c^T x : x \in F \}.$$

Boundedness is equivalent to $c \in R(A^T)$ as the following lemma shows.

LEMMA 4. Let $a,b \in R^m$; $c \in R^n$; $A \in R^{m \times n}$ be such that IP(a,b,c,A) is consistent. Then IP(a,b,c,A) is bounded if and only if

$$c \in N(A)^\perp. \tag{123}$$

PROOF. From (122), $F = F + N(A)$. Therefore

$$\max \{ c^T x : x \in F \} = \max \{ c^T x : x \in F + N(A) \}$$

$$= \max \{ (P_{R(A^T)} c + P_{N(A)} c)^T x : x \in F + N(A) \}, \quad \text{by (49),}$$

$$= \max \{ c^T P_{R(A^T)} x : x \in F \} + \max \{ c^T x : x \in N(A) \},$$

where the first term

$$\max \{ c^T P_{R(A^T)} x : x \in F \} = \max \{ c^T A^\dagger A x : a \leqslant Ax \leqslant b \}$$

is finite, and the second term

$$\max \{ c^T x : x \in N(A) \}$$

is finite if and only if $c \in N(A)^\perp$. ∎

We introduce now a function $\eta : R^m \times R^m \times R^m \to R^m$, defined for $u, v, w \in R^m$ by

$$\eta(u,v,w) = [\eta_i] \qquad (i = 1, \ldots, m)$$

where

$$\eta_i = \begin{cases} u_i, & \text{if} \quad w_i < 0 \\ v_i, & \text{if} \quad w_i > 0 \\ \lambda_i u_i + (1 - \lambda_i) v_i & \text{where} \quad 0 \leqslant \lambda_i \leqslant 1, \quad \text{if} \quad w_i = 0 \end{cases} \qquad (124)$$

A component of $\eta(u,v,w)$ is equal to the corresponding component of u or v, if the corresponding component of w is negative or positive, respectively. If a component of w is zero, then the corresponding component of $\eta(u,v,w)$ is the closed interval with the corresponding components of u and v as endpoints. Thus η maps points in $R^m \times R^m \times R^m$ into sets in R^m, and any statement below about $\eta(u,v,w)$ is meant for all values of $\eta(u,v,w)$, unless otherwise specified.

The next result gives all the optimal solutions of IP(a,b,c,A) with A of full row rank.

THEOREM 15. (Ben-Israel and Charnes [3]). Let $a,b \in R^m$; $c \in R^n$; $A \in R_m^{m \times n}$ be such that IP(a,b,c,A) is consistent and bounded, and let $A^{(1)}$ be any $\{1\}$-inverse of A. Then the general optimal solution of IP(a,b,c,A) is

$$x = A^{(1)} \eta(a,b,A^{(1)T}c) + y, \qquad y \in N(A). \qquad (125)$$

PROOF.   From $A \in R_m^{m \times n}$ it follows that $R(A) = R^m$, so that any $u \in R^m$ can be written as

$$u = Ax \qquad (126)$$

where

$$x = A^{(1)}u + y, \qquad y \in N(A), \qquad \text{by Corollary 2.} \qquad (127)$$

Substituting (126) and (127) in (121), we get, by using (123), the equivalent IP

$$\text{maximize} \; \{ c^T A^{(1)} u : a \leqslant u \leqslant b \}$$

whose general optimal solution is, by the definition (124) of η,

$$u = \eta(a,b,A^{(1)T}c)$$

which gives (125) by using (127).   ∎

EXERCISES

89. Let $a,b \in R^m$; $c \in R^n$; $A \in R^{m \times n}$ be such that $IP(a,b,c,A)$ is consistent and bounded. Let $A^{(1)} \in A\{1\}$ and let $z_0 \in N(A^T)$ satisfy

$$z^T \eta_0 \leqslant 0$$

for some $\eta_0 \in \eta(a,b,(A^{(1)}P_{R(A)})^T c + z_0)$. Then

$$x_0 = A^{(1)}P_{R(A)}\eta_0 + y, \qquad y \in N(A)$$

is an optimal solution of IP (a,b,c,A) if and only if it is a feasible solution (Zlobec and Ben-Israel [2]).

90. Let $b \in R^m$; $A \in R^{m \times n}$; $c \in R^n$ and let $u \in R^n$ be a positive vector such that the problem

$$\text{maximize } \{ c^T x : Ax = b, 0 \leqslant x \leqslant u \} \tag{128}$$

is consistent. Let $z_0 \in R(A^T)$ satisfy

$$z^T \eta_0 \leqslant z^T A^\dagger b$$

for some $\eta_0 \in \eta(0,u,P_{N(A)}c + z_0)$. Then

$$x_0 = A^\dagger b + P_{N(A)}\eta_0$$

is an optimal solution of (128) if and only if it is a feasible solution (Zlobec and Ben-Israel [2]).

11. A {1,2}-INVERSE FOR THE INTEGRAL SOLUTION OF LINEAR EQUATIONS

Let K denote the *ring of integers* $0, \pm 1, \pm 2, \ldots$ and let:

K^m be the *m dimensional vector space over K,*
$K^{m \times n}$ be the *$m \times n$ matrices over K,*
$K_r^{m \times n}$ be the same with rank r.

Any vector in K^m will be called an *integral vector*. Similarly, any element of $K^{m \times n}$ will be called an *integral matrix*.

Let $A \in K^{m \times n}$, $b \in K^m$ and let the linear equation

$$Ax = b \tag{5}$$

be consistent. In many applications one has to determine if (5) has integral solutions, in which case one has to find some or all of them. If A is nonsingular and its inverse is also integral, then (5) has the unique integral solution $x = A^{-1}b$ for any integral b. A nonsingular matrix $A \in K^{n \times n}$ whose inverse A^{-1} is also in $K^{n \times n}$ is called a *unit matrix*; e.g. Marcus and Minc [1], p. 42.

In this section, which is based on the work of Hurt and Waid [1], we study the integral solution of (5) for any $A \in K^{m \times n}$ and $b \in K^m$. Using the Smith normal form of A (Theorem 16 below), a $\{1,2\}$-inverse is found (Corollary 12) which can be used to determine the existence of integral solutions, and to list all of them if they exist (Corollaries 13 and 14).

Two matrices $A, S \in K^{m \times n}$ are said to be *equivalent over K* if there exist two unit matrices $P \in K^{m \times m}$ and $Q \in K^{n \times n}$ such that

$$PAQ = S. \tag{129}$$

The following theorem states that any integral matrix is equivalent over K to a diagonal integral matrix.

THEOREM 16. Let $A \in K_r^{m \times n}$. Then A is equivalent over K to a matrix $S = [s_{ij}] \in K_r^{m \times n}$ such that:

    (a) $s_{ii} \neq 0$, $i = 1, \ldots, r$
    (b) $s_{ij} = 0$ otherwise
and
    (c) s_{ii} divides $s_{i+1,i+1}$ for $i = 1, \ldots, r-1$.

Remark. S is called the *Smith normal form* of A, and its nonzero elements s_{ii} $(i = 1, \ldots, r)$ are the *invariant factors* of A; see, e.g., Marcus and Minc [1] pp. 42–44.

PROOF. The proof given in Marcus and Minc [1], p. 44, is constructive and describes an algorithm to

    (i) find the greatest common divisor of the elements of A,
    (ii) bring it to position $(1,1)$, and
    (iii) make zeros of all other elements in the first row and column.

This is done, in an obvious way, by using a sequence of elementary row and column operations consisting of:

$$\text{interchanging two rows [columns]} \tag{130}$$

$$\text{subtracting an integer multiple of one row [column]}$$

$$\text{from another row [column].} \tag{131}$$

The matrix $B = [b_{ij}]$ so obtained is equivalent over K to A, and

$$b_{11} \text{ divides } b_{ij} \quad (i > 1, j > 1),$$

$$b_{i1} = b_{1j} = 0 \quad (i > 1, j > 1).$$

Setting $s_{11} = b_{11}$, one repeats the algorithm for the $(m-1) \times (n-1)$ matrix $[b_{ij}]$ $(i > 1, j > 1)$, etc.

The algorithm is repeated r times and stops when the bottom right $(m-r) \times (n-r)$ submatrix is zero, giving the Smith normal form S.

The unit matrix $P[Q]$ in (129) is the product of all the elementary row [column] operators, in the right order. ∎

Using the Smith normal form, a {1,2}-inverse with special integral properties can now be given:

COROLLARY 12 (Hurt and Waid [1]). Let $A \in K^{m \times n}$. Then there is an $n \times m$ matrix X satisfying

$$AXA = A, \tag{1.1}$$

$$XAX = X, \tag{1.2}$$

$$AX \in K^{m \times m}, \quad XA \in K^{n \times n}. \tag{132}$$

PROOF. Let

$$PAQ = S \tag{129}$$

be the Smith normal form of A, and let

$$\hat{A} = QS^{\dagger}P. \tag{133}$$

Then

$$PAQ = S = SS^{\dagger}S = PAQS^{\dagger}PAQ = PA\hat{A}AQ,$$

proving $A = A\hat{A}A$. $\hat{A}A\hat{A} = \hat{A}$ is similarly proved. The integrality of $A\hat{A}$ and $\hat{A}A$ follows from that of $PA\hat{A} = SS^{\dagger}P$ and $\hat{A}AQ = QS^{\dagger}S$, respectively. ∎

In the rest of this section we denote by $\hat{A}, \hat{B}$ the {1,2}-inverses of A, B as given in Corollary 12.

COROLLARY 13 (Hurt and Waid [1]). Let A, B, D be integral matrices, and let the matrix equation

$$AXB = D \tag{1}$$

be consistent. Then (1) has an integral solution if and only if the matrix

$$\hat{A}D\hat{B}$$

is integral, in which case the general integral solution of (1) is

$$X = \hat{A}D\hat{B} + Y - \hat{A}AYB\hat{B}, \quad Y \in K^{n \times m}.$$

PROOF.   Follows from Corollary 12 and Theorem 1.

COROLLARY 14   (Hurt and Waid [1]).   Let A and b be integral, and let the vector equation

$$Ax = b \qquad\qquad (5)$$

be consistent. Then (5) has an integral solution if and only if the vector

$$\hat{A}b$$

is integral, in which case the general integral solution of (5) is

$$x = \hat{A}b + (I - \hat{A}A)y, \qquad y \in K^n.$$

EXERCISES

91.   Two matrices $A, B \in K^{m \times n}$ are equivalent over K if and only if B can be obtained from A by a sequence of   elementary row and column operations (130), (131).

Hint.   Use Ex. 1.3.

92.   Describe in detail the algorithm mentioned in the proof of Theorem 16.

93.   Use the results of Sections 10 and 11 to find the integral optimal solutions of the interval program

$$\max\{ c^T x : a \leqslant Ax \leqslant b \}$$

where a, b, c, and A are integral.

94.   If K is the *ring of polynomials with real coefficients*, or the *ring of polynomials with complex coefficients*, the results of this section hold; see, e.g., Marcus and Minc [1], p. 40. Interpret Corollaries 12 and 14 in these two cases.

12. AN APPLICATION OF THE BOTT–DUFFIN INVERSE TO ELECTRICAL NETWORKS

In this section, which is based on the work of Bott and Duffin [1], we keep the discussion of electrical networks at the minimum sufficient to illustrate the application of the Bott–Duffin inverse studied in Section 9. The reader is referred to the original work of Bott and Duffin for further information.

An electrical network is described topologically in terms of its graph consisting of *nodes* (also *vertices, junctions*, etc.) and *branches* (also *edges*), and electrically in terms of its (branch) *currents* and *voltages*.

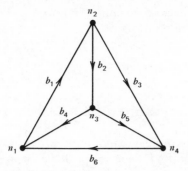

Figure 1. AN EXAMPLE OF A NETWORK.

Let the network consist of m elements called *nodes* denoted by n_i, $i = 1,\ldots,m$ (which, in the present limited discussion, can be represented by m points in the plane), and n ordered pairs of nodes called *branches* denoted by $b_j, j = 1,\ldots,n$ (represented here by directed segments joining the paired nodes). For example, the network represented by Fig. 1 has four nodes denoted by n_i ($i = 1,\ldots,4$) and six branches, $\{n_1, n_2\}$, $\{n_2, n_3\}$, $\{n_2, n_4\}$, $\{n_3, n_1\}$, $\{n_3, n_4\}$, and $\{n_4, n_1\}$, denoted by b_1, b_2, b_3, b_4, b_5, and b_6, respectively.

A network with m nodes and n branches can be represented by an $m \times n$ matrix, called the *(node-branch) incidence matrix of the network*, denoted here by $M = [m_{ij}]$ and defined as follows:

(i) The ith row of M corresponds to the node n_i, $i = 1,\ldots,m$.

(ii) The jth column of M corresponds to the branch b_j, $j = 1,\ldots,n$.

(iii) If $b_j = \{n_k, n_l\}$, then

$$m_{ij} = \begin{cases} 1, & i = k \\ -1, & i = l \\ 0, & i \neq k, l, \end{cases} \quad j = 1,\ldots,n.$$

For example, the incidence matrix of the network in Fig. 1 is

$$M = \begin{bmatrix} 1 & 0 & 0 & -1 & 0 & -1 \\ -1 & 1 & 1 & 0 & 0 & 0 \\ 0 & -1 & 0 & 1 & 1 & 0 \\ 0 & 0 & -1 & 0 & -1 & 1 \end{bmatrix}.$$

Two nodes n_k and n_l (or the two corresponding rows of M) are called *directly connected* if either $\{n_k, n_l\}$ or $\{n_l, n_k\}$ is a branch, i.e., if there is a column in M having its nonzero entries in rows k and l. Two nodes n_k or n_l (or the two corresponding rows of M) are called *connected* if there is a sequence of nodes,

$$\{n_k, n_p, \ldots, n_q, n_l\}$$

in which every two adjacent nodes are directly connected. Finally, a network (or its incidence matrix) is called *connected* if every two nodes are connected.

In this section we consider only *direct current networks*, referring the reader to Bott and Duffin [1] and to Ex. 96 below, for alternating current networks. A direct current network is described electrically in terms of two real-valued functions, the *current* and the *potential*, defined on the sets of branches and nodes, respectively.

For $j = 1, \ldots, m$, the *current in branch* b_j, denoted by y_j, is the current (measured in amperes) flowing in b_j. A positive [negative] sign of y_j denotes that the directions of b_j and y_j are identical [opposite].

For $i = 1, \ldots, m$, the *potential at node* n_i, denoted by p_i, is the voltage difference (measured in volts) between n_i and some reference point, which can be taken as one of the nodes. A related function which is more often used, is the *voltage*, defined on the set of branches. For $j = 1, \ldots, n$, the *voltage across branch* $b_j = \{n_k, n_l\}$, denoted by x_j, is defined as the potential difference

$$x_j = p_k - p_l.$$

From the definition of the incidence matrix M it is clear that the vector of branch voltages $x = [x_j]$, $j = 1, \ldots, n$, and the vector of node potentials $p = [p_i]$, $i = 1, \ldots, m$, are related by

$$x = M^T p \tag{134}$$

The currents and voltages are assumed to satisfy Kirchhoff laws. The *Kirchhoff current law* is a conservation theorem for the currents (or electrical charges), stating that for each node, the net current entering the node is zero, i.e., the sum of incoming currents equals the sum of outgoing currents. From the definition of the incidence matrix M it follows that the Kirchhoff current law can be written as

$$My = 0, \tag{135}$$

where $y = [y_j]$, $j = 1, \ldots, n$, is the vector of branch currents.

The *Kirchhoff voltage law* states that the potential function is single valued. This statement usually assumes the equivalent form that the sum of the branch voltages directed around any closed circuit is zero.

From (134), (135), and (49), it follows that the Kirchhoff current and voltage laws define two complementary orthogonal subspaces:

$N(M)$, the currents satisfying Kirchhoff current law;
$R(M^T)$, the voltages satisfying Kirchhoff voltage law.

Each branch b_j, $j = 1, \ldots, n$, of the network will be regarded as having a *series voltage generator* of v_j volts and a *parallel current generator* of w_j amperes. These are related to the branch currents and voltages by Ohm's law

$$a_j(x_j - v_j) + (y_j - w_j) = 0, \qquad j = 1, \ldots, n, \qquad (136)$$

where $a_j > 0$ is the *conductivity* of the branch b_j, measured in mhos.

Thus, the branch currents y and voltages x are found by solving the following constrained system:

$$Ax + y = Av + w, \qquad x \in R(M^T), \qquad y \in N(M), \qquad (137)$$

where $A = [\operatorname{diag} a_j]$ is the diagonal matrix of branch conductivities, v and w are the given vectors of generated voltages and currents, respectively, and M is the incidence matrix. It can be shown that the Bott–Duffin inverse of A with respect to $R(M^T)$, $A^{(-1)}_{(R(M^T))}$, exists; see, e.g., Ex. 95 below. Therefore, by Theorem 14, the unique solution of (137) is

$$x = A^{(-1)}_{(R(M^T))}(Av + w), \qquad (138)$$

$$y = \left(I - AA^{(-1)}_{(R(M^T))}\right)(Av + w). \qquad (139)$$

The physical significance of the matrix $A^{(-1)}_{(R(M^T))}$ should be clear from (138). The (i,j)th entry of $A^{(-1)}_{(R(M^T))}$ is the voltage across branch b_i as a result of inserting a current source of one ampere in branch b_j; $i,j = 1, \ldots, n$. Because of this property, $A^{(-1)}_{(R(M^T))}$ is called the *transfer matrix* of the network.

Since the conductivity matrix A is nonsingular, the network equations (137) can be rewritten as

$$A^{-1}y + x = A^{-1}w + v, \qquad y \in N(M), \qquad x \in R(M^T). \qquad (140)$$

By Exs. 95 and 85, the unique solution of (140) is

$$y = (A^{-1})^{(-1)}_{(N(M))}(A^{-1}w + v), \qquad (141)$$

$$x = \left(I - A^{-1}(A^{-1})^{(-1)}_{(N(M))}\right)(A^{-1}w + v) \qquad (142)$$

The matrix $(A^{-1})_{(N(M))}^{(-1)}$ is called the *dual transfer matrix*, its (i,j)th entry being the current in branch b_i as a result of inserting a one-volt generator parallel to branch b_j. Comparing the corresponding equations in (138)–(139) and in (141)–(142), we prove that the transfer matrices $A_{(R(M^T))}^{(-1)}$ and $(A^{-1})_{(N(M))}^{(-1)}$ satisfy

$$A^{-1}(A^{-1})_{(N(M))}^{(-1)} + A_{(R(M^T))}^{(-1)}A = I, \qquad (143)$$

which can also be proved directly from Ex. 86(b).

The correspondence between results like (138)–(139) and (141)–(142) is called *electrical duality*; see, e.g., the discussion in Bott and Duffin [1], Duffin [1], and Sharpe and Styan [1], [2], [3], for further results on duality and on applications of generalized inverses in electrical networks.

EXERCISES

95. Let $A \in C^{n \times n}$ be such that $(Ax, x) \neq 0$ for every nonzero vector x in L, a subspace of C^n. Then $A_{(L)}^{(-1)}$ exists, i.e., $(AP_L + P_{L^\perp})$ is nonsingular.

PROOF. If $Ax + y = 0$ for some $x \in L$ and $y \in L^\perp$, then $Ax \in L^\perp$ and therefore $(Ax, x) = 0$. ∎

See also Exs. 87 and 80(b) above.

96. In alternating current networks without mutual coupling, equations (136) still hold for the branches, by using complex, instead of real, constants and variables. The complex a_j is then the admittance of branch b_j. Alternating current networks with mutual coupling due to transformers, are still represented by (137), where the *admittance matrix* A is symmetric, its off-diagonal elements giving the mutual couplings; see, e.g., Bott and Duffin [1].

97. *Incidence matrices.* Let M be a connected $m \times n$ incidence matrix. Then, for any $M^{(1,3)} \in M\{1,3\}$,

$$I - MM^{(1,3)} = \frac{1}{m}ee^T,$$

where ee^T is the $m \times m$ matrix whose elements are all 1 (see also Ijiri [1]).

PROOF. From $(I - MM^{(1,3)})M = O$ it follows for any two directly connected nodes n_i and n_j (i.e., for any column of M having its $+1$ and -1 in rows i and j), that the ith and jth columns of $I - MM^{(1,3)}$ are identical. Since M is connected, all columns of $I - MM^{(1,3)}$ are identical. Since $I - MM^{(1,3)}$ is symmetric, all rows of $I - MM^{(1,3)}$ are also identical. Therefore, all the elements of $I - MM^{(1,3)}$ are identical, say

$$I - MM^{(1,3)} = \alpha ee^T$$

for some real α. Now $I - MM^{(1,3)}$ is idempotent, proving that $\alpha = 1/m$. ∎

98. Let M be a connected $m \times n$ incidence matrix. Then rank $M = m - 1$.

PROOF.

$$P_{N(M^T)} = I - P_{R(M)}, \qquad \text{by (50)},$$

$$= I - MM^{(1,3)}, \qquad \text{by Ex. 1.10 and Lemma 3},$$

$$= \frac{1}{m} ee^T, \qquad \text{by Ex. 97},$$

proving that $\dim N(M^T) = \text{rank } P_{N(M^T)} = 1$, and therefore

$$\text{rank } M = \dim R(M) = m - \dim N(M^T) = m - 1. \qquad ∎$$

99. *Trees.* Let a connected network consist of m nodes and n branches, and let M be its incidence matrix. A *tree* is defined as consisting of the m nodes, and any $m - 1$ branches which correspond to linearly independent columns of M. Show that:

(a) A tree is a connected network which contains no closed circuit.

(b) Any column of M not among the $m - 1$ columns corresponding to a given tree, can be expressed uniquely as a linear combination of those $m - 1$ columns, using only the coefficients $0, + 1$, and $- 1$.

(c) Any branch not in a given tree, lies in a unique closed circuit whose other branches, or the branches obtained from them by reversing their directions, belong to the tree.

100. Let $A = [\text{diag} a_j]$, $a_j \neq 0$, $j = 1, \ldots, n$, and let M be a connected $m \times n$ incidence matrix. Show that the discriminant (see Ex. 84)

$$d_{A, R(M^T)} = \det(AP_{R(M^T)} + P_{N(M)})$$

is the sum, over all trees $\{b_{j_1}, b_{j_2}, \ldots, b_{j_{m-1}}\}$ in the network, of the products

$$a_{j_1} a_{j_2} \cdots a_{j_{m-1}} \qquad (\text{Bott and Duffin } [1]).$$

SUGGESTED FURTHER READING

Section 1. Bjerhammar [3], Hearon [4], Jones [2], Morris and Odell [1], Sheffield [2].

Section 4. Afriat [1], Chipman and Rao [1], Graybill and Marsaglia [1], Greville [8], Wedderburn [1].

Section 5. Ward, Boullion, and Lewis [1].

Section 6.   Afriat [1], Anderson and Duffin [1], Ben-Israel [7], Chipman and Rao [1], Glazman and Ljubich [1], Greville [8], Petryshyn [1], Stewart [2].

Section 9.   Rao and Mitra [2].

Section 10.   For applications of generalized inverses in mathematical programming see also Beltrami [1], Ben-Israel ([8], [11], [14]), Ben-Israel and Kirby [1], Charnes and Cooper [1], Charnes, Cooper, and Thompson [1], Charnes and Kirby [1], Kirby [1], Nelson, Lewis, and Boullion [1], Rosen ([1], [2]), Zlobec [3].

Section 11.   Bowman and Burdet [1], and Charnes and Granot [1].

3
MINIMAL PROPERTIES
OF GENERALIZED
INVERSES

1. LEAST-SQUARES SOLUTIONS OF INCONSISTENT LINEAR SYSTEMS

For given $A \in C^{m \times n}$ and $b \in C^m$, the linear system

$$Ax = b \tag{1}$$

is *consistent*, i.e., has a solution for x, if and only if $b \in R(A)$. Otherwise, the *residual* vector

$$r = b - Ax \tag{2}$$

is nonzero for all $x \in C^n$, and it may be desired to find an *approximate solution* of (1), by which is meant a vector x making the residual vector (2) "closest" to zero in some sense, i.e., minimizing some norm of (2). An approximate solution that is often used, expecially in statistical applications, is the *least-squares solution* of (1), defined as a vector x minimizing the Euclidean norm of the residual vector, i.e., minimizing the sum of squares of moduli of the residuals

$$\sum_{i=1}^{m} |r_i|^2 = \sum_{i=1}^{m} \left| b_i - \sum_{j=1}^{n} a_{ij} x_j \right|^2 = \|b - Ax\|^2. \tag{3}$$

In this section the Euclidean vector norm—see, e.g., Ex. 1.31—is denoted simply by $\| \; \|$.

The following theorem shows that $\|Ax - b\|$ is minimized by choosing $x = Xb$, where $X \in A\{1,3\}$, thus establishing a relation between the $\{1,3\}$-

103

inverses of A and the least-squares solutions of $Ax = b$, characterizing each of these two concepts in terms of the other.

THEOREM 1. Let $A \in C^{m \times n}$, $b \in C^m$. Then $\|Ax - b\|$ is smallest when $x = A^{(1,3)}b$, where $A^{(1,3)} \in A\{1,3\}$. Conversely, if $X \in C^{n \times m}$ has the property that, for all b, $\|Ax - b\|$ is smallest when $x = Xb$, then $X \in A\{1,3\}$.

PROOF. Writing

$$Ax - b = (Ax - P_{R(A)}b) + (P_{R(A)}b - b)$$

it follows from Ex. 2.38 that

$$\|Ax - b\|^2 = \|Ax - P_{R(A)}b\|^2 + \|P_{R(A)}b - b\|^2 \tag{4}$$

since $(Ax - P_{R(A)}b) \in R(A)$ and $(P_{R(A)}b - b) = -(I - P_{R(A)})b \in R(A)^{\perp}$. Evidently, (4) assumes its minimum value if and only if

$$Ax = P_{R(A)}b, \tag{5}$$

which holds if $x = A^{(1,3)}b$ for any $A^{(1,3)} \in A\{1,3\}$, since by Theorem 2.8, (2.30), and Lemma 2.3

$$AA^{(1,3)} = P_{R(A)}. \tag{6}$$

Conversely, if X is such that for all b, $\|Ax - b\|$ is smallest when $x = Xb$, (5) gives $AXb = P_{R(A)}b$ for all b, and therefore

$$AX = P_{R(A)}.$$

Thus, by Theorem 2.3, $X \in A\{1,3\}$. ∎

COROLLARY 1. A vector x is a least-squares solution of $Ax = b$ if and only if

$$Ax = P_{R(A)}b = AA^{(1,3)}b.$$

Thus, the general least-squares solution is

$$x = A^{(1,3)}b + (I_n - A^{(1,3)}A)y, \tag{7}$$

with $A^{(1,3)} \in A\{1,3\}$ and arbitrary $y \in C^n$.

It will be noted that the least-squares solution is unique only when A is of full column rank (the most frequent case in statistical applications). Otherwise, (7) is an infinite set of such solutions.

EXERCISES, EXAMPLES, AND SUPPLEMENTARY NOTES

1. *Normal equation.* Show that a vector x is a least-squares solution of $Ax = b$ if and only if x is a solution of

$$A^*Ax = A^*b, \tag{8}$$

often called the *normal equation* of $Ax = b$.

Solution. Writing

$$b = P_{R(A)}b + P_{N(A^*)}b \qquad \text{by (2.49),}$$

it follows from (5) that x is a least-squares solution if and only if

$$Ax - b \in N(A^*),$$

which is (8).

Alternative solution. A necessary condition for the vector x^0 to be a least-squares solution of $Ax = b$ is that the partial derivatives $\partial f / \partial x_j$ of the function

$$f(x) = \|Ax - b\|^2 = \sum_{i=1}^{m} \left(\sum_{j=1}^{n} a_{ij} x_j - b_i \right)^* \left(\sum_{j=1}^{n} a_{ij} x_j - b_i \right) \tag{9}$$

vanish at x^0, i.e., that $\nabla f(x^0), = 0$, where

$$\nabla f(x^0) = \left[\frac{\partial f}{\partial x_j}(x^0) \right],$$

is the *gradient of f at x^0*. Now it can be verified that the gradient of (9) at x^0 is

$$\nabla f(x^0) = 2A^*(Ax^0 - b),$$

proving the necessity of (8). The sufficiency follows from the identity

$$(Ax - b)^*(Ax - b) - (Ax^0 - b)^*(Ax^0 - b)$$

$$= (x - x^0)^* A^* A (x - x^0) + 2 \operatorname{Re}(x - x^0)^* A^*(Ax^0 - b),$$

which holds for all $x, x^0 \in C^n$, where Re denotes the real part.

2. For any $A \in C^{m \times n}$ and $b \in C^m$, the normal equation (8) is consistent.

3. *Ill-conditioning.* The linear equation $Ax = b$, and the matrix A, are said to be *ill-conditioned* (or badly conditioned) if the solutions are very sensitive to small changes in the data; see, e.g., Noble [2], Chapter 8, and

Wilkinson [2]. The use of the normal equations (8) in finding least-squares solutions is limited by the fact that the matrix A^*A is frequently ill-conditioned and very sensitive to roundoff errors; see, e.g., Taussky [1] and Ex. 6.15. Methods for computing least-squares solutions which take account of this difficulty have been studied by several authors. We mention in particular Björk [1], [2], and [3], Björk and Golub [1], Businger and Golub [1] and [2], Golub and Wilkinson [1], and Noble [2]. Three such methods are mentioned in Exs. 6, 10, and 11 below. These methods can be used, with slight modifications, to compute the generalized inverse. The reader who is not interested in numerical methods may skip Exs. 4 through 11.

4. The following example illustrates the ill-conditioning of the normal equation. Let

$$A = \begin{bmatrix} 1 & 1 \\ \epsilon & 0 \\ 0 & \epsilon \end{bmatrix}$$

and let the elements of

$$A^TA = \begin{bmatrix} 1+\epsilon^2 & 1 \\ 1 & 1+\epsilon^2 \end{bmatrix}$$

be computed using double-precision and then rounded to single-precision with t binary digits. If $|\epsilon| < \sqrt{2^{-t}}$ then the rounded A^TA is

$$fl(A^TA) = \begin{bmatrix} 1 & 1 \\ 1 & 1 \end{bmatrix} \qquad (\textit{fl denotes floating point})$$

which is of rank 1, whereas A is of rank 2. Thus, for any $b \in R^3$, the computed normal equation

$$fl(A^TA)x = fl(A^Tb)$$

may be inconsistent or may have solutions which are not least-squares solutions of $Ax = b$.

5. *Factorization methods.* Let A be factorized as

$$A = FG \qquad (10)$$

where G is of full row-rank. Show that the normal equation (8) is equivalent to

$$F^*Ax = F^*b. \tag{11}$$

The factorization (10) is useful if the system (11) is not ill-conditioned, or at least not worse-conditioned than the system (1). Two such factorizations are given in Exs. 6 and 10 below.

6. QR factorization. Let $A \in C_n^{m \times n}$ (where full column-rank is assumed for convenience; the modifications necessary for the general case are the subject of Ex. 9). Then the QR factorization of A is

$$A = QR = \tilde{Q}\tilde{R} \tag{12}$$

where $Q \in C^{m \times m}$ is unitary (i.e., $Q^*Q = I$), $R = \begin{bmatrix} \tilde{R} \\ O \end{bmatrix}$ where $\tilde{R}$ is an $n \times n$ upper triangular nonsingular matrix, and $\tilde{Q}$ consists of the first n columns of Q. The columns of the unitary matrix Q form an orthogonal basis for C^m, and it is clear from (12) that the columns of $\tilde{Q}$ (and the upper triangular matrix $\tilde{R}$) may be obtained by orthogonalizing the columns of A. (It also follows from (12) that each column of $\tilde{Q}$ and each row of $\tilde{R}$ is determined uniquely up to a scalar factor of modulus one.)

The two principal ways of computing the QR-factorization are:

(1) Using a Gram–Schmidt type of orthogonalization; see, e.g., Rice [1] and Björk [1] where a detailed error analysis is given for least-squares solutions.

(2) Using Householder transformations; see, e.g., Wilkinson [1], Parlett [1], and Golub [3].

These two ways are compared in Golub [3].

Given the QR-factorization (12), it follows from Ex. 5 that the normal equation (8) is equivalent to

$$\tilde{Q}^*Ax = \tilde{Q}^*b$$

or to

$$\tilde{R}x = \tilde{Q}^*b, \qquad \text{since } \tilde{Q}^*\tilde{Q} = I_n. \tag{13}$$

Now $\tilde{R}$ is upper triangular, and thus (13) is solved by backward substitution.

7. Using the notations of Ex. 6, let

$$Q^*b = c = [c_i] \qquad (i = 1, \dots, m).$$

Show that the minimum value of $\|Ax - b\|^2$ is $\sum_{i=n+1}^{m} |c_i|^2$.

(*Hint*: $\|Ax - b\|^2 = \|Q^*(Ax - b)\|^2$ since Q is unitary.)

8. Show that the $\tilde{Q}\tilde{R}$-factorization for the matrix of Ex. 4, computed and rounded as in Ex. 4, is

$$
A = \begin{bmatrix} 1 & 1 \\ \epsilon & 0 \\ 0 & \epsilon \end{bmatrix} \approx fl(\tilde{Q}) fl(\tilde{R}) = \begin{bmatrix} 1 & \dfrac{\epsilon}{\sqrt{2}} \\ \epsilon & \dfrac{-1}{\sqrt{2}} \\ 0 & \dfrac{1}{\sqrt{2}} \end{bmatrix} \begin{bmatrix} 1 & 1 \\ 0 & \epsilon\sqrt{2} \end{bmatrix}.
$$

Use this to compute the least-squares solution of

$$
\begin{bmatrix} 1 & 1 \\ \epsilon & 0 \\ 0 & \epsilon \end{bmatrix} \begin{bmatrix} x_1 \\ x_2 \end{bmatrix} = \begin{bmatrix} 1 \\ \epsilon \\ 2\epsilon \end{bmatrix}.
$$

Answer. The (rounded) least-squares solution obtained by using (13) with the rounded matrices $fl(\tilde{Q})$ and $fl(\tilde{R})$ is

$$
x_1 = 0, \qquad x_2 = 1.
$$

The exact least-squares solution is

$$
x_1 = \frac{\epsilon^2}{2 + \epsilon^2}, \qquad x_2 = \frac{2(1 + \epsilon^2)}{2 + \epsilon^2}.
$$

9. Modify the results of Exs. 6 and 7 for the case $A \in C_r^{m \times n}$, $r < n$.

10. *Noble's method.* Let again $A \in C_n^{m \times n}$ and assume that A is partitioned as

$$
A = \begin{bmatrix} A_1 \\ A_2 \end{bmatrix} \qquad \text{where } A_1 \in C_n^{n \times n}.
$$

Then A may be factorized as

$$A = \begin{bmatrix} I \\ S \end{bmatrix} A_1 \quad \text{where } S = A_2 A_1^{-1} \in C^{(m-n) \times n}. \tag{14}$$

Let now $b \in C^m$ be partitioned as $b = \begin{bmatrix} b_1 \\ b_2 \end{bmatrix}$, $b_1 \in C^n$. Then by Ex. 5, the normal equation reduces to

$$(I + S^*S)A_1 x = b_1 + S^* b_2 \tag{15}$$

(which reduces further to $A_1 x = b_1$ if and only if $Ax = b$ is consistent).

The matrix S can be obtained by applying Gauss–Jordan elimination to the matrix

$$\begin{bmatrix} A_1 & b_1 & I \\ A_2 & b_2 & O \end{bmatrix},$$

transforming it into

$$\begin{bmatrix} I & A_1^{-1}b_1 & A_1^{-1} \\ O & b_2 - Sb_1 & -S \end{bmatrix},$$

from which S can be read. (See Noble [2], pp. 262–265.)

11. *Iterative refinement of solutions.* Let $x^{(0)}$ be an approximate solution of the consistent equation $Ax = b$, and let $\hat{x}$ be an exact solution. Then the error $\delta x = \hat{x} - x^{(0)}$ satisfies

$$A\delta x = A\hat{x} - Ax^{(0)}$$

$$= b - Ax^{(0)}$$

$$= r^{(0)}, \quad \text{the residual corresponding to } x^{(0)}.$$

This suggests the following *iterative refinement of solutions*, due to Wilkinson [1] (see also Moler [1]):

The initial approximation: $x^{(0)}$, given.
The kth residual: $r^{(k)} = b - Ax^{(k)}$.

The kth correction, $\delta x^{(k)}$, is obtained by solving: $A\delta x^{(k)} = r^{(k)}$.
The $(k+1)$st approximation: $x^{(k+1)} = x^{(k)} + \delta x^{(k)}$.
Double precision is used in computing the residuals, but not elsewhere.
The iteration is stopped if $\|\delta x^{(k)}\|/\|x^{(k)}\|$ falls below a prescribed number.
If the sequence $\{x^{(k)}: k = 0, 1, \ldots\}$ converges, it converges to a solution of $Ax = b$.

The use of this method to solve linear equations which are equivalent to the normal equation, such as (13) or (16), has been successful in finding, or improving, least-squares solutions. The reader is referred to Golub and Wilkinson [1], Björk [2] and [3], and Björk and Golub [1].

12. Show that the vector x is a least-squares solution of $Ax = b$ if and only if there is a vector r such that the vector $\begin{bmatrix} r \\ x \end{bmatrix}$ is a solution of

$$\begin{bmatrix} I & A \\ A^* & O \end{bmatrix} \begin{bmatrix} r \\ x \end{bmatrix} = \begin{bmatrix} b \\ 0 \end{bmatrix}. \tag{16}$$

13. Let $A \in C^{m \times n}$ and let $b_1, b_2, \ldots, b_k \in C^m$. Show that a vector x minimizes

$$\sum_{i=1}^{k} \|Ax - b_i\|^2$$

if and only if x is a least-squares solution of

$$Ax = \frac{1}{k} \sum_{i=1}^{k} b_i.$$

14. Let $A_i \in C^{m \times n}$, $b_i \in C^m$ $(i = 1, \ldots, k)$. Show that a vector x minimizes

$$\sum_{i=1}^{k} \|A_i x - b_i\|^2 \tag{17}$$

if and only if x is a solution of

$$\left(\sum_{i=1}^{k} A_i^* A_i \right) x = \sum_{i=1}^{k} A_i^* b_i. \tag{18}$$

Solution. x minimizes (17) if and only if x is a least-squares solution of
the system

$$\begin{bmatrix} A_1 \\ A_2 \\ \cdot \\ \cdot \\ A_k \end{bmatrix} x = \begin{bmatrix} b_1 \\ b_2 \\ \cdot \\ \cdot \\ b_k \end{bmatrix},$$

whose normal equation is (18).

15.   Let $A \in C^{m \times n}$, $b \in C^m$, and let α^2 be a positive real number. Show
that the function

$$\|Ax - b\|^2 + \alpha^2 \|x\|^2 \tag{19}$$

has a unique minimizer x_{α^2} given by

$$x_{\alpha^2} = (A^*A + \alpha^2 I)^{-1} A^* b \tag{20}$$

whose norm $\|x_{\alpha^2}\|$ is a monotone decreasing function of α^2.

Solution.   (19) is a special case of (17) with $k = 2$, $A_1 = A$, $A_2 = \alpha I$, $b_1 = b$,
and $b_2 = 0$. Substituting these values in (18) we get

$$(A^*A + \alpha^2 I)x = A^* b,$$

which has the unique solution (20), since $(A^*A + \alpha^2 I)$ is nonsingular.

Using (2.49) or Lemma 1 below, it is possible to write b (uniquely) as

$$b = Au + v, \quad u \in R(A^*), \quad v \in N(A^*). \tag{21}$$

Substituting this in (20) gives

$$x_{\alpha^2} = (A^*A + \alpha^2 I)^{-1} A^* Au. \tag{22}$$

Now let $\{u_1, u_2, \ldots, u_r\}$ be an orthonormal basis of $R(A^*)$ consisting of
eigenvectors of A^*A corresponding to nonzero eigenvalues, say

$$A^* A u_j = \lambda_j u_j, \quad \lambda_j > 0 \quad (j = 1, \ldots, r).$$

If $u = \sum_{j=1}^r \beta_j u_j$ is the representation of u in terms of the above basis, then

(22) gives

$$x_{\alpha^2} = \sum_{j=1}^{r} \frac{\lambda_i \beta_j}{\lambda_j + \alpha^2} u_j$$

whose norm squared is

$$\|x_{\alpha^2}\|^2 = \sum_{j=1}^{r} \left(\frac{\lambda_j}{\lambda_j + \alpha^2} \right)^2 |\beta_j|^2,$$

a monotone decreasing function of α^2. ∎

Problems of minimizing expressions like (19) in infinite-dimensional spaces and subject to given constraints arise often in control theory. The reader is referred to Porter [1], especially to Section 4.4 and pp. 353–354 where additional references are given.

16. *Constrained least-squares solutions.* A vector x is said to be a *constrained least-squares solution* of $Ax = b$ subject to given constraints, if x is a solution of the constrained minimization problem: Minimize $\|Ax - b\|$ subject to the given constraints. Let $A_1 \in C^{m_1 \times n}$, $b_1 \in C^{m_1}$, $A_2 \in C^{m_2 \times n}$, $b_2 \in R(A_2)$. Characterize the solutions of the problem:

$$\text{minimize } \|A_1 x - b_1\|^2 \text{ subject to } A_2 x = b_2. \tag{23}$$

Solution. The general solution of $A_2 x = b_2$ is

$$x = A_2^{(1)} b_2 + (I - A_2^{(1)} A_2) y, \tag{24}$$

where $A_2^{(1)} \in A_2\{1\}$ and y ranges over C^n. Substituting (24) in $A_1 x = b_1$ gives the equation

$$A_1(I - A_2^{(1)} A_2) y = b_1 - A_1 A_2^{(1)} b_2. \tag{25}$$

Therefore x is a solution of (23) if and only if x is given by (24) where y is a least-squares solution of (25).

17. Show that a vector $x \in C^n$ is a solution of (23) if and only if there is a vector $y \in C^{m_2}$ such that the vector $\begin{bmatrix} x \\ y \end{bmatrix}$ is a solution of

$$\begin{bmatrix} A_1^* A_1 & A_2^* \\ A_2 & O \end{bmatrix} \begin{bmatrix} x \\ y \end{bmatrix} = \begin{bmatrix} A_1^* b_1 \\ b_2 \end{bmatrix}. \tag{26}$$

Compare this with Ex. 1. Similarly, find a characterization analogous to that given in Ex. 12. See also Björk and Golub [1].

18. Let $A \in C^{m \times n}$, $b \in C^m$, and let p be positive real number. Show that the problem

$$\text{minimize } \|Ax - b\| \text{ subject to } \|x\| = p \tag{27}$$

has the unique solution

$$x = (A^*A + \alpha^2 I)^{-1}A^*b$$

where α is (uniquely) determined by

$$\|(A^*A + \alpha^2 I)^{-1}A^*b\| = p.$$

Hint. Use Ex. 15.

See also Forsythe and Golub ([1], Section 7), and Forsythe [1].

2. SOLUTIONS OF MINIMUM NORM

When the system (1) has a multiplicity of solutions for x, there is a unique solution of minimum norm. This follows from Ex. 2.74, restated here as,

LEMMA 1. Let $A \in C^{m \times n}$. Then A is a one-to-one mapping of $R(A^*)$ onto $R(A)$. ∎

COROLLARY 2. Let $A \in C^{m \times n}$, $b \in R(A)$. Then there is a unique minimum norm solution of

$$Ax = b \tag{1}$$

given as the unique solution of (1) which lies in $R(A^*)$.

PROOF. By Lemma 1, Eq. (1) has a unique solution x_0 in $R(A^*)$. Now the general solution is given as

$$x = x_0 + y, \quad y \in N(A),$$

and by Ex. 2.38

$$\|x\|^2 = \|x_0\|^2 + \|y\|^2$$

proving that $\|x\| > \|x_0\|$ unless $x = x_0$. ∎

The following theorem relates minimum-norm solutions of $Ax = b$ and {1,4}-inverses of A, characterizing each of these two concepts in terms of the other.

THEOREM 2. Let $A \in C^{m \times n}$, $b \in C^m$. If $Ax = b$ has a solution for x, the unique solution for which $\|x\|$ is smallest is given by

$$x = A^{(1,4)}b,$$

where $A^{(1,4)} \in A\{1,4\}$. Conversely, if $X \in C^{n \times m}$ is such that, whenever $Ax = b$ has a solution, $x = Xb$ is the solution of minimum norm, then $X \in A\{1,4\}$.

PROOF. If $Ax = b$ is consistent, then for any $A^{(1,4)} \in A\{1,4\}$, $x = A^{(1,4)}b$ is a solution (by Corollary 2.2), lies in $R(A^*)$ (by Ex. 1.10) and thus, by Lemma 1, is the unique solution in $R(A^*)$, and thus the unique minimum-norm solution by Corollary 2.

Conversely, let X be such that, for all $b \in R(A)$, $x = Xb$ is the solution of $Ax = b$ of minimum norm. Setting b equal to each column of A, in turn, we conclude that

$$XA = A^{(1,4)}A$$

and $X \in A\{1,4\}$, by Theorem 2.4. ∎

The unique minimum-norm least-squares solution of $Ax = b$, and the generalized inverse $A^\dagger$ of A, are related as follows.

COROLLARY 3 (Penrose [2]). Let $A \in C^{m \times n}$, $b \in C^m$. Then, among the least-squares solutions of $Ax = b$, $A^\dagger b$ is the one of minimum norm. Conversely, if $X \in C^{n \times m}$ has the property that, for all b, Xb is the minimum-norm least-squares solution of $Ax = b$, then $X = A^\dagger$.

PROOF. By Corollary 1, the least-squares solutions of $Ax = b$ coincide with the solutions of

$$Ax = AA^{(1,3)}b \tag{5}$$

Thus the minimum-norm least-squares solution of $Ax = b$ is the minimum norm solution of (5). But by Theorem 2, the latter is

$$x = A^{(1,4)}AA^{(1,3)}b$$

$$= A^\dagger b$$

by Theorem 1.4.

A matrix X having the properties stated in the last sentence of the theorem must satisfy $Xb = A^\dagger b$ for all $b \in C^m$, and therefore $X = A^\dagger$. ∎

The *minimum-norm least-squares solution*, $x_0 = A^\dagger b$, (also called the *best approximate solution*; e.g., Penrose [2]) of $Ax = b$, can thus be characterized by the following two inequalities:

$$\|Ax_0 - b\| \leqslant \|Ax - b\| \qquad \text{for all } x \qquad (28)$$

and

$$\|x_0\| < \|x\| \qquad (29)$$

for any $x \neq x_0$ which gives equality in (28).

EXERCISES, EXAMPLES, AND SUPPLEMENTARY NOTES

19. Let A be given as in Ex. 1.6 and let

$$b = \begin{bmatrix} -i \\ 1 \\ 1 \end{bmatrix}.$$

Show that the general least-squares solution of $Ax = b$ is

$$x = \frac{1}{19} \begin{bmatrix} 0 \\ 1 \\ 0 \\ -4 \\ 0 \\ 0 \end{bmatrix} + \begin{bmatrix} 1 & 0 & 0 & 0 & 0 & 0 \\ 0 & 0 & -\frac{1}{2} & 0 & -1+2i & \frac{1}{2}i \\ 0 & 0 & 1 & 0 & 0 & 0 \\ 0 & 0 & 0 & 0 & -2 & -1-i \\ 0 & 0 & 0 & 0 & 1 & 0 \\ 0 & 0 & 0 & 0 & 0 & 1 \end{bmatrix} \begin{bmatrix} y_1 \\ y_2 \\ y_3 \\ y_4 \\ y_5 \\ y_6 \end{bmatrix},$$

where $y_1, y_2, \ldots, y_6$ are arbitrary, while the residual vector for the least-squares solution is

$$\frac{1}{19} \begin{bmatrix} 2i \\ 12 \\ -2 \end{bmatrix}.$$

20. In Ex. 19 show that the minimum-norm least-squares solution is

$$x = \frac{1}{874} \begin{bmatrix} 0 \\ 26-36i \\ 13-18i \\ -55-9i \\ -12-2i \\ -46+59i \end{bmatrix}.$$

21. Let $A \in C^{m \times n}$, $b \in C^m$, and $a \in C^n$. Show that if $Ax = b$ has a solution for x, then the unique solution for which $\|x - a\|$ is smallest is given by

$$x = A^{(1,4)}b + (I - A^{(1,4)}A)a$$

$$= A^{(1,4)}b + P_{N(A)}a.$$

22. Show that for any $A \in C^{m \times n}$, as $\lambda \to 0$ through any neighborhood of 0 in C, the following limit exists and

$$\lim_{\lambda \to 0} (A^*A + \lambda I)^{-1}A^* = A^\dagger \tag{30}$$

(den Broeder and Charnes [1], Ben-Israel [10]).

Solution. We must show that

$$\lim_{\lambda \to 0} (A^*A + \lambda I)^{-1}A^*y = A^\dagger y \tag{31}$$

for all $y \in C^m$. Since $N(A^*) = N(A^\dagger)$, by Ex. 2.32, (31) holds trivially for $y \in N(A^*)$. Therefore it suffices to prove (31) for $y \in N(A^*)^\perp = R(A)$. By Lemma 1, for any $y \in R(A)$ there is a unique $x \in R(A^*)$ such that $y = Ax$. Proving (31) thus amounts to proving for all $x \in R(A^*)$

$$\lim_{\lambda \to 0} (A^*A + \lambda I)^{-1}A^*Ax = A^\dagger Ax$$

$$= x, \qquad \text{since } A^\dagger A = P_{R(A^*)}. \tag{32}$$

Let $\{u_1, \ldots, u_r\}$ be a basis for $R(A^*)$ consisting of eigenvectors of A^*A, say

$$A^*Au_j = \lambda_j u_j \qquad (\lambda_j > 0, j = 1, \ldots, r).$$

Writing $x \in R(A^*)$ in terms of this basis

$$x = \sum_{j=1}^{r} \beta_j u_j,$$

we verify that for all $\lambda \neq -\lambda_1, -\lambda_2, \ldots, -\lambda_r$

$$(A^*A + \lambda I)^{-1} A^* A x = \sum_{j=1}^{r} \frac{\lambda_j \beta_j}{\lambda_j + \lambda} u_j,$$

which tends, as $\lambda \to 0$, to $\sum_{j=1}^{r} \beta_j u_j = x$.

Alternative solution. Following the last solution up to (32), it suffices to show that

$$\lim_{\lambda \to 0} (A^*A + \lambda I_n)^{-1} A^* A = A^\dagger A = P_{R(A^*)}.$$

Now let $A^*A = FF^*$, $F \in C_r^{n \times r}$ be a full-rank factorization. Then

$$(A^*A + \lambda I_n)^{-1} A^* A = (FF^* + \lambda I_n)^{-1} FF^*$$

for any λ for which the inverses exist. We now use the identity

$$(FF^* + \lambda I_n)^{-1} FF^* = F(F^*F + \lambda I_r)^{-1} F^*$$

and note that F^*F is nonsingular so that $\lim_{\lambda \to 0} (F^*F + \lambda I_r)^{-1} = (F^*F)^{-1}$. Collecting these facts we conclude that

$$\lim_{\lambda \to 0} (A^*A + \lambda I_n)^{-1} A^* A = F(F^*F)^{-1} F^*$$

$$= FF^\dagger$$

$$= P_{R(A^*)},$$

since the columns of F are a basis for $R(A^*A) = R(A^*)$.

Still another proof is given in Ex. 4.40.

23. Use Exs. 15 and 22 to conclude that the solutions $\{x_{\alpha^2}\}$ of the minimization problems:

$$\text{minimize } \left\{ \|Ax - b\|^2 + \alpha^2 \|x\|^2 \right\}$$

converge to $A^\dagger b$ as $\alpha \to 0$. Explain this result in view of Corollary 3.

24. For a given $A \in C^{m \times n}$, $b \in C^m$ and a positive real number p, solve the problem

$$\text{minimize } \|Ax - b\| \text{ subject to } \|x\| \leqslant p. \tag{33}$$

Solution. If

$$\|A^\dagger b\| \leqslant \|p\| \tag{34}$$

then $x = A^\dagger b$ is a solution of (33), and is the unique solution if and only if (34) is an equality.

If (34) does not hold, then (33) has the unique solution, given in Ex. 18. (See also Balakrishnan [1], Theorem 2.3.)

25. *Matrix spaces.* For any $A, B \in C^{m \times n}$ define

$$R(A, B) = \{ Y = AXB \in C^{m \times n} : X \in C^{n \times m} \} \tag{35}$$

and

$$N(A, B) = \{ X \in C^{n \times m} : AXB = O \} \tag{36}$$

which we shall call the *range* and *null space* of (A, B), respectively. Let $C^{m \times n}$ be endowed with the inner product

$$(U, W) = \text{trace } W^* U = \sum_{i=1}^m \sum_{j=1}^n u_{ij} \overline{w}_{ij}. \tag{37}$$

Then for every $A, B \in C^{m \times n}$ the sets $R(A, B)$ and $N(A^*, B^*)$ are complementary orthogonal subspaces of $C^{m \times n}$.

Solution. As in Ex. 2.2 we use the one-to-one correspondence

$$v_{n(i-1)+j} = u_{ij} \qquad (i = 1, \ldots, m; j = 1, \ldots, n) \tag{38}$$

between the matrices $U = [u_{ij}] \in C^{m \times n}$ and the vectors $v(U) = [v_k] \in C^{mn}$. The correspondence (38) is a nonsingular linear transformation mapping $C^{m \times n}$ onto C^{mn}. Linear subspaces of $C^{m \times n}$ and C^{mn} thus correspond under (38).

It follows from (38) that the inner product (37) is equal to the standard inner product of the corresponding vectors $v(U)$ and $v(W)$. Thus, $(U, W) = (v(U), v(W)) = v(W)^* v(U)$. Also, from (2.10) we deduce that under (38), $R(A, B)$ and $N(A^*, B^*)$ correspond to $R(A \otimes B^T)$ and $N(A^* \otimes B^{*T})$, respectively. By (2.9), the latter is the same as $N((A \otimes B^T)^*)$, which by (2.50) is the orthogonal complement of $R(A \otimes B^T)$ in C^{mn}. Therefore, $R(A, B)$ and $N(A^*, B^*)$ are orthogonal complements in $C^{m \times n}$.

26. *Characterization of* $\{1,3\}$-, $\{1,4\}$-, *and* $\{1,2,3,4\}$-*inverses.* Let the norm used in $C^{m \times n}$ be

$$\|U\| = \sqrt{\text{trace } U^* U} ; \tag{39}$$

see, e.g., (1.53), which is the Euclidean norm of the vector $v(U)$. Show that for every $A \in C^{m \times n}$:

(a) $X \in A\{1,3\}$ if and only if X is a least-squares solution of

$$AX = I_m, \tag{40}$$

i.e., minimizing $\|Ax - I\|$ in the norm (39).

(b) $X \in A\{1,4\}$ if and only if X is a least-squares solution of

$$XA = I_n. \tag{41}$$

(c) $A^\dagger$ is the minimum-norm least-squares solution of both (40) and (41).

Solution. These results are based on the fact that the norm $\|U\|$ defined by (39) is merely the Euclidean norm of the corresponding vector $v(U)$.

(a) Writing the equation (40) as

$$(A \otimes I)v(X) = v(I), \tag{42}$$

it follows from Corollary 1 that the general least-squares solution of (42) is

$$v(X) = (A \otimes I)^{(1,3)} v(I) + \left(I - (A \otimes I)^{(1,3)} (A \otimes I) \right) y, \tag{43}$$

where y is an arbitrary element of C^{mn}. From (2.8) and (2.9) it follows that every $\{1,3\}$-inverse of $A \otimes I$ is equal to $(A^{(1,3)} \otimes I)$ for some $A^{(1,3)} \in A\{1,3\}$. Therefore the general least-squares solution of (40) is the matrix corresponding to (43), namely

$$X = A^{(1,3)} + (I - A^{(1,3)} A) Y, \qquad Y \in C^{n \times m},$$

which is the general $\{1,3\}$-inverse of A by Corollary 2.3.

(b) Taking the conjugate transpose of (41), we get

$$A^* X^* = I_n.$$

The set of least-squares solutions of the last equation is by (a)

$$A^* \{1,3\},$$

which coincides with $A\{1,4\}$.

(c) This is left to the reader.

27. Let A, B, D be complex matrices having dimensions consistent with the matrix equation

$$AXB = D.$$

Show that the minimum-norm least-squares solution of the last equation is

$$X = A^\dagger D B^\dagger \qquad (\text{Penrose [2]}).$$

28. Let $A \in C^{m \times n}$ and let X be a $\{1\}$-inverse of A; i.e., let X satisfy

$$AXA = A. \qquad (1.1)$$

Then the following are equivalent:

(a) $X = A^\dagger$,

(b) $X \in R(A^*, A^*)$,

(c) X is the minimum-norm solution of (1.1).

(Ben-Israel [12])

PROOF.   The general solution of (1.1) is by Theorem 2.1

$$X = A^\dagger A A^\dagger + Y - A^\dagger A Y A A^\dagger, \qquad Y \in C^{n \times m}$$
$$= A^\dagger + Y - A^\dagger A Y A A^\dagger \qquad (44)$$

Now it is easy to verify that

$$A^\dagger \in R(A^*, A^*), \qquad Y - A^\dagger A Y A A^\dagger \in N(A, A),$$

and using the norm (39) it follows from Ex. 25 that X of (44) satisfies

$$\|X\|^2 = \|A^\dagger\|^2 + \|Y - A^\dagger A Y A A^\dagger\|^2,$$

and the equivalence of (a), (b), and (c) is obvious. ∎

29. *Restricted generalized inverses.*   Let the matrix $A \in C^{m \times n}$ and the subspace $S \subset C^n$ be given. Then for any $b \in C^m$, the point $Xb \in S$ minimizes $\|Ax - b\|$ in S, if and only if $X = P_S(AP_S)^{(1,3)}$ is any S-restricted $\{1,3\}$-inverse of A.

PROOF.   Follows from Section 2.8 and Theorem 1.

30. Let A, S be as in Ex. 29. Then for any $b \in C^m$ for which the system

$$Ax = b, \qquad x \in S \qquad (2.97)$$

is consistent, Xb is the minimum norm solution of (2.97) if and only if $X = P_S(AP_S)^{(1,4)}$ is any S-restricted $\{1,4\}$-inverse of A.

PROOF.   Follows from Section 2.8 and Theorem 2.

31. Let A, S be as above. Then for any $b \in C^m$, Xb is the minimum-norm least-squares solution of (2.97) if and only if $X = P_S(AP_S)^\dagger$, the S-restricted Moore–Penrose inverse of A (Minamide and Nakamura [2]).

3. WEIGHTED GENERALIZED INVERSES

It may be desired to give different weights to the different squared residuals of the linear system $Ax = b$. This is a more general problem than the one solved by the $\{1,3\}$-inverses. A still further generalization which, however, presents no greater mathematical difficulty, is the minimizing of a given positive definite quadratic form in the residuals, or, in other words, the minimizing of

$$\|Ax - b\|_W^2 = (Ax - b)^* W(Ax - b), \qquad (45)$$

where W is a given positive definite matrix.

When A is not of full column rank, this problem does not have a unique solution for x, and we may choose from the class of "generalized least-squares solutions" the one for which

$$\|x\|_U^2 = x^* Ux \qquad (46)$$

is smallest, where U is a second positive definite matrix. If $A \in C^{m \times n}$, W is of order m and U is of order n.

Since every inner product in C^n can be represented as $x^* Uy$ for some positive definite matrix U (see Ex. 32), it follows that the problem of minimizing (45), and the problem of minimizing (46) among all the minimizers of (45), differ from the problems treated in Sections 1 and 2 only in the different choices of inner products and their associated norms in C^m and C^n. These seemingly more general problems can be reduced by a simple transformation to the "unweighted" problems considered in Sections 1 and 2. Every positive definite matrix H has a unique positive definite *square root*: that is, a positive definite K such that $K^2 = H$ (see, e.g., Ex. 6.34 below). Let us denote this K by $H^{1/2}$, and its inverse by $H^{-1/2}$.

We shall now introduce the transformations

$$\tilde{A} = W^{1/2}AU^{-1/2}, \qquad \tilde{x} = U^{1/2}x, \qquad \tilde{b} = W^{1/2}b, \qquad (47)$$

and it is easily verified that

$$\|Ax - b\|_W = \|\tilde{A}\tilde{x} - \tilde{b}\| \qquad (48)$$

and

$$\|x\|_U = \|\tilde{x}\|, \qquad (49)$$

expressing the norms $\| \quad \|_W$ and $\| \quad \|_U$ in terms of the Euclidean norms of the transformed vectors. This observation leads to the following two theorems.

THEOREM 3. Let $A \in C^{m \times n}$, $b \in C^m$, and let $W \in C^{m \times m}$ be positive definite. Then $\|Ax - b\|_W$ is smallest when $x = Xb$, where X satisfies

$$AXA = A, \quad (WAX)^* = WAX. \tag{50}$$

Conversely, if $X \in C^{m \times n}$ has the property that, for all b, $\|Ax - b\|_W$ is smallest when $x = Xb$, then X satisfies (50).

PROOF. In view of (48), it follows from Theorem 1 that $\|Ax - b\|_W$ is smallest when $\tilde{x} = Y\tilde{b}$, where Y satisfies

$$\tilde{A}Y\tilde{A} = \tilde{A}, \quad (\tilde{A}Y)^* = \tilde{A}Y, \tag{51}$$

and also if $Y \in C^{n \times m}$ has the property that, for all b, $\|Ax - b\|_W$ is smallest when $\tilde{x} = Y\tilde{b}$, then Y satisfies (51).

Now let

$$X = U^{-1/2}YW^{1/2} \tag{52}$$

so that

$$Y = U^{1/2}XW^{-1/2}. \tag{53}$$

Then it is easily verified by means of (47) and (53) that

$$\tilde{x} = Y\tilde{b} \Leftrightarrow x = Xb, \tag{54}$$

$$\tilde{A}Y\tilde{A} = \tilde{A} \Leftrightarrow AXA = A, \tag{55}$$

$$(\tilde{A}Y)^* = \tilde{A}Y \Leftrightarrow (WAX)^* = WAX. \quad \blacksquare$$

See also Ex. 35.

THEOREM 4. Let $A \in C^{m \times n}$, $b \in C^m$, and let $U \in C^{n \times n}$ be positive definite. If $Ax = b$ has a solution for x, the unique solution for which $\|x\|_U$ is smallest is given by

$$x = Xb,$$

where X satisfies

$$AXA = A, \quad (UXA)^* = UXA. \tag{56}$$

Conversely, if $X \in C^{n \times m}$ is such that, whenever $Ax = b$ has a solution, $x = Xb$ is the solution for which $\|x\|_U$ is smallest, then X satisfies (56).

PROOF. In view of (47),

$$Ax = b \Leftrightarrow \tilde{A}\tilde{x} = \tilde{b}.$$

Then it follows from (49) and Theorem 2 that, if $Ax = b$ has a solution for x, the unique solution for which $\|x\|_U$ is smallest is given by $\tilde{x} = Y\tilde{b}$, where Y satisfies

$$\tilde{A} Y \tilde{A} = \tilde{A}, \qquad (Y\tilde{A})^* = Y\tilde{A}, \tag{57}$$

and furthermore if $Y \in C^{n \times m}$ has the property that, whenever $Ax = b$ has a solution, $\|x\|_U$ is smallest when $\tilde{x} = Y\tilde{b}$, then Y satisfies (57).

As in the proof of Theorem 3, let X be given by (52), so that (53) holds. Then we have, in addition to (54) and (55),

$$(Y\tilde{A})^* = Y\tilde{A} \Leftrightarrow (UXA)^* = UXA. \quad \blacksquare$$

See also Ex. 37.

From Theorems 3 and 4 and Corollary 3, we can easily deduce:

COROLLARY 4. Let $A \in C^{m \times n}$, $b \in C^m$, and let $W \in C^{m \times m}$ and $U \in C^{n \times n}$ be positive definite. Then, there is a unique matrix

$$X = A^{(1,2)}_{(W,U)} \in A\{1,2\}$$

satisfying

$$(WAX)^* = WAX, \qquad (UXA)^* = UXA. \tag{58}$$

Moreover, $\|Ax - b\|_W$ assumes its minimum value for $x = Xb$, and in the set of vectors x for which this minimum value is assumed, $x = Xb$ is the one for which $\|x\|_U$ is smallest.

If $Y \in C^{n \times m}$ has the property that, for all b, $x = Yb$ is the vector of C^m for which $\|x\|_U$ is smallest among those for which $\|Ax - b\|_W$ assumes its minimum value, then $Y = A^{(1,2)}_{(W,U)}$. $\quad \blacksquare$

See also Exs. 38–45.

EXERCISES

32. *Inner products in C^n.* A function $f: C^n \times C^n \to C$ is called an *inner product* if for every $x, y, z \in C^n$ and $\alpha \in C$:

(a) $f(\alpha x + y, z) = \alpha f(x, z) + f(y, z)$ (*linearity*)

(b) $f(x, y) = \overline{f(y, x)}$ (*Hermitian symmetry*)

(c) $f(x, x) \geqslant 0$, $f(x, x) = 0 \Leftrightarrow x = 0$ (*positivity*).

Show that to every inner product $f: C^n \times C^n \to C$, there corresponds a unique positive definite $U \in C^{n \times n}$ such that

$$f(x, y) = y^* U x \qquad \text{for every } x, y \in C^n. \tag{59}$$

Solution. The inner product f and the positive definite matrix $U = [u_{ij}]$ satisfying (59), completely determine each other by

$$f(e_j, e_i) = u_{ij} \qquad (i, j = 1, \ldots, n),$$

where e_i is the ith unit vector.

33. *Square root.* Let H be Hermitian positive definite with the spectral decomposition

$$H = \sum_{i=1}^{k} \lambda_i E_i. \qquad (2.72)$$

Then

$$H^{1/2} = \sum_{i=1}^{k} \lambda_i^{1/2} E_i.$$

34. *Cholesky factorization.* Let H be Hermitian positive definite. Then H can be factorized as

$$H = R_H^* R_H, \qquad (60)$$

where R_H is an upper triangular matrix. (60) is called the *Cholesky factorization* of H; see, e.g., Wilkinson [1].

Show that the results of Section 3 can be derived by using the Cholesky factorization

$$U = R_U^* R_U \quad \text{and} \quad W = R_W^* R_W \qquad (61)$$

of U and W, respectively, instead of their square-root factorizations.

Hint. Instead of (47) use

$$\tilde{A} = R_W A R_U^{-1}, \qquad \tilde{x} = R_U x, \qquad \tilde{b} = R_W b.$$

35. Let A, b, and W be as in Theorem 3. Show that a vector $x \in C^n$ minimizes $\|Ax - b\|_W$ if and only if x is a solution of

$$A^* W A x = A^* W b,$$

and compare with Ex. 1.

36. Let $A_1 \in C^{m_1 \times n}$, $b_1 \in C^{m_1}$, $A_2 \in C^{m_2 \times n}$, $b_2 \in R(A_2)$, and let $W \in C^{m_1 \times m_1}$ be positive definite. Consider the problem

$$\text{minimize } \|A_1 x - b_1\|_W \text{ subject to } A_2 x = b_2. \qquad (62)$$

Show that a vector $x \in C^n$ is a minimizer of (62) if and only if there is a vector $y \in C^{m_2}$ such that the vector $\begin{bmatrix} x \\ y \end{bmatrix}$ is a solution of

$$\begin{bmatrix} A_1^* W A_1 & A_2^* \\ A_2 & O \end{bmatrix} \begin{bmatrix} x \\ y \end{bmatrix} = \begin{bmatrix} A_1^* W b_1 \\ b_2 \end{bmatrix}.$$

Compare with Ex. 17.

37. Let $A \in C^{m \times n}$, $b \in R(A)$, and let $U \in C^{n \times n}$ be positive definite. Show that the problem

$$\text{minimize } \|x\|_U \text{ subject to } Ax = b \tag{63}$$

has the unique minimizer

$$x = U^{-1} A^* (A U^{-1} A^*)^{(1)} b$$

and the minimum value

$$b^* (A U^{-1} A^*)^{(1)} b,$$

where $(A U^{-1} A^*)^{(1)}$ is any $\{1\}$-inverse of $A U^{-1} A^*$ (Rao [2], p. 49).

Outline of solution. Equation 63 is equivalent to the problem

$$\text{minimize } \|\tilde{x}\| \text{ subject to } \tilde{A}\tilde{x} = \tilde{b}$$

where $\tilde{x} = U^{1/2} x$, $\tilde{A} = A U^{-1/2}$, $\tilde{b} = b$. The unique minimizer of the last problem is, by Theorem 2,

$$\tilde{x} = Y\tilde{b} \quad \text{for any } Y \in \tilde{A}\{1,4\}.$$

Therefore, the unique minimizer of (63) is

$$x = U^{-1/2} X b \quad \text{for any } X \in (A U^{-1/2})\{1,4\}.$$

Complete the proof by choosing

$$X = U^{-1/2} A^* (A U^{-1} A^*)^{(1)}$$

which by Theorem 1.3 is a $\{1,2,4\}$-inverse of $A U^{-1/2}$.

38. *The weighted inverse* $A_{(W,U)}^{(1,2)}$. Chipman [2] first called attention to the unique $\{1,2\}$-inverse given by Corollary 4. However, instead of the second equation of (58), he used

$$(XAV)^* = XAV.$$

Show that these two relations are equivalent. How are U and V related?

39. Use Theorems 3 and 4 to show that

$$A^{(1,2)}_{\{W,U\}} = U^{-1/2}(W^{1/2}AU^{-1/2})^\dagger W^{1/2},$$

or equivalently, using (61),

$$A^{(1,2)}_{\{W,U\}} = R_U^{-1}(R_W A R_U^{-1})^\dagger R_W.$$

40. Use Exs. 35 and 37 to show that

$$A^{(1,2)}_{\{W,U\}} = U^{-1}A^* WA(A^* WAU^{-1}A^* WA)^{(1)}A^* W.$$

41. For a given A and an arbitrary $X \in A\{1,2\}$, do there exist positive definite matrices W and U such that $X = A^{(1,2)}_{\{W,U\}}$? Show that this question reduces to the following simpler one. Given an idempotent E, is there a positive definite V, such that VE is Hermitian? Show that such a V is given by

$$V = E^* HE + (I - E^*)K(I - E),$$

where H and K are arbitrary positive definite matrices. (This slightly generalizes a result of Ward, Boullion, and Lewis [1], who took $H = K = I$.)

Solution. Since H and K are positive definite, $x^* Vx = 0$ only if both the equations

$$Ex = 0, \qquad (I - E)x = 0 \tag{64}$$

hold. But addition of the two equations (63) gives $x = 0$. Therefore V is positive definite. Moreover,

$$VE = E^* HE$$

is clearly Hermitian.

42. As a particular illustration, let

$$E = \begin{bmatrix} 1 & 1 \\ 0 & 0 \end{bmatrix},$$

and show that V can be taken as any matrix of the form

$$V = \begin{bmatrix} a & a \\ a & b \end{bmatrix}, \tag{65}$$

where $b > a > 0$. Show that (65) can be written in the form

$$V = aE^*E + c(I - E^*)(I - E),$$

where a and c are arbitrary positive scalars.

43. Use Ex. 41 to prove that if X is an arbitrary element of $A\{1,2\}$, there exist positive definite W and U such that $X = A_{(W,U)}^{(1,2)}$ (Ward, Boullion, and Lewis [1]).

44. Show that

$$A_{(W,U)}^{(1,2)} = A_{T,S}^{(1,2)}$$

(see Theorem 2.10(c)), where the subspaces T, S and the positive definite matrices W, U are related by

$$T = U^{-1}N(A)^{\perp} \tag{66}$$

and

$$S = W^{-1}R(A)^{\perp} \tag{67}$$

or equivalently, by

$$U = P_{N(A),T}^* U_1 P_{N(A),T} + P_{T,N(A)}^* U_2 P_{T,N(A)} \tag{68}$$

and

$$W = P_{R(A),S}^* W_1 P_{R(A),S} + P_{S,R(A)}^* W_2 P_{S,R(A)} \tag{69}$$

where U_1, U_2, W_1, and W_2 are arbitrary positive definite matrices of appropriate dimensions.

Solution. From (58), we have

$$XA = U^{-1}A^*X^*U,$$

and therefore

$$R(X) = R(XA) = U^{-1}R(A^*) = U^{-1}N(A)^{\perp}$$

by Corollary 2.7 and (2.49). Also,

$$AX = W^{-1}X^*A^*W,$$

and therefore

$$N(X) = N(AX) = N(A^*W) = W^{-1}N(A^*) = W^{-1}R(A)^{\perp}$$

by Corollary 2.7 and (2.50). Finally, from Exs. 41 and 2.26 it follows that the general positive definite matrix U mapping T onto $N(A)^{\perp}$ is given by (68). Equation (69) is similarily proved.

45. Let $A = FG$ be a full-rank factorization. Use Ex. 44 and Theorem 2.11(d) to show that

$$A^{(1,2)}_{(W,U)} = U^{-1}G^*(F^*WAU^{-1}G^*)^{-1}F^*W.$$

Compare with Ex. 39.

4*. ESSENTIALLY STRICTLY CONVEX NORMS AND THE ASSOCIATED PROJECTORS AND GENERALIZED INVERSES

In the previous sections various generalized inverses were characterized and studied in terms of their minimization properties with respect to the class of *ellipsoidal* (or *weighted Euclidean*) *norms*

$$\|x\|_U = (x^*Ux)^{1/2}, \tag{46}$$

where U is positive definite.

Given any two ellipsoidal norms $\|\ \|_W$ and $\|\ \|_U$ on C^m and C^n, respectively, (defined by (46) and two given positive definite matrices $W \in C^{m \times m}$ and $U \in C^{n \times n}$), it was shown in Corollary 4 that every $A \in C^{m \times n}$ has a unique $\{1,2\}$-inverse $A^{(1,2)}_{(W,U)}$ with the following minimization property:

For any $b \in C^m$, the vector $A^{(1,2)}_{(W,U)}b$ satisfies

$$\|AA^{(1,2)}_{(W,U)}b - b\|_W \leqslant \|Ax - b\|_W, \qquad \text{for all } x \in C^n, \tag{70}$$

and

$$\|A^{(1,2)}_{(W,U)}b\|_U < \|x\|_U \tag{71}$$

for any $A^{(1,2)}_{(W,U)}b \neq x \in C^n$ which gives equality in (70). In particular, for $W = I_m$ and $U = I_n$ the inverse mentioned above is the Moore–Penrose inverse

$$A^{(1,2)}_{(I_m, I_n)} = A^\dagger \qquad \text{for every } A \in C^{m \times n}.$$

In this section, which is based on Erdelsky [1], Newman and Odell [1], and Holmes [1], similar minimizations are attempted for norms in the more general class of essentially strictly convex norms. The resulting projectors and generalized inverses are, in general, not even linear transformations, but they still retain many useful properties that justify their study.

*This section requires familiarity with the basic properties of convex functions and convex sets in finite-dimensional spaces; see, e.g., Rockafellar [1].

In this section we denote by $\alpha, \beta, \varphi, \ldots$ various vector norms on finite-dimensional spaces; see, e.g., Ex. 1.29.

Let φ be a norm on C^n and let L be a subspace of C^n. Then for any point $x \in C^n$ there is a point $y \in L$ which is "closest" to x in the norm φ, i.e., a point $y \in L$ satisfying

$$\varphi(y - x) = \inf \{ \varphi(l - x) : l \in L \}; \qquad (72)$$

see, e.g., Ex. 46 below. Generally, the closest point is not unique; see, e.g., Ex. 47. However, Lemma 1 below guarantees the uniqueness of closest points, for the special class of essentially strictly convex norms.

From the definition of a vector norm (see Ex. 1.29), it is obvious that every norm φ on C^n is a *convex function*, i.e., for any $x, y \in C^n$ and $0 \leqslant \lambda \leqslant 1$,

$$\varphi(\lambda x + (1 - \lambda)y) \leqslant \lambda \varphi(x) + (1 - \lambda)\varphi(y).$$

A function $\varphi : C^n \to R$ is called *strictly convex* if for all $x \neq y$ in C^n and $0 < \lambda < 1$,

$$\varphi(\lambda x + (1 - \lambda)y) < \lambda \varphi(x) + (1 - \lambda)\varphi(y). \qquad (73)$$

If $\varphi : C^n \to R$ is a norm, then (73) is clearly violated for $y = \mu x$, $\mu \geqslant 0$. Thus a norm φ on C^n is not strictly convex. Following Holmes [1], a norm φ on C^n is called here an *essentially strictly convex* (abbreviated e.s.c.) *norm* if φ satisfies (73) for all $x \neq 0$ and $y \notin \{ \mu x : \mu \geqslant 0 \}$. Equivalently, a norm φ on C^n is e.s.c. if

$$\left. \begin{array}{c} x \neq y \in C^n, \quad \varphi(x) = \varphi(y) \\ 0 < \lambda < 1 \end{array} \right\} \Rightarrow \varphi(\lambda x + (1 - \lambda)y) < \varphi(x). \qquad (74)$$

The following lemma is a special case of a result in Clarkson [1].

LEMMA 2. Let φ be an e.s.c. norm on C^n. Then for any subspace $L \subset C^n$ and any point $x \in C^n$, there is a unique point $y \in L$ closest to x, i.e.,

$$\varphi(y - x) = \inf \{ \varphi(l - x) : l \in L \}. \qquad (72)$$

PROOF. If $y_1, y_2 \in L$ satisfy (72) and $y_1 \neq y_2$, then for any $0 < \lambda < 1$

$$\varphi(\lambda y_1 + (1 - \lambda)y_2 - x) < \varphi(y_1 - x), \qquad \text{by (74)},$$

showing that the point $\lambda y_1 + (1 - \lambda)y_2$, which is in L, is closer to x than y_1, a contradiction. ∎

DEFINITION 1. Let φ be an e.s.c. norm on C^n and let L be a subspace of C^n. Then the φ-*metric projector on* L, denoted by $P_{L,\varphi}$ is the mapping $P_{L,\varphi} : C^n \rightarrow L$ assigning to each point in C^n its (unique) closest point in L, i.e.,

$$P_{L,\varphi}(x) \in L$$

and

$$\varphi\big(P_{L,\varphi}(x) - x\big) \leqslant \varphi(l - x), \qquad \text{for all } x \in C^n \text{ and } l \in L. \tag{75}$$

If φ is a general norm, then the projector $P_{L,\varphi}$ defined as above is a point-to-set mapping, since the closest point $P_{L,\varphi}(x)$ need not be unique for all $x \in C^n$ and $L \subset C^n$. An excellent survey of metric projectors in normed linear spaces, is given in Holmes [1], Section 32; see also Exs. 66–74 below.

Some properties of $P_{L,\varphi}$ in the e.s.c. case are collected in the following theorem, a special case of results by Aronszajn and Smith and Hirschfeld; see also Singer [1], p. 140, Theorem 6.1.

THEOREM 5.   Let φ be an e.s.c. norm on C^n. Then for any subspace $L \subset C^n$ and every point $x \in C^n$:

(a) $P_{L,\varphi}(x) = x$ if and only if $x \in L$,

(b) $P_{L,\varphi}^2(x) = P_{L,\varphi}(x)$,

(c) $P_{L,\varphi}(\lambda x) = \lambda P_{L,\varphi}(x)$ for all $\lambda \in C$,

(d) $P_{L,\varphi}(x + y) = P_{L,\varphi}(x) + y$ for all $y \in L$,

(e) $P_{L,\varphi}\big(x - P_{L,\varphi}(x)\big) = 0$,

(f) $\big|\varphi\big(x - P_{L,\varphi}(x)\big) - \varphi\big(y - P_{L,\varphi}(y)\big)\big| \leqslant \varphi(x - y)$ for all $y \in C^n$,

(g) $\varphi\big(x - P_{L,\varphi}(x)\big) \leqslant \varphi(x)$,

(h) $\varphi\big(P_{L,\varphi}(x)\big) \leqslant 2\varphi(x)$,

(i) $P_{L,\varphi}$ is continuous on C^n.

PROOF.   (a) Follows from (72) and (75) since the infimum in (72) is zero if and only if $x \in L$.

(b) $P_{L,\varphi}^2(x) = P_{L,\varphi}(P_{L,\varphi}(x))$

$$= P_{L,\varphi}(x) \qquad \text{by (a), since } P_{L,\varphi}(x) \in L.$$

(c) For any $z \in L$ and $\lambda \neq 0$

$$\varphi(\lambda x - z) = \varphi\left(\lambda x - \lambda\frac{z}{\lambda}\right)$$

$$= |\lambda|\varphi\left(x - \frac{z}{\lambda}\right)$$

$$\geqslant |\lambda|\varphi(x - P_{L,\varphi}(x)) \qquad \text{by (75)}$$

$$= \varphi(\lambda x - \lambda P_{L,\varphi}(x)),$$

which proves (c) for $\lambda \neq 0$. For $\lambda = 0$, (c) is obvious.

(d) From (75) it follows that for all $z \in L$

$$\varphi(P_{L,\varphi}(x) + y - (x+y)) \leqslant \varphi(z + y - (x+y)),$$

proving (d).

(e) Follows from (d).

(f) For all $x, y \in C^n$

$$\varphi(x - P_{L,\varphi}(x)) \leqslant \varphi(x - P_{L,\varphi}(y)) \leqslant \varphi(x-y) + \varphi(y - P_{L,\varphi}(y))$$

and thus

$$\varphi(x - P_{L,\varphi}(x)) - \varphi(y - P_{L,\varphi}(y)) \leqslant \varphi(x-y),$$

from which (f) follows by interchanging x and y.

(g) Follows from (f) by taking $y = 0$.

(h) $\varphi(P_{L,\varphi}(x)) \leqslant \varphi(P_{L,\varphi}(x) - x) + \varphi(x)$

$$\leqslant 2\varphi(x) \qquad \text{by (g).}$$

(i) Let $\{x_k\} \subset C^n$ be a sequence converging to x:

$$\lim_{k \to \infty} x_k = x.$$

Then the sequence $\{P_{L,\varphi}(x_k)\}$ is bounded, by (h), and hence contains a convergent subsequence, also denoted by $\{P_{L,\varphi}(x_k)\}$. Let

$$\lim_{k \to \infty} P_{L,\varphi}(x_k) = y.$$

Then

$$\varphi\big(P_{L,\varphi}(x_k)-x_k\big)\leqslant\varphi\big(P_{L,\varphi}(x)-x_k\big)$$

for $k=1,2,\ldots$, and in the limit,

$$\varphi(y-x)\leqslant\varphi\big(P_{L,\varphi}(x)-x\big)$$

proving that $y=P_{L,\varphi}(x)$. ∎

The function $P_{L,\varphi}$ is homogeneous by Theorem 5(c), but in general it is not additive; i.e., it does not necessarily satisfy

$$P_{L,\varphi}(x+y)=P_{L,\varphi}(x)+P_{L,\varphi}(y),\qquad\text{for all }x,y\in C^n.$$

Thus, in general, $P_{L,\varphi}$ is not a linear transformation. The following three corollaries deal with cases where $P_{L,\varphi}$ is linear.

For any $l\in L$ we define the *inverse image of l under $P_{L,\varphi}$*, denoted by $P_{L,\varphi}^{-1}(l)$, as

$$P_{L,\varphi}^{-1}(l)=\big\{x\in C^n:P_{L,\varphi}(x)=l\big\}.$$

We recall that a *linear manifold* (also *affine set, flat, linear variety*) in C^n is a set of the form

$$x+L=\{x+l:l\in L\},$$

where x and L are a given point and subspace, respectively, in C^n.

The following result is a special case of Theorem 6.4 in Singer [1], p. 144.

COROLLARY 5.   Let φ be an e.s.c. norm on C^n and let L be a subspace of C^n. Then the following statements are equivalent.

  (a) $P_{L,\varphi}$ is additive.
  (b) $P_{L,\varphi}^{-1}(0)$ is a linear subspace.
  (c) $P_{L,\varphi}^{-1}(l)$ is a linear manifold for any $l\in L$.

PROOF.   First we show that

$$P_{L,\varphi}^{-1}(0)=\big\{x-P_{L,\varphi}(x):x\in C^n\big\}.\tag{76}$$

From Theorem 5(f) it follows that

$$P_{L,\varphi}^{-1}(0)\supset\big\{x-P_{L,\varphi}(x):x\in C^n\big\}.$$

The reverse containment follows by writing each $x\in P_{L,\varphi}^{-1}(0)$ as

$$x=x-P_{L,\varphi}(x).$$

The equivalence of (a) and (b) is obvious from (76). The equivalence of (b) and (c) follows from

$$P_{L,\varphi}^{-1}(l) = l + P_{L,\varphi}^{-1}(0), \qquad \text{for all } l \in L, \tag{77}$$

which is a result of Theorem 5 (d) and (e). ∎

COROLLARY 6. Let L be a hyperplane of C^n, i.e., an $(n-1)$-dimensional subspace of C^n. Then $P_{L,\varphi}$ is additive for any e.s.c. norm φ on C^n.

PROOF. Let u be a vector not contained in L. Then any $x \in C^n$ is uniquely represented as

$$x = \lambda u + l, \qquad \text{where } \lambda \in C, l \in L.$$

Therefore, by (76),

$$P_{L,\varphi}^{-1}(0) = \left\{ \lambda u + \left(l - P_{L,\varphi}(\lambda u + l) \right) : \lambda \in C, l \in L \right\}$$

$$= \left\{ \lambda u + P_{L,\varphi}(-\lambda u) : \lambda \in C \right\}, \qquad \text{by Theorem 5(d)}$$

$$= \left\{ \lambda \left(u - P_{L,\varphi}(u) \right) : \lambda \in C \right\}, \qquad \text{by Theorem 5(c)}$$

is a line, proving that $P_{L,\varphi}$ is additive, by Corollary 5. ∎

COROLLARY 7 (Erdelsky [1]). Let φ be an e.s.c. norm on C^n and let r be an integer, $1 \leqslant r < n$. If $P_{L,\varphi}$ is additive for all r-dimensional subspaces of C^n, then it is additive for all subspaces of higher dimension.

PROOF. Let L be a subspace with $\dim L > r$, and assume that $P_{L,\varphi}$ is not additive. Then by Corollary 5, $P_{L,\varphi}^{-1}(0)$ is not a subspace; i.e., there exist $x_1, x_2 \in P_{L,\varphi}^{-1}(0)$ such that $P_{L,\varphi}(x_1 + x_2) = y \neq 0$. Let now M be an r-dimensional subspace of L which contains y. Then $x_1, x_2 \in P_{M,\varphi}^{-1}(0)$, but $P_{M,\varphi}(x_1 + x_2) = y \neq 0$, a contradiction of the hypothesis that $P_{M,\varphi}$ is additive. ∎

See also Exs. 69–72 for additional results on the linearity of the projectors $P_{L,\varphi}$.

Following Boullion and Odell ([2] pp. 43–44) we define generalized inverses associated with pairs of e.s.c. norms as follows.

DEFINITION 2. Let α and β be e.s.c. norms on C^m and C^n, respectively. For any $A \in C^{m \times n}$ we define the *generalized inverse associated with α and β*, (also called the α-β *generalized inverse*; see, e.g., Boullion and Odell [2] p. 44), denoted by $A_{\alpha,\beta}^{(-1)}$, as

$$A_{\alpha,\beta}^{(-1)} = (I - P_{N(A),\beta}) A^{(1)} P_{R(A),\alpha}, \tag{78}$$

where $A^{(1)}$ is any $\{1\}$-inverse of A.

The right-hand side of (78) means that the three transformations

$$P_{R(A),\alpha}: C^m \to R(A),$$

$$A^{(1)}: C^m \to C^n,$$

and

$$I - P_{N(A),\beta}: C^n \to P_{N(A),\beta}^{-1}(0),$$

see, e.g., (76), are performed in this order. We show now that $A_{\alpha,\beta}^{(-1)}$ is a single-valued transformation which does not depend on the particular $\{1\}$-inverse used in its definition. For any $y \in C^m$, the set

$$\left\{ A^{(1)} P_{R(A),\alpha}(y) : A^{(1)} \in A\{1\} \right\}$$

obtained as $A^{(1)}$ ranges over $A\{1\}$, is, by Theorem 1.2, the set of solutions of the linear equation

$$Ax = P_{R(A),\alpha}(y) \,,$$

a set which can be written as

$$A^\dagger P_{R(A),\alpha}(y) + \{ z : z \in N(A) \}.$$

Now, for any $z \in N(A)$, it follows from Theorem 5 (a) and (d) that

$$(I - P_{N(A),\beta})(A^\dagger P_{R(A),\alpha}(y) + z) = (I - P_{N(A),\beta})A^\dagger P_{R(A),\alpha}(y)$$

proving that

$$A_{\alpha,\beta}^{(-1)}(y) = (I - P_{N(A),\beta})A^\dagger P_{R(A),\alpha}(y), \qquad \text{for all } y \in C^n, \quad (79)$$

independently of the $\{1\}$-inverse $A^{(1)}$ used in the definition (78).

If the norms α and β are Euclidean, then $P_{R(A),\alpha}$ and $P_{N(A),\beta}$ reduce to the orthogonal projectors $P_{R(A)}$ and $P_{N(A)}$, respectively, and $A_{\alpha,\beta}^{(-1)}$ is, by (79), just the Moore–Penrose inverse $A^\dagger$; see also Exs. 67–70 and 74 below. Thus many properties of $A^\dagger$ are specializations of the corresponding properties of $A_{\alpha,\beta}^{(-1)}$, some of which are collected in the following Theorem. In particular, the minimization properties of $A^\dagger$ are special cases of statements (i) and (j) below.

THEOREM 6 (Erdelsky [1], Newman and Odell [1]). Let α and β be e.s.c. norms on C^m and C^n respectively. Then, for any $A \in C^{m \times n}$:

(a) $A_{\alpha,\beta}^{-1}: C^m \to C^n$ is a homogeneous transformation.

(b) $A_{\alpha,\beta}^{-1}$ is additive (hence linear) if $P_{R(A),\alpha}$ and $P_{N(A),\beta}$ are additive.

(c) $N\left(A_{\alpha,\beta}^{(-1)}\right) = P_{R(A),\alpha}^{-1}(0)$,

(d) $R\left(A_{\alpha,\beta}^{(-1)}\right) = P_{N(A),\beta}^{-1}(0)$,

where, as in the case of linear transformations, we denote

$$N\left(A_{\alpha,\beta}^{(-1)}\right) = \left\{ y \in C^m : A_{\alpha,\beta}^{(-1)}(y) = 0 \right\},$$

$$R\left(A_{\alpha,\beta}^{(-1)}\right) = \left\{ A_{\alpha,\beta}^{(-1)}(y) : y \in C^m \right\}.$$

(e) $AA_{\alpha,\beta}^{(-1)} = P_{R(A),\alpha}$.

(f) $A_{\alpha,\beta}^{(-1)}A = I - P_{N(A),\beta}$.

(g) $AA_{\alpha,\beta}^{(-1)}A = A$.

(h) $A_{\alpha,\beta}^{(-1)}AA_{\alpha,\beta}^{(-1)} = A_{\alpha,\beta}^{(-1)}$.

(i) For any $b \in C^m$, an α-*approximate solution* of

$$Ax = b \tag{1}$$

is defined as any vector $x \in C^n$ minimizing $\alpha(Ax - b)$. Then x is an α-approximate solution of (1) if and only if

$$Ax = AA_{\alpha,\beta}^{(-1)}(b). \tag{80}$$

(j) For any $b \in C^m$, the equation

$$Ax = b \tag{1}$$

has a unique α-approximate solution of minimal β-norm, given by $A_{\alpha,\beta}^{(-1)}(b)$; that is, for every $b \in C^m$,

$$\alpha\left(AA_{\alpha,\beta}^{(-1)}(b) - b\right) \leqslant \alpha(Ax - b), \qquad \text{for all } x \in C^n, \tag{81}$$

and

$$\beta\left(A_{\alpha,\beta}^{(-1)}(b)\right) < \beta(x) \tag{82}$$

for any $x \neq A_{\alpha,\beta}^{(-1)}(b)$ with equality in (81).

PROOF.   (a) Follows from the definition and Theorem 5(c).

   (b) Obvious from definition (78).

   (c) From (78) it is obvious that

$$N\left(A_{\alpha,\beta}^{(-1)}\right) \supset P_{R(A),\alpha}^{-1}(0).$$

Conversely, if $y \notin P_{R(A),\alpha}^{-1}(0)$; i.e., if $P_{R(A),\alpha}(y) \neq 0$, then $A^{\dagger}P_{R(A),\alpha}(y) \neq 0$ since $(A^{\dagger})_{[R(A)]}$ is nonsingular (see Ex. 2.76), and consequently

$$(I - P_{N(A),\beta})A^{\dagger}P_{R(A),\alpha}(y) \neq 0, \qquad \text{by Theorem 5(a).}$$

   (d) From (76) and the definition (78) it is obvious that

$$R\left(A_{\alpha,\beta}^{(-1)}\right) \subset P_{N(A),\beta}^{-1}(0).$$

Conversely, let $x \in P_{N(A),\beta}^{-1}(0)$. Then, by (76),

$$x = (I - P_{N(A),\beta})z, \qquad \text{for some } z \in C^n$$

$$= (I - P_{N(A),\beta})P_{R(A^*)}z, \qquad \text{by Theorem 5(d)}$$

$$= (I - P_{N(A),\beta})A^{\dagger}Az$$

$$= (I - P_{N(A),\beta})A^{\dagger}P_{R(A),\alpha}(Az)$$

$$= A_{\alpha,\beta}^{(-1)}(Az). \qquad (83)$$

   (e) Obvious from (79).

   (f) For any $z \in C^n$ it follows from (83) that

$$(I - P_{N(A),\beta})z = A_{\alpha,\beta}^{(-1)}(Az).$$

   (g) Obvious from (e) and Theorem 5(a).

   (h) Obvious from (f) and (d).

   (i) A vector $x \in C^n$ is an α-approximate solution of (1) if and only if

$$\alpha(Ax - b) \leqslant \alpha(y - b), \qquad \text{for all } y \in R(A),$$

or equivalently

$$Ax = P_{R(A),\alpha}(b), \qquad \text{by (75)}$$

$$= AA_{\alpha,\beta}^{(-1)}(b), \qquad \text{by (e).}$$

(j) From (80) it follows that x is an α-approximate solution of (1) if and only if

$$x = A^\dagger A A_{\alpha,\beta}^{(-1)}(b) + z, \qquad z \in N(A) \tag{84}$$

$$= A^\dagger P_{R(A),\alpha}(b) + z, \qquad z \in N(A), \qquad \text{by (e)}.$$

Now, by Lemma 2 and Definition 1, the β-norm of

$$A^\dagger P_{R(A),\alpha}(b) + z, \qquad z \in N(A)$$

is minimized uniquely at

$$z = -P_{N(A),\beta} A^\dagger P_{R(A),\alpha}(b),$$

which, substituted in (84), gives

$$x = (I - P_{N(A),\beta}) A^\dagger P_{R(A),\alpha}(b)$$

$$= A_{\alpha,\beta}^{(-1)}(b). \qquad \blacksquare$$

See Exs. 74–77 for additional results on the generalized inverses $A_{\alpha,\beta}^{(-1)}$.

EXERCISES AND EXAMPLES

46. *Closest points.* Let φ be a norm on C^n and let L be a nonempty closed set in C^n. Then, for any $x \in C^n$, the infimum

$$\inf\{\varphi(l - x) : l \in L\}$$

is attained at some point $y \in L$, called φ-*closest to x in L.*

PROOF. Let $z \in L$. Then the set

$$K = L \cap \{l \in C^n : \varphi(l - x) \leqslant \varphi(z - x)\}$$

is closed (being the intersection of two closed sets) and bounded, hence compact. The continuous function $\varphi(l - x)$ attains its infimum at some $l \in K$, but by definition of K,

$$\inf\{\varphi(l - x) : l \in K\} = \inf\{\varphi(l - x) : l \in L\}.$$

47. Let φ be the l_1-norm on R^2,

$$\varphi(x) = \varphi\left(\begin{bmatrix} x_1 \\ x_2 \end{bmatrix}\right) = |x_1| + |x_2|$$

see, e.g., Ex. 1.31, and let $L=\{x\in R^2:x_1+x_2=0\}$. Then the set of φ-closest points in L to $\begin{bmatrix} 1 \\ 1 \end{bmatrix}$ is

$$\left\{\begin{bmatrix} \alpha \\ -\alpha \end{bmatrix}:-1\leqslant\alpha\leqslant 1\right\}.$$

48. Let $\|\ \ \|$ be the Euclidean norm on C^n, let $S\subset C^n$ be a convex set and let x,y be two points in $C^n:x\notin S$ and $y\in S$. Then the following statements are equivalent:

(a) y is $\|\ \ \|$-closest to x in S.

(b) $s\in S\Rightarrow\mathrm{Re}(y-x,s-y)\geqslant 0.$

PROOF (adapted from Goldstein [1], p. 99).
 (a)$\Rightarrow$(b)   For any $0\leqslant\lambda\leqslant 1$ and $s\in S$,

$$y+\lambda(s-y)\in S.$$

Now

$$0\leqslant\|x-y-\lambda(s-y)\|^2-\|x-y\|^2$$

$$=2\lambda\,\mathrm{Re}(y-x,s-y)+\lambda^2\|s-y\|^2$$

$$<0\quad\text{if }\mathrm{Re}(y-x,s-y)<0\quad\text{and}\quad 0<\lambda<-\frac{2\,\mathrm{Re}(y-x,s-y)}{\|s-y\|^2},$$

a contradiction to (a).
 (b)$\Rightarrow$(a).   For any $s\in S$,

$$\|x-s\|^2-\|x-y\|^2=\|s\|^2-2\,\mathrm{Re}(s,x)+2\,\mathrm{Re}(y,x)-\|y\|^2$$

$$=\|s-y\|^2+2\,\mathrm{Re}(y-x,s-y)$$

$$\geqslant 0\quad\text{if (b).}\quad\blacksquare$$

49. *A hyperplane separation theorem.* Let S be a nonempty closed convex set in C^n, x a point not in S. Then there is a real hyperplane

$$\{z\in C^n:\mathrm{Re}(u,z)=\alpha\},\qquad\text{for some }0\neq u\in C^n,\alpha\in R$$

which separates S and x, in the sense that

$$s \in S \Rightarrow \mathrm{Re}(u,s) \geqslant \alpha,$$

and

$$\mathrm{Re}(u,x) < \alpha.$$

PROOF. Let x_S be the $\| \ \|$-closest point to x in S, where $\| \ \|$ is the Euclidean norm. The point x_S is unique, by the same proof as in Lemma 2, since $\| \ \|$ is e.s.c. Then, for any $s \in S$,

$$\mathrm{Re}(x_S - x, s) \geqslant \mathrm{Re}(x_S - x, x_S), \qquad \text{by Ex. 48,}$$

$$> \mathrm{Re}(x_S - x, x),$$

since

$$\mathrm{Re}(x_S - x, x_S - x) = \|x_S - x\|^2 > 0.$$

The proof is completed by choosing

$$u = x_S - x, \qquad \alpha = \mathrm{Re}(x_S - x, x_S). \qquad \blacksquare$$

50. Gauge functions and their duals. A function $\varphi : C^n \to R$ is called a *gauge function* (also a *Minkowski functional*) if for all $x \in C^n$

(G1) φ is continuous

(G2) $\varphi(x) \geqslant 0$ and $\varphi(x) = 0$ only if $x = 0$

(G3) $\varphi(\alpha x) = \alpha \varphi(x)$ for all $\alpha \geqslant 0$

and

(G4) $\varphi(x + y) \leqslant \varphi(x) + \varphi(y)$, for all $x, y \in C^n$.

A gauge function $\varphi : C^n \to R$ is called *symmetric* if for all

$$x = [x_1, x_2, \ldots, x_n]^T \in C^n,$$

(G5) $\varphi(x) = \varphi(x_1, x_2, \ldots, x_n) = \varphi(x_{\pi(1)}, x_{\pi(2)}, \ldots, x_{\pi(n)})$

for every permutation $\{\pi(1), \pi(2), \ldots, \pi(n)\}$ of $\{1, 2, \ldots, n\}$, and

(G6) $\varphi(x) = \varphi(x_1, x_2, \ldots, x_n) = \varphi(\lambda_1 x_1, \lambda_2 x_2, \ldots, \lambda_n x_n)$

for every scalar sequence $\{\lambda_1, \lambda_2, \ldots, \lambda_n\}$ satisfying

$$\begin{cases} |\lambda_i| = 1 & \text{if } \varphi : C^n \to R, \\ \lambda_i = \pm 1 & \text{if } \varphi : R^n \to R, \end{cases} \quad i = 1, \ldots, n.$$

Let $\varphi: C^n \to R$ satisfy (G1)–(G3). The *dual* (also *conjugate**) *function* of φ is the function $\varphi_D : C^n \to R$ defined by

$$\varphi_D(y) = \sup_{x \neq 0} \frac{\mathrm{Re}(y,x)}{\varphi(x)}. \tag{85}$$

Then:
  (a) The supremum in (85) is attained, and

$$\varphi_D(y) = \max_{x \in S_i} \frac{\mathrm{Re}(y,x)}{\varphi(x)}, \qquad i = 1 \text{ or } \varphi, \tag{86}$$

where

$$S_1 = \left\{ x \in C^n : \|x\|_1 = \sum_{i=1}^n |x_i| = 1 \right\} \tag{87}$$

and

$$S_\varphi = \left\{ x \in C^n : \varphi(x) = 1 \right\}. \tag{88}$$

  (b) φ_D is a gauge function.
  (c) φ_D satisfies (G5) [(G6)] if φ does.
  (d) If φ is a gauge function (i.e., if φ also satisfies (G4)), then φ is the conjugate of φ_D (Bonnesen and Fenchel [1], von Neumann [3]).

PROOF. (a). From (G3) it follows that the constraint $x \neq 0$ in (85) can be replaced by $x \in S_1$, or alternatively, by $x \in S_\varphi$. The supremum is attained since S_1 is compact.
  (b),(c). The continuity of φ_D follows from (G1), (86), and the compactness of S_1. It is easy to show that φ shares with φ_D each of the properties (G2), (G3), (G5), and (G6), while (G4) holds for φ_D, by definition (85), without requiring that it hold for φ.
  (d) From (85) it follows that

$$\mathrm{Re}(y,x) \leqslant \varphi(x)\varphi_D(y), \qquad \text{for all } x,y \in C^n, \tag{89}$$

and hence

$$\varphi(x) \geqslant \sup_{y \neq 0} \frac{\mathrm{Re}(y,x)}{\varphi_D(y)}. \tag{90}$$

To show equality in (90) we note that the set

$$B = \left\{ z : \varphi(z) \leqslant 1 \right\}$$

---

*Originally, φ_D was called the *conjugate* of φ by Bonnesen and Fenchel [1] and von Neumann [3]. However, in the modern convexity literature, the word *conjugate function* has a different meaning; see, e.g., Rockafellar [1]. Therefore we use here *dual* rather than *conjugate*.

is a closed convex set in C^n, an easy consequence of the definition of a gauge function. From the hyperplane separation theorem (see e.g. Ex. 49 above) we conclude:

If a point x is contained in every closed half-space $\{z:\mathrm{Re}(u,z)<1\}$

which contains B, then $x\in B$, i. e. $\varphi(x)\leqslant 1$. $\hspace{2cm}$ (91)

From (86) and (88), it follows that

$$B\subset\{z:\mathrm{Re}(y,z)\leqslant 1\}$$

is equivalent to

$$\varphi_D(y)\leqslant 1.$$

Statement (91) is thus equivalent to

$$\{\varphi_D(y)\leqslant 1\Rightarrow\mathrm{Re}(y,x)\leqslant 1\}\Rightarrow\varphi(x)\leqslant 1$$

which proves equality in (90). ∎

51. *Convex bodies and gauge functions.* A *convex body* in C^n is a closed bounded convex set with nonempty interior.

Let $B\subset C^n$ be a convex body and let $0\in\mathrm{int}\,B$, where $\mathrm{int}\,B$ denotes the *interior* of B. The *gauge function* (or *Minkowski functional*) *of* B is the function $\varphi^B:C^n\rightarrow R$ defined by

$$\varphi^B(x)=\inf\{\lambda>0:x\in\lambda B\}. \hspace{2cm} (92)$$

Then

(a) φ^B is a gauge function, i.e., it satisfies (G1)–(G4) of Ex. 50.

(b) $B=\{x\in C^n:\varphi^B(x)\leqslant 1\}$.

(c) $\mathrm{int}\,B=\{x\in C^n:\varphi^B(x)<1\}$.

Conversely, if $\varphi:C^n\rightarrow R$ is any gauge function, then φ is the gauge function φ^B of the convex body B defined by

$$B=\{x\in C^n:\varphi(x)\leqslant 1\}, \hspace{2cm} (93)$$

which has 0 as an interior point.

Thus (92) and (93) establish a one-to-one correspondence between all gauge functions $\varphi:C^n\rightarrow R$ and all convex bodies $B\subset C^n$ with $0\in\mathrm{int}\,B$.

52. A set $B \subset C^n$ is called *equilibrated* if

$$x \in B, |\lambda| \leqslant 1 \Rightarrow \lambda x \in B.$$

Clearly, 0 is an interior point of any equilibrated convex body.

Let B be a convex body, $0 \in \mathrm{int}\, B$. Then B is equilibrated if and only if its gauge function φ^B satisfies

$$\varphi^B(\lambda x) = |\lambda| \varphi^B(x) \qquad \text{for all } \lambda \in C, \, x \in C^n. \tag{94}$$

53. *Vector norms.* From the definitions of a *vector norm* (Ex. 1.29) and a *gauge function* (Ex. 50) it follows that a function $\varphi : C^n \to R$ is a norm if and only if φ is a gauge function satisfying (94).

Thus (92) and (93) establish a one-to-one correspondence between all norms $\varphi : C^n \to R$ and all equilibrated convex bodies $B \subset C^n$ (Householder [1], Chapter 2).

54. If a norm $\varphi : C^n \to R$ is *unitarily invariant* (i.e., if $\varphi(Ux) = \varphi(x)$ for all $x \in C^n$ and any unitary $U \in C^{n \times n}$) then $\varphi : C^n \to R$ is a symmetric gauge function (see Ex. 50). Is the converse true?

55. *Dual norms.* The *dual* (also *polar*) of a nonempty set $B \subset C^n$ is the set B_D defined by

$$B_D = \{ y \in C^n : x \in B \Rightarrow \mathrm{Re}(y,x) \leqslant 1 \}. \tag{95}$$

Let $B \subset C^n$ be an equilibrated convex body. Then

  (a) B_D is an equilibrated convex body.

  (b) $(B_D)_D = B$, i.e., B is the dual of its dual.

  (c) Let φ^B be the norm corresponding to B via (92). Then the dual of φ^B, computed by (85),

$$\varphi_D^B(y) = \sup_{x \neq 0} \frac{\mathrm{Re}(y,x)}{\varphi^B(x)}, \tag{96}$$

is the norm corresponding to B_D. The norm φ_D^B, defined by (96), is called the *dual* of φ^B.

  (d) $(\varphi_D^B)_D = \varphi^B$, i.e., φ^B is the dual of its dual. Such pairs $\{\varphi^B, \varphi_D^B\}$ are called *dual norms* (Householder [1], Chapter 2).

56. l_p-*norms.* If φ is an l_p-norm, $p \geqslant 1$, (see Exs. 1.30–1.31), then its dual φ_D is an l_q-norm where q is determined by

$$\frac{1}{p} + \frac{1}{q} = 1.$$

In particular, the l_1 and l_∞ norms are dual, while the Euclidean norm (the l_2-norm) is self-dual.

57. *The generalized Cauchy inequality.* Let $\{\varphi, \varphi_D\}$ be dual norms on C^n. Then

$$\text{Re}(y, x) \leqslant \varphi(x)\varphi_D(y), \qquad \text{for all } x, y \in C^n, \tag{89}$$

and for any $x \neq 0$ [$y \neq 0$] there exists a $y \neq 0$ [$x \neq 0$] giving equality in (89). Such pairs $\{x, y\}$ are called *dual vectors* (*with respect to the norm* φ).

If φ is the Euclidean norm, then (89) reduces to the classical Cauchy inequality (Householder [1]).

58. *A Tchebycheff solution of* $Ax = b$, $A \in C_n^{(n+1) \times n}$. A *Tchebycheff approximate solution* of the system

$$Ax = b \tag{1}$$

is, by the definition in Theorem 6(i), a vector x minimizing the Tchebycheff norm

$$\|r\|_\infty = \max_{i=1,\ldots,m} \{|r_i|\}$$

of the residual vector

$$r(x) = b - Ax. \tag{2}$$

Let $A \in C_n^{(n+1) \times n}$ and $b \in C^{n+1}$ be given such that (1) is inconsistent. Then (1) has a unique Tchebycheff approximate solution given by

$$x = A^\dagger (b + r), \tag{97}$$

where the residual $r = [r_i]$ is

$$r_i = \frac{\sum_{j=1}^{n+1} |(P_{N(A^*)}b)_j|^2}{\sum_{j=1}^{n+1} |(P_{N(A^*)}b)_j|} \frac{(P_{N(A^*)}b)_i}{|(P_{N(A^*)}b)_i|}, \qquad i = 1, \ldots, n+1. \tag{98}$$

(The real case appeared in Cheney [1], p. 41, and Meicler [1].)

PROOF. From

$$r(x) - b = -Ax \in R(A)$$

it follows that any residual r satisfies

$$P_{N(A^*)}r = P_{N(A^*)}b$$

or equivalently

$$(P_{N(A^*)}b, r) = (b, P_{N(A^*)}b), \tag{99}$$

since $\dim N(A^*) = 1$ and $b \notin R(A)$. (Equation (99) represents the hyperplane of residuals; see, e.g., Cheney [1], Lemma, p. 40). A routine computation now shows, that among all vectors r satisfying (99) there is a unique vector of minimum Tchebycheff norm given by (98), from which (97) follows since $N(A) = \{0\}$. ∎

59. Let $A \in C_n^{(n+1) \times n}$ and $b \in C^{n+1}$ be given such that (1) is inconsistent. Then, for any norm φ on C^n, a φ-approximate solution of (1) is given by

$$x = A^\dagger(b + r),$$

where the residual r is a dual vector of $P_{N(A^*)}b$ with respect to the norm φ, and the error of approximation is

$$\varphi(r) = \frac{(b, P_{N(A^*)}b)}{\varphi_D(P_{N(A^*)}b)} \, .$$

PROOF. Follows from (99) and Ex. 57.

60. Let $\{\varphi, \varphi_D\}$ be dual norms with unit balls $B = \{x : \varphi(x) \leqslant 1\}$ and $B_D = \{y : \varphi_D(y) \leqslant 1\}$, respectively, and let $\{x_0, y_0\}$ be dual vectors of norm one, i.e., $\varphi(x_0) = 1$, $\varphi_D(y_0) = 1$, and

$$(x_0, y_0) = \varphi(x_0)\varphi_D(y_0).$$

Then

(a) The hyperplane

$$H = \{ x : \mathrm{Re}(y_0, x) = \varphi(x_0)\varphi_D(y_0) \}$$

supports B at x_0, that is, $x_0 \in H$ and B lies on one side of H, i.e.,

$$x \in B \Rightarrow \mathrm{Re}(y_0, x) \leqslant \mathrm{Re}(y_0, x_0) = \varphi(x_0)\varphi_D(y_0).$$

(b) The hyperplane

$$\{ y : \mathrm{Re}(x_0, y) = \varphi(x_0)\varphi_D(y_0) \}$$

supports B_D at y_0.

PROOF. Follows from (89).

61. A closed convex set B is called *rotund* if its boundary contains no line segments, or equivalently, if each one of its boundary points is an extreme point.

A closed convex set is called *smooth* if it has, at each boundary point, a unique supporting hyperplane.

Show that an equilibrated convex body B is rotund if and only if its dual B_D is smooth.

PROOF. *If.* If B is not rotund then its boundary contains two points $x_1 \neq x_2$ and the line segment $\{\lambda x_1 + (1-\lambda) x_2 : 0 \leqslant \lambda \leqslant 1\}$ joining them; that is,

$$\varphi(\lambda x_1 + (1-\lambda) x_2) = 1, \qquad 0 \leqslant \lambda \leqslant 1,$$

where φ is the gauge function of B.

For any $0 < \lambda < 1$ let y_λ be a dual vector of $\lambda x_1 + (1-\lambda) x_2$ with $\varphi_D(y_\lambda) = 1$. Then

$$\mathrm{Re}(\lambda x_1 + (1-\lambda) x_2, y_\lambda) = 1$$

and, by (89)

$$\mathrm{Re}(x_1, y_\lambda) = \mathrm{Re}(x_2, y_\lambda) = 1,$$

showing that y_λ is a dual vector of both x_1 and x_2, and by Ex. 60(b), both hyperplanes

$$\{y : \mathrm{Re}(x_i, y) = 1\}$$

support B_D at y_λ.

Only if. Follows by reversing the above steps. ∎

For additional results and references on rotundity in linear spaces see the survey of Cudia [1].

62. Let φ be a norm on C^n and let B be its unit ball,

$$B = \{x : \varphi(x) \leqslant 1\}.$$

Then

(a) φ is e.s.c. if and only if B is rotund.

(b) φ is Gateaux differentiable; that is the limit

$$\varphi'(x; y) = \lim_{t \to 0} \frac{\varphi(x + ty) - \varphi(x)}{t}$$

exists for all $x, y \in C^n$, if and only if B is smooth.

63. Give an example of dual norms $\{\varphi, \varphi_D\}$ such that φ is e.s.c. but φ_D is not.

Solution. Let

$$B = \left\{ \begin{bmatrix} x_1 \\ x_2 \end{bmatrix} \in R^2 : x_1 \geqslant \tfrac{1}{2}(x_2+1)^2 - 1,\ x_2 \geqslant \tfrac{1}{2}(x_1+1)^2 - 1 \right\}.$$

Then B is an equilibrated convex body. B is rotund but not smooth (the points $\begin{bmatrix} 1 \\ 1 \end{bmatrix}$ and $\begin{bmatrix} -1 \\ -1 \end{bmatrix}$ are "corners" of B), so, by Ex. 59, the dual set B_D is not rotund. Hence, by Ex. 62(a), the gauge function φ^B is an e.s.c. norm but its dual φ_D^B is not.

64. *Norms of homogeneous transformations.* Let α and β be norms on C^n and C^m, respectively. Let $A : C^n \rightarrow C^m$ be a continuous transformation which is homogeneous; that is

$$A(\lambda x) = \lambda A(x), \qquad \text{for all } \lambda \in C, x \in C^n.$$

The *norm* (also *least upper bound*) *of A corresponding to* $\{\alpha, \beta\}$, denoted by $\|A\|_{\alpha, \beta}$ (also by $\text{lub}_{\alpha, \beta}(A)$) is defined as

$$\|A\|_{\alpha, \beta} = \sup_{x \neq 0} \frac{\alpha(Ax)}{\beta(x)}$$

$$= \max_{\beta(x)=1} \alpha(Ax), \tag{100}$$

since A is continuous and homogeneous.

Then for any A, A_1, A_2 as above:

(a) $\quad \|A\|_{\alpha, \beta} \geqslant 0 \quad$ with equality if and only if A

is the zero transformation,

(b) $\quad \|\lambda A\|_{\alpha, \beta} = |\lambda| \|A\|_{\alpha, \beta} \quad$ for all $\lambda \in C$,

(c) $\quad \|A_1 + A_2\|_{\alpha, \beta} \leqslant \|A_1\|_{\alpha, \beta} + \|A_2\|_{\alpha, \beta}.$

(d) If B_α, B_β are the unit balls of α, β, respectively, then

$$\|A\|_{\alpha, \beta} = \inf \{ \lambda > 0 : AB_\beta \subset \lambda B_\alpha \}.$$

(e) If

$$A_1 : C^n \to C^m \quad \text{and} \quad A_2 : C^m \to C^p$$

are continuous homogeneous transformations and if α, β, and γ are norms on C^n, C^m, and C^p, respectively, then

$$\|A_2 A_1\|_{\alpha,\gamma} \leqslant \|A_1\|_{\alpha,\beta} \|A_2\|_{\beta,\gamma}$$

(Bauer [2], Householder [1]).

65. If $A : C^n \to C^m$ is a linear transformation and if $\alpha = \beta$, i.e., if the same norm is used in C^n and C^m, then definition (100) reduces to that given in Ex. 1.34.

Let α and β be norms on C^n and C^m, respectively. Then for any $A \in C^{m \times n}$

$$\|A\|_{\alpha,\beta} = \|A^*\|_{\beta_D, \alpha_D}. \tag{101}$$

PROOF. From (89) and (100) it follows that for all $x \in C^n$, $y \in C^m$

$$\text{Re}(Ax,y) \leqslant \alpha(Ax)\alpha_D(y) \leqslant \|A\|_{\alpha,\beta}\beta(x)\alpha_D(y),$$

with equality for at least one pair $x \neq 0, y \neq 0$. The dual inequalities

$$\text{Re}(x,A^*y) \leqslant \beta(x)\beta_D(A^*x) \leqslant \|A^*\|_{\beta_D,\alpha_D}\alpha_D(y)\beta(x)$$

then show that

$$\|A\|_{\alpha,\beta} \leqslant \|A^*\|_{\beta_D,\alpha_D},$$

from which (101) follows by reversing the roles of A and A^* and by using Ex. 55(d). ∎

66. *Projective bounds.* Let α be an e.s.c. norm on C^n. The *projective bound* of α, denoted by $Q(\alpha)$, is defined as

$$Q(\alpha) = \sup_L \|P_{L,\alpha}\|_{\alpha,\alpha}, \tag{102}$$

where the supremum is taken over all subspaces L with dimension $1 \leqslant \dim L \leqslant n-1$. (The α-metric projector $P_{L,\alpha}$ is continuous and homogeneous, by Theorem 5(c) and (i), allowing the use of (100) to define $\|P_{L,\alpha}\|_{\alpha,\alpha}$).

Then

(a) The supremum in (102) is finite and is attained for a k-dimensional subspace, for each $k = 1, 2, \ldots, n-1$.

(b) The projective bound satisfies

$$1 \leqslant Q(\alpha) < 2 \tag{103}$$

and the upper limit is approached arbitrarily closely by e.s.c. norms (Erdelsky [1]).

PROOF. (a) It can be shown that the $n-1$ sets of real numbers

$$S_j = \{\alpha(P_{L,\alpha}(x)) : \alpha(x) = 1, \quad L \text{ is } j\text{-dimensional}\}, \qquad j = 1, 2, \ldots, n-1,$$

are identical, bounded and contain the supremum $Q(\alpha)$.

(b) From Theorem 5(a) and (h) it follows that

$$1 \leqslant Q(\alpha) \leqslant 2.$$

Let x be such that $\|P_{L,\alpha}\|_{\alpha,\alpha} = \alpha(P_{L,\alpha}(x))$ and $\alpha(x) = 1$. Then $P_{L,\alpha}(x) \neq 0$ and consequently

$$1 = \alpha(x) = \alpha(0-x) > \alpha(P_{L,\alpha}(x) - x)$$

and

$$\|P_{L,\alpha}\|_{\alpha,\alpha} = \alpha(P_{L,\alpha}(x)) \leqslant \alpha(P_{L,\alpha}(x) - x) + \alpha(x) < 2,$$

proving (103). Let $\{B_k\}$ be a sequence of rotund equilibrated convex bodies in R^2 satisfying

$$B_{k+1} \subset B_k, \qquad k = 1, 2, \ldots$$

and "converging" to

$$B = \left\{ \begin{bmatrix} x_1 \\ x_2 \end{bmatrix} \in R^2 : |x_1| \leqslant 1, |x_2| \leqslant 1 \right\}.$$

Then the corresponding norms $\{\varphi^{B_k}\}$ are e.s.c., by Ex. 62(a), and "approximate" φ^B, which is the l_∞-norm on R^2,

$$\varphi^B\left(\begin{bmatrix} x_1 \\ x_2 \end{bmatrix} \right) = \max\{|x_1|, |x_2|\}.$$

Finally, by (92)

$$\varphi^{B_k}(x) \leqslant \varphi^{B_{k+1}}(x), \qquad k = 1, 2, \ldots$$

and

$$\sup_k Q(\varphi^{B_k}) = 2. \quad \blacksquare$$

67. *Projective norms.* An e.s.c. norm α on C^n for which the projective bound

$$Q(\alpha) = 1$$

is called a *projective norm.* All ellipsoidal norms

$$\|x\|_U = (x^* U x)^{1/2}, \qquad U \text{ positive definite}, \tag{46}$$

are projective.

Conversely, for spaces of dimension ≥ 3, all projective norms are ellipsoidal, both in the real case (Kakutani [1]) and in the complex case (Bohnenblust [1]). An example of a nonellipsoidal projective norm on R^2 is

$$\alpha\left(\begin{bmatrix} x_1 \\ x_2 \end{bmatrix}\right) = \begin{cases} (|x_1|^p + |x_2|^p)^{1/p} & \text{if } x_1 x_2 \geq 0 \\ (|x_1|^q + |x_2|^q)^{1/q} & \text{if } x_1 x_2 < 0 \end{cases}$$

where $(1/p) + (1/q) = 1$, $1 < p \neq 2$ (Erdelsky [1]).

68. If α is a projective norm, L is a subspace for which the α-metric projector $P_{L,\alpha}$ is linear, and N denotes

$$N = P_{L,\alpha}^{-1}(0), \tag{104}$$

then

$$L = P_{N,\alpha}^{-1}(0) \qquad \text{(Erdelsky [1])}.$$

PROOF. $L \subset P_{N,\alpha}^{-1}(0)$. If $x \in L$ and $y \in N$ then

$$P_{L,\alpha}(x+y) = x,$$

by Theorem 5(a) and consequently,

$$\alpha(x) \leq \|P_{L,\alpha}\|_{\alpha,\alpha} \alpha(x+y)$$

$$\leq Q(\alpha)\alpha(x+y)$$

$$= \alpha(x+y)$$

for all $y \in N$, proving that $P_{N,\alpha}(x) = 0$.

$P_{N,\alpha}^{-1}(0) \subset L$. If $x \in P_{N,\alpha}^{-1}(0)$, then, by (76), it can be written as

$$x = x_1 + x_2, \qquad x_1 \in L, \qquad x_2 \in N.$$

Therefore,

$$0 = P_{N,\alpha}(x) = P_{N,\alpha}(x_1) + x_2, \qquad \text{by Theorem 5(d)}$$

$$= x_2, \qquad \text{since } L \subset P_{N,\alpha}^{-1}(0),$$

proving that

$$x = x_1 \in L. \qquad \blacksquare$$

Projective Norms and the Linearity of Metric Projectors

The following four exercises probe the relations between the linearity of the α-metric projector $P_{L,\alpha}$ and the projectivity of the norm α. Exercise 69 shows that

$$\alpha \text{ projective} \Rightarrow P_{L,\alpha} \text{ linear for all } L,$$

and a partial converse is proved in Ex. 71.

69. If α is a projective norm on C^n, then $P_{L,\alpha}$ is linear for all subspaces L of C^n (Erdelsky [1]).

PROOF. By Corollary 7 it suffices to prove linearity of $P_{L,\alpha}$ for all one-dimensional subspaces L.

Let $\dim L = 1$, $l \in L$, $\alpha(l) = 1$, and let $l + N$ be a supporting hyperplane of $B_\alpha = \{x : \alpha(x) \leqslant 1\}$ at l. Since

$$\alpha(l) \leqslant \alpha(x), \qquad \text{for all } x \in l + N,$$

it follows from Definition 1 that

$$P_{N,\alpha}(l) = 0$$

and hence

$$L \subset P_{N,\alpha}^{-1}(0).$$

Now $P_{N,\alpha}$ is linear by Corollary 6, since $\dim N = n - 1$, which also shows that $P_{N,\alpha}^{-1}(0)$ is a 1-dimensional subspace, by (76), and hence

$$L = P_{N,\alpha}^{-1}(0).$$

From Ex. 68 it follows then that

$$N = P_{L,\alpha}^{-1}(0),$$

and the linearity of $P_{L,\alpha}$ is established by Corollary 5(b). ∎

70. If α is an e.s.c. norm on C^n, L is a subspace for which $P_{L,\alpha}$ is linear, and N denotes

$$N = P_{L,\alpha}^{-1}(0). \tag{104}$$

Then

$$L = P_{N,\alpha}^{-1}(0) \tag{105}$$

if, and only if,

$$P_{L,\alpha} + P_{N,\alpha} = I \quad \text{(Erdelsky [1]).}$$

PROOF. Follows from (76).

71. Let α be an e.s.c. norm on C^n and let $1 \leqslant k \leqslant n-1$ be an integer such that, for every k-dimensional subspace L of C^n:

$$P_{L,\alpha} \text{ is linear}$$

and

$$L = P_{N,\alpha}^{-1}(0), \tag{105}$$

where N is given by (104). Then α is projective (Erdelsky [1]).

PROOF. Let α be nonprojective; i.e., let $Q(\alpha) > 1$. Then there is a k-dimensional subspace L and two points x, y in C^n such that

$$y = P_{L,\alpha}(x) \tag{106}$$

and

$$\alpha(y) = \|P_{L,\alpha}\|_{\alpha,\alpha}\alpha(x) = Q(\alpha)\alpha(x) > \alpha(x). \tag{107}$$

Let $N = P_{L,\alpha}^{-1}(0)$. Then

$$0 \neq y - x \in N, \quad \text{by (107), (106), and (76)} \tag{108}$$

and

$$\alpha(x) = \alpha(y - (y - x)) < \alpha(y). \tag{109}$$

Now

$$y = P_{L,\alpha}(y) + P_{N,\alpha}(y), \quad \text{by (105) and Ex. 70}$$

$$= y + P_{N,\alpha}(y), \quad \text{by (106) and Theorem 5(a),}$$

proving that

$$P_{N,\alpha}(y) = 0,$$

which, by (108) and (75), contradicts (109).   ∎

72. Let φ_p be the l_p-norm, $1 < p < \infty$, on C^n. Then P_{L,φ_p} is linear for every subspace L if and only if $p = 2$ (Newman and Odell [1]).

73. *Essentially strictly convex norms.* Let α be an e.s.c. norm on C^n, $0 \neq x \in C^n$ and L a subspace of C^n. Then

$$x \in P_{L,\alpha}^{-1}(0)$$

if, and only if, there is a dual y of x with respect to α (i.e., a vector $y \neq 0$ satisfying $(y,x) = \alpha(x)\alpha_D(y)$), such that $y \in L^\perp$ (Erdelsky [1]).

74. If α and α_D are both e.s.c. norms on C^n, L is a subspace of C^n for which $P_{L,\alpha}$ is linear, and $N = P_{L,\alpha}^{-1}(0)$, then

(a) $L^\perp = P_{N^\perp,\alpha_D}^{-1}(0)$,

(b) $P_{N^\perp,\alpha_D} = (P_{L,\alpha})^*$ (Erdelsky [1]).

PROOF.   (a) Since both α and α_D are e.s.c., it follows from Exs. 62(a), 61, and 60 that every $0 \neq x$ has a dual $0 \neq y$ with respect to α, and x is a dual of y. Now

$$y \in P_{N^\perp,\alpha_D}^{-1}(0) \Leftrightarrow x \in N^{\perp\perp} = N,$$

by Ex. 73, which also shows that

$$x \in N \Leftrightarrow y \in L^\perp,$$

proving (a).

(b) By (a) and Corollary 5(b), $P_{N^\perp,\alpha_D}$ is linear. Let x and y be arbitrary vectors, written as

$$x = x_1 + x_2, \qquad x_1 \in L, \qquad x_2 \in N, \qquad \text{by (76)}$$

and

$$y = y_1 + y_2, \qquad y_1 \in N^\perp, \qquad y_2 \in L^\perp, \qquad \text{by (a) and (76).}$$

Then

$$(P_{L,\alpha}(x), y) = (x_1, y_1) = (x, P_{N^\perp,\alpha_D}(y)).$$

75. *Dual norms.*   Let α and α_D be dual norms on C^n. Then:

(a) If α and α_D are both e.s.c., then $Q(\alpha) = Q(\alpha_D)$.

(b) If α is projective, then α_D is e.s.c.

(c) If α is projective, then so is α_D (Erdelsky [1]).

α-β Generalized Inverses

76. Let α and β be e.s.c. norms on C^m and C^n, respectively, and let $A \in C^{m \times n}$.

If $B \in C^{n \times m}$ satisfies

$$AB = P_{R(A), \alpha}, \tag{110}$$

$$BA = I - P_{N(A), \beta}, \tag{111}$$

$$\text{rank } B = \text{rank } A, \tag{112}$$

then

$$B = A_{\alpha, \beta}^{(-1)} \quad \text{(Erdelsky [1]).}$$

Thus, if the α-β generalized inverse of A is linear, it can be defined by (110)–(112).

77. Let α and β be e.s.c. norms on C^m and C^n, respectively. Then

$$\left(A_{\alpha, \beta}^{(-1)} \right)_{\beta, \alpha}^{(-1)} = A \quad \text{for all } A \in C^{m \times n} \tag{113}$$

if and only if α and β are projective norms (Erdelsky [1]).

PROOF. If α and β are projective, then $A_{\alpha, \beta}^{(-1)}$ is linear for any A, by Theorem 6(b) and Ex. 69. Let $R' = R(A_{\alpha, \beta}^{(-1)})$ and $N' = N(A_{\alpha, \beta}^{(-1)})$. Then by Exs. 68. 69. 73 and Theorem 6(c), (d), (e), (g), and (h),

$$A_{\alpha, \beta}^{(-1)} A = I - P_{N(A), \beta} = P_{R', \beta},$$

$$A A_{\alpha, \beta}^{(-1)} = P_{R(A), \alpha} = I - P_{N', \alpha},$$

$$\text{rank} A_{\alpha, \beta}^{(-1)} = \text{rank} A,$$

and (113) follows from Ex. 76.

Only if. If (113) holds for all $A \in C^{m \times n}$ then

$$I - P_{N(A), \beta} = A_{\alpha, \beta}^{(-1)} A = A_{\alpha, \beta}^{(-1)} \left(A_{\alpha, \beta}^{(-1)} \right)_{\beta, \alpha}^{(-1)} = P_{R', \beta},$$

$$P_{R(A), \alpha} = A A_{\alpha, \beta}^{(-1)} = \left(A_{\alpha, \beta}^{(-1)} \right)_{\beta, \alpha}^{(-1)} A_{\alpha, \beta}^{(-1)} = I - P_{N', \alpha},$$

and α and β are projective by Ex. 71. ∎

78. If α and β are projective norms on C^m and C^n, respectively, then

$$\left(A_{\alpha, \beta}^{(-1)} \right)^* = \left(A^* \right)_{\beta_D, \alpha_D}^{(-1)}, \quad \text{for all } A \in C^{m \times n} \quad \text{(Erdelsky [1]).} \tag{114}$$

PROOF.   From Theorem 6(d) and (f) and Exs. 68, 69, and 70

$$AA_{\alpha,\beta}^{(-1)} = P_{R(A),\alpha} = I - P_{N,\alpha}, \qquad N = P_{R(A),\alpha}^{-1}(0),$$

$$A_{\alpha,\beta}^{(-1)}A = I - P_{N(A),\beta} = P_{M,\beta}, \qquad M = P_{N(A),\beta}^{-1}(0),$$

and

$$R(A) = P_{N,\alpha}^{(-1)}(0),$$

$$N(A) = P_{M,\alpha}^{(-1)}(0).$$

Since α_D and β_D are e.s.c. norms, by Ex. 75(b), it follows from Ex. 74(b) that

$$AA_{\alpha,\beta}^{(-1)} = I - (P_{R(A)^{\perp},\alpha_D})^* = I - (P_{N(A^*),\alpha_D})^*,$$

$$A_{\alpha,\beta}^{(-1)}A = (P_{N(A)^{\perp},\beta_D})^* = (P_{R(A^*),\beta_D})^*,$$

and hence

$$(A_{\alpha,\beta}^{(-1)})^*A^* = I - P_{N(A^*),\alpha_D},$$

$$A^*(A_{\alpha,\beta}^{(-1)})^* = P_{R(A^*),\beta_D},$$

from which (114) follows by using Ex. 76. ∎

79. If α and β are e.s.c. norms on C^m and C^n, respectively, then for any $O \neq A \in C^{m \times n}$

$$\frac{1}{\|A_{\alpha,\beta}^{(-1)}\|_{\beta,\alpha}} \leqslant \inf\{\|X\|_{\alpha,\beta} : X \in C^{m \times n}, \text{rank }(A+X) < \text{rank } A\}$$

$$\leqslant \frac{q}{\|A_{\alpha,\beta}^{(-1)}\|_{\beta,\alpha}}, \tag{115}$$

where

$$q = 1 \qquad \text{if rank } A = m,$$

and

$$q = Q(\alpha) \qquad \text{otherwise.}$$

In particular, if α is projective,

$$\frac{1}{\|A_{\alpha,\beta}^{(-1)}\|_{\beta,\alpha}} = \inf\left\{ \|X\|_{\alpha,\beta} : X \in C^{m\times n}, \text{ rank } (A+X) < \text{rank } A \right\}$$

(Erdelsky [1]). (116)

A special case of (116) is given in Ex. 6.23 below.

5. AN EXTREMAL PROPERTY OF THE BOTT–DUFFIN INVERSE WITH APPLICATION TO ELECTRICAL NETWORKS

An important extremal property of the Bott–Duffin inverse, studied in Sections 2.9 and 2.12, is stated in the following theorem.

THEOREM 7 (Bott and Duffin [1]).   Let $A \in C^{n\times n}$ be Hermitian, and let L be a subspace of C^n such that $A_{(L)}^{(-1)}$ exists.* Then, for any two vectors $v, w \in C^n$, the quadratic function

$$q(x) = \tfrac{1}{2}(x-v)^*A(x-v) - w^*x \tag{117}$$

has a unique stationary value in L, when

$$x = A_{(L)}^{(-1)}(Av + w). \tag{118}$$

Conversely, if the Hermitian matrix A and the subspace L are such that for any two vectors $v, w \in C^n$, the quadratic function (117) has a stationary value in L, then $A_{(L)}^{(-1)}$ exists and the stationary point is unique for any v, w and given by (118).

PROOF. *A stationary point of q in L* is a point $x \in L$ at which the gradient

$$\nabla q(x) = \left[\frac{\partial}{\partial x_j} q(x) \right] \qquad (j = 1, \ldots, n)$$

is orthogonal to L, i.e., $\nabla q(x) \in L^\perp$. The value of q at a stationary point is called a *stationary value* of q.

Differentiating (117) we see that the sought stationary point $x \in L$ satisfies

$$\nabla q(x) = A(x-v) - w \in L^\perp,$$

*See Ex. 2.80 for conditions equivalent to the existence of $A_{(L)}^{(-1)}$.

and by taking $y = -\nabla q(x)$ we conclude that x is a stationary point of q in L if and only if x is a solution of

$$Ax + y = Av + w, \qquad x \in L, y \in L^{\perp}. \tag{119}$$

Thus the existence of a stationary value of q for any v, w is equivalent to the consistency of (119) for any v, w, i.e., to the existence of $A_{(L)}^{(-1)}$, in which case (118) is the unique stationary point in L. ■

COROLLARY 8. Let $A \in C^{n \times n}$ be Hermitian positive definite and let L be a subspace of C^n. Then for any $v, w \in C^n$ the function

$$q(x) = \tfrac{1}{2}(x - v)^* A(x - v) - w^* x \tag{117}$$

has a unique minimum in L, when

$$x = A_{(L)}^{(-1)}(Av + w). \tag{118}$$

PROOF. Follows from Theorem 7, since $A_{(L)}^{(-1)}$ exists, by Ex. 2.95, and the stationary value of q is actually a minimum since A is positive definite. ■

We return now to the direct current electrical network of Section 2.12, consisting of m nodes $\{n_i : i = 1, \ldots, m\}$ and n branches $\{b_j : j = 1, \ldots, n\}$, with

$a_j > 0$, the *conductance* of b_j,

$A = [\operatorname{diag} a_j]$, the *conductance matrix*,

x_j, the *voltage* across b_j,

y_j the *current* in b_j,

v_j, the *voltage* generated by the *source* in series with b_j,

w_j, the *current* generated by the *source* in parallel with b_j

and

M, the (node-branch) *incidence matrix*.

We recall that the branch voltages x and currents y are uniquely determined by the following three physical laws:

$$Ax + y = Av + w \qquad (\textit{Ohm's Law}), \tag{120}$$

$$y \in N(M) \qquad (\textit{Kirchhoff's current law}), \tag{121}$$

$$x \in R(M^T) \qquad (\textit{Kirchhoff's voltage law}), \tag{122}$$

and that x, y are given by

$$x = A^{(-1)}_{(R(M^T))}(Av + w), \tag{2.138}$$

$$y = (I - AA^{(-1)}_{(R(M^T))})(Av + w), \tag{2.139}$$

or dually, by (2.142) and (2.141).

A classical variational principle of Kelvin (Thomson [1]) and Maxwell ([1], pp. 903–908), states that the voltages x and the currents y are such that the rate of energy dissipation in the network is minimized. This variational principle is given in the following corollary.

COROLLARY 9.   Let A, M, x, y, v, w be as above. Then
(a) The vector x_0 of branch voltages is the unique minimizer of

$$q(x) = \tfrac{1}{2}(x - v)^* A(x - v) - w^* x \tag{117}$$

in $R(M^T)$, and the vector y_0 of branch currents is

$$y_0 = -\nabla q(x_0) = -A(x_0 - v) + w \in R(M^T)^{\perp} = N(M). \tag{123}$$

(b) The vector y_0 is unique minimizer of

$$p(y) = \tfrac{1}{2}(y - w)^* A^{-1}(y - w) - v^* y \tag{124}$$

in $N(M)$, and the vector x_0 is

$$x_0 = -\nabla p(y_0) = -A^{-1}(y_0 - w) + v \in N(M)^{\perp} = R(M^T). \tag{125}$$

PROOF.   Since the conductance matrix A is positive definite, it follows by comparing (118) and (2.138) that x_0 is the unique minimizer of (117) in $R(M^T)$, and the argument used in the proof of Theorem 7 shows that $y_0 = -\nabla q(x_0)$ as given in (123). Part (b) follows from the dual derivation (2.141) and (2.142) of y_0 and x_0, respectively, as solutions of the dual network equations (2.140). ∎

Corollary 9 shows that the voltage x is uniquely determined by the function (117) to be minimized subject to Kirchhoff's voltage law (122). Kirchhoff's current law (121) and Ohm's law (120) are then consequences of (123).

Dually, the current y is uniquely determined by the function (124) to be minimized subject to Kirchhoff's current law (121), and the other two laws (120) and (122) then follow from (125).

Further references on the extremal properties of the network functions and solutions are Dennis [1], Stern [1] and [2], and Guillemin [1]. Corollary 9 is a special case of the Duality Theory of Convex Programming; see, e.g., Rockafellar [1].

EXERCISES

80. Let $A \in C^{n \times n}$ be Hermitian positive semidefinite, and let the subspace $L \subset C^n$ and the vector $w \in C^n$ be given. Then the quadratic function

$$\tfrac{1}{2} x^* A x - w^* x \tag{126}$$

has a minimum in L if and only if the system

$$A x - w \in L^\perp, \quad x \in L \tag{127}$$

is consistent, in which case the solutions x of (127) are the minimizers of (126) in L.

81. Show that the consistency of (127) is equivalent to the condition

$$x \in L, \, A x = 0 \Rightarrow w^* x = 0,$$

which is obviously equivalent to the boundedness from below of (126) in L, hence to the existence of a minimizer in L.

82. Show that $A_{(L)}^{(-1)}$ exists if and only if the system (127) has a unique solution for any $w \in C^n$, in which case this solution is

$$x = A_{(L)}^{(-1)} w.$$

83. Give the general solution of (127) in case it is consistent but $A_{(L)}^{(-1)}$ does not exist.

SUGGESTED FURTHER READING

Section 1.   Desoer and Whalen [1], Erdelyi and Ben-Israel [1], Levinge and Wedin [1], Osborne [1], Peters and Wilkinson [1], and the references on applications to statistics given at the end of the Introduction.

For various applications in control theory and in system theory, see Balakrishnan [1], Barnett [1], Ho and Kalman [1], Kalman ([1], [2], [3], [4]), Kalman, Ho, and Narenda [1], Kishi [1], Kuo and Kazda [1], Minamide and Nakamura ([1], [2]), Porter ([1], [2]), Porter and Williams ([1], [2]), Wahba and Nashed [1], and Zadeh and Desoer [1].

Section 2.   Erdelyi and Ben-Israel [1], Osborne [2], Rosen [3].

4

SPECTRAL GENERALIZED INVERSES

1. INTRODUCTION

In this chapter we shall study generalized inverses having some of the spectral properties (i.e., properties relating to eigenvalues and eigenvectors) of the inverse of a nonsingular matrix. Only square matrices are considered, since only they have eigenvalues and eigenvectors.

The four Penrose equations of Chapter 1,

$$AXA = A, \tag{1}$$

$$XAX = X, \tag{2}$$

$$(AX)^* = AX, \tag{3}$$

$$(XA)^* = XA, \tag{4}$$

will now be supplemented further by the following equations applicable only to square matrices:

$$A^k XA = A^k, \tag{1^k}$$

$$AX = XA, \tag{5}$$

$$A^k X = XA^k, \tag{5^k}$$

$$AX^k = X^k A. \tag{6^k}$$

In these equations k is a given positive integer. For example, we shall have occasion to refer to a $\{1^k, 2, 5\}$-inverse of A.

159

2. SPECTRAL PROPERTIES OF A NONSINGULAR MATRIX

If A is nonsingular it is easy to see that every eigenvector of A associated with the eigenvalue λ is also an eigenvector of A^{-1} associated with the eigenvalue λ^{-1}. (A nonsingular matrix does not have 0 as an eigenvalue.)

A matrix $A \in C^{n \times n}$ that is not diagonable does not have n linearly independent eigenvectors (see Ex. 2.25). However, it does have n linearly independent principal vectors. Following Wilkinson [1], we define a *principal vector* of A of *grade p* associated with the eigenvalue λ as a vector x such that

$$(A - \lambda I)^p x = 0, \qquad (A - \lambda I)^{p-1} x \neq 0.$$

Here p is some positive integer.

Evidently principal vectors are a generalization of eigenvectors. In fact, an eigenvector is a principal vector of grade 1. We shall find it convenient to abbreviate "principal vector of grade p associated with the eigenvalue λ" to "λ-vector of A of grade p."

It is not difficult to show (see Ex. 3) that, if A is nonsingular, a vector x is a λ^{-1}-vector of A^{-1} of grade p if and only if it is a λ-vector of A of grade p. In the remainder of this chapter, we shall explore the extent to which singular square matrices have generalized inverses with comparable spectral properties.

EXERCISES

1. A square matrix A is diagonable if and only if all its principal vectors are eigenvectors.

2. For a given eigenvalue λ, the maximal grade of the λ-vectors of A is the multiplicity of λ as a zero of the minimum polynomial of A.

3. If A is nonsingular, x is a λ^{-1}-vector of A^{-1} of grade p if and only if it is a λ-vector of A of grade p. [*Hint*: Show that $A^{-p}(A - \lambda I)^p = (-\lambda)^p (A^{-1} - \lambda^{-1} I)^p$. Using this and the analogous relation obtained by replacing A by A^{-1}, show that $(A - \lambda I)^r x = 0$ if and only if $(A^{-1} - \lambda^{-1} I)^r x = 0$ for $r = 0, 1, \ldots$.]

4. If A is nonsingular and diagonable, A^{-1} is the *only* matrix related to A by the property stated in Ex. 3.

5. If A is nonsingular and not diagonable, there are matrices other than A^{-1} having the spectral relationship to A described in Ex. 3. For example, consider

$$A = \begin{bmatrix} \lambda & 1 \\ 0 & \lambda \end{bmatrix}, \qquad X = \begin{bmatrix} \lambda^{-1} & c \\ 0 & \lambda^{-1} \end{bmatrix} \qquad (\lambda, c \neq 0).$$

Show that, for $p = 1, 2$, x is a λ^{-1}-vector of X of grade p if and only if it is a λ-vector of A of grade p. (Note that $X = A^{-1}$ for $c = -\lambda^{-2}$.)

3. SPECTRAL INVERSE OF A DIAGONABLE MATRIX

In investigating the existence of generalized inverses of a singular square matrix, we shall begin with diagonable matrices, because they are the easiest to deal with. Evidently some extension must be made of the spectral property enjoyed by nonsingular matrices, because a singular square matrix has 0 as one of its eigenvalues. Given a diagonable matrix $A \in C^{n \times n}$, let us seek a matrix X such that every eigenvector of A associated with the eigenvalue λ (for every λ in the spectrum of A) is also an eigenvector of X associated with the eigenvalue $\lambda^{\dagger}$, where $\lambda^{\dagger}$ is as defined in Section 1.3.

Since A has n linearly independent eigenvectors, there is a nonsingular matrix P, having such a set of eigenvectors as columns, such that

$$AP = PJ, \tag{7}$$

where

$$J = \mathrm{diag}(\lambda_1, \lambda_2, \ldots, \lambda_n)$$

is a Jordan form of A. We shall need the diagonal matrix obtained from J by replacing each diagonal element λ_i by $\lambda_i^{\dagger}$. By Ex. 1.21, this is, in fact, the Moore–Penrose inverse of J; that is,

$$J^{\dagger} = \mathrm{diag}(\lambda_1^{\dagger}, \lambda_2^{\dagger}, \ldots, \lambda_n^{\dagger}).$$

Because of the spectral requirement imposed on X, we must have

$$XP = PJ^{\dagger}. \tag{8}$$

Solving (7) and (8) for A and X gives

$$A = PJP^{-1}, \qquad X = PJ^{\dagger}P^{-1}. \tag{9}$$

Since J and $J^{\dagger}$ are both diagonal, they commute with each other. As a result, it follows from (9) that $X \in A\{1, 2, 5\}$.

We do not wish to limit our consideration to diagonable matrices. We began with them because they are easier to work with. The result just obtained suggests that we should examine the existence and properties (especially spectral properties) of $\{1, 2, 5\}$-inverses for square matrices in general.

4. THE GROUP INVERSE

It follows from (5) and from Corollary 2.7 that a $\{1,2,5\}$-inverse of A, if it exists, is a $\{1,2\}$-inverse X such that $R(X)=R(A)$ and $N(X)=N(A)$. By Theorem 2.10, there is at most one such inverse.

This unique $\{1,2,5\}$-inverse is called the *group inverse*, and we shall denote it by $A^{\#}$. The name "group inverse" was given by I. Erdelyi [4], because the positive and negative powers of a given matrix A (the latter being interpreted as powers of $A^{\#}$), together with the projector $AA^{\#}$ as the unit element, constitute an Abelian group. Both he and Englefield [1] (who called it the "commuting reciprocal inverse") drew attention to the spectral properties of the group inverse. As we shall see later, however, the group inverse is a particular case of the Drazin [1] pseudoinverse, or $\{1^{k},2,5\}$-inverse, which had been known for some time before the publication of Englefield [1] and Erdelyi [4].

The group inverse is not restricted to diagonable matrices; however, it does not exist for all square matrices. By Section 2.5, and Theorem 2.10, such an inverse exists if and only if $R(A)$ and $N(A)$ are complementary subspaces.

In Section 6 below we shall define the index of a square matrix A as the smallest positive integer k such that rank $A^{k}=$ rank A^{k+1}. There we shall discuss the properties of the index in some detail. Here we mention only that $R(A)$ and $N(A)$ are complementary subspaces if and only if A has index 1. This is shown in Ex. 6 at the end of this section. We have, therefore, the following theorem.

THEOREM 1.   A square matrix A has a group inverse if and only if its index is 1, or, in other words, if and only if

$$\operatorname{rank} A = \operatorname{rank} A^{2}. \tag{10}$$

When the group inverse exists, it is unique. ■

An alternative proof of uniqueness is as follows. Let $X, Y \in A\{1,2,5\}$, $E=AX=XA$, and $F=AY=YA$. Then $E=F$, since

$$E = AX = AYAX = FE,$$

$$F = YA = YAXA = FE.$$

Therefore,

$$X = EX = FX = YE = YF = Y.$$

The following theorem gives an equivalent condition for the existence of $A^{\#}$ that is often more convenient in numerical work, and also an explicit formula for $A^{\#}$.

THEOREM 2 (Cline [3]). Let a square matrix A have the full-rank factorization

$$A = FG. \tag{11}$$

Then A has a group inverse if and only if GF is nonsingular, in which case

$$A^{\#} = F(GF)^{-2}G. \tag{12}$$

PROOF. Let $r = \operatorname{rank} A$. Then $GF \in C^{r \times r}$. Now,

$$A^2 = FGFG,$$

and so

$$\operatorname{rank} A^2 = \operatorname{rank} GF$$

by Ex. 1.8. Therefore (10) holds if and only if GF is nonsingular, and the first part of the theorem is established. It is easily verified that (1), (2) and (5) hold with A given by (11) and X by the right member of (12). Formula (12) then follows from the uniqueness of the group inverse. ∎

For an important class of matrices, the group inverse and the Moore–Penrose inverse are the same. We shall call a square matrix A *range-Hermitian** if

$$R(A^*) = R(A).$$

It follows from (2.49) that a range-Hermitian matrix A also has the property

$$N(A^*) = N(A).$$

Using the notation of Theorem 2.10, the preceding discussion shows that

$$A^{\#} = A^{(1,2)}_{R(A), N(A)},$$

while Ex. 2.32 establishes that

$$A^{\dagger} = A^{(1,2)}_{R(A^*), N(A^*)}.$$

The two inverses are equal, therefore, if and only if $R(A) = R(A^*)$ and $N(A) = N(A^*)$. But this is true if and only if A is range-Hermitian. Thus we have proved:

---

*Schwerdtfeger [1], Pearl [3], and some other writers call such a matrix an EP_r or EP matrix.

THEOREM 3. $A^{\#} = A^{\dagger}$ if and only if A is range-Hermitian.

The approach of (9) can be extended from diagonable matrices to all square matrices of index 1. To do this we shall need the following lemma.

LEMMA 1. Let J be a square matrix in Jordan form. Then J is range-Hermitian if and only if it has index 1.

PROOF. *Only if:* Follows from Ex. 7.

If: If J is nonsingular, rank J = rank J^2 and J is range-Hermitian by Ex. 15. If J has only 0 as an eigenvalue, it is nilpotent. In this case, it follows easily from the structure of the Jordan form that rank $J^2 <$ rank J unless rank $J = 0$, i.e., J is a null matrix, in which case it is trivially range-Hermitian.

If J has both zero and nonzero eigenvalues, it can be partitioned in the form

$$J = \begin{bmatrix} J_1 & O \\ O & J_2 \end{bmatrix},$$

where J_1 is nonsingular and has as eigenvalues the nonzero eigenvalues of J, while J_2 is nilpotent. By the same reasoning employed in the preceding paragraph, rank J = rank J^2 implies $J_2 = O$. It then follows from Ex. 15 that J is range-Hermitian. ∎

THEOREM 4 (Erdelyi). Let A have index 1 and let

$$A = PJP^{-1},$$

where P is nonsingular and J is a Jordan normal form of A. Then

$$A^{\#} = PJ^{\dagger}P^{-1}. \tag{13}$$

PROOF. It is easily verified that relations (1), (2), (5), and (10) are similarity invariants. Therefore,

$$J^{\#} = P^{-1}A^{\#}P \tag{14}$$

and also rank J = rank J^2. It then follows from Lemma 1 and Theorem 3 that

$$J^{\#} = J^{\dagger}, \tag{15}$$

and (13) follows from (14) and (15). ∎

EXERCISES

6. Let $A \in C^{n \times n}$. Then $R(A)$ and $N(A)$ are complementary subspaces if and only if A has index 1.

PROOF. If A is nonsingular, $R(A) = C^n$ and $N(A) = \{0\}$. Thus $R(A)$ and $N(A)$ are trivially complementary. Since a nonsingular matrix has index 1, it remains to prove the statement for singular A. Since

$$\dim R(A) + \dim N(A) = \operatorname{rank} A + \operatorname{null} A = n,$$

if follows from statement (c) of Ex. 2.20 that $R(A)$ and $N(A)$ are complementary if and only if

$$R(A) \cap N(A) = \{0\}, \tag{16}$$

and we need only prove that (16) is equivalent to (10), which follows since rank $A^2 <$ rank A is equivalent to the existence of a linear relation among the columns of A^2 that does not hold for the corresponding columns of A, or, in other words, a vector x such that $A^2 x = 0$, $Ax \neq 0$. But this is tantamount to the existence of a nonzero vector $Ax \in R(A) \cap N(A)$. ∎

7. Every range-Hermitian matrix has index 1.

PROOF. If A is range-Hermitian, then by (2.49), $N(A) = R(A)^{\perp}$. Thus $R(A)$ and $N(A)$ are complementary subspaces and Ex. 6 applies.

8. If A is nonsingular, $A^{\#} = A^{-1}$.

9. $A^{\#\#} = A$.

10. $A^{*\#} = A^{\#*}$.

11. $A^{T\#} = A^{\#T}$.

12. $(A^l)^{\#} = (A^{\#})^l$ for every positive integer l.

13. Let A have index 1 and denote $(A^{\#})^j$ by A^{-j} for $j = 1, 2, \ldots$ Also denote $AA^{\#}$ by A^0. Then show that

$$A^l A^m = A^{l+m}$$

for *all* integers l and m. (Thus, the "powers" of A, positive, negative, and zero, constitute an Abelian group under matrix multiplication.)

14. Show that

$$A^{\#} = A(A^3)^{(1)}A, \tag{17}$$

where $(A^3)^{(1)}$ is an arbitrary element of $A^3\{1\}$.

15. Show that a nonsingular matrix is range-Hermitian.

16. Show that a normal matrix is range-Hermitian. (*Hint*: Use Corollary 1.2.)

Remark. It follows from Exs. 7 and 16 that

$$\{ \text{matrices of index } 1 \} \supset \{ \text{range-Hermitian matrices} \}$$

$$\supset \{ \text{normal matrices} \} \supset \{ \text{Hermitian matrices} \}.$$

17. A square matrix A is range-Hermitian if and only if A commutes with $A^{\dagger}$.

18. A square matrix A is range-Hermitian if and only if there is a matrix Y such that $A^* = YA$ (Katz [1]).

5. SPECTRAL PROPERTIES OF THE GROUP INVERSE

Even when A is not diagonable, the group inverse has spectral properties comparable to those of the inverse of a nonsingular matrix. However, in this case, $A^{\#}$ is not the only matrix having such properties. This has already been illustrated in the case of a nonsingular matrix (see Ex. 5).

We note that if a square matrix A has index 1, its 0-vectors are all of grade 1, i.e., null vectors of A. This follows from the fact that (10) implies $N(A^2) = N(A)$ by Ex. 1.11.

The following two lemmas are needed in order to establish the spectral properties of the group inverse. The second is stated in greater generality than is required for the immediate purpose because it will be used in connection with spectral generalized inverses other than the group inverse.

LEMMA 2. Let x be a λ-vector of A with $\lambda \neq 0$. Then $x \in R(A^l)$, where l is an arbitrary positive integer.

PROOF. We have

$$(A - \lambda I)^p x = 0$$

for some positive integer p. Expanding the left member by the binomial theorem, transposing the final term, and dividing by its coefficient $(-\lambda)^{p-1} \neq 0$ gives

$$x = c_1 A x + c_2 A^2 x + \cdots + c_p A^p x, \tag{18}$$

where

$$c_i = (-1)^{i-1} \lambda^{-i} \binom{p}{i}.$$

Successive multiplication of (18) by A gives

$$Ax = c_1 A^2 x + c_2 A^3 x + \cdots + c_p A^{p+1} x$$

$$A^2 x = c_1 A^3 x + c_2 A^4 x + \cdots + c_p A^{p+2} x$$

$$\cdots \cdots \cdots \cdots \cdots \cdots \cdots \cdots \cdots$$

(19)

$$A^{l-1} x = c_1 A^l x + c_2 A^{l+1} x + \cdots + c_p A^{p+l-1} x.$$

Successive substitution of equations (19) in the right member of (18) gives eventually

$$x = A^l q(A) x,$$

where q is some polynomial. ∎

LEMMA 3. Let A be a square matrix and let

$$XA^{l+1} = A^l \tag{20}$$

for some positive integer l. Then every λ-vector of A of grade p for $\lambda \neq 0$ is a λ^{-1}-vector of X of grade p.

PROOF. The proof will be by induction on the grade p. Let $\lambda \neq 0$ and $Ax = \lambda x$. Then $A^{l+1} x = \lambda^{l+1} x$, and therefore $x = \lambda^{-l-1} A^{l+1} x$. Accordingly,

$$Xx = \lambda^{-l-1} XA^{l+1} x = \lambda^{-l-1} A^l x = \lambda^{-1} x.$$

Thus the lemma is true for $p = 1$.

Now, suppose it is true for $p = 1, 2, \ldots, r$, and let x be a λ-vector of A of grade $r + 1$. Then, by Lemma 2,

$$x = A^l y$$

for some y. Thus,

$$(X - \lambda^{-1} I) x = (X - \lambda^{-1} I) A^l y = X(A^l - \lambda^{-1} A^{l+1}) y$$

$$= X(I - \lambda^{-1} A) A^l y = -\lambda^{-1} X(A - \lambda I) x.$$

By the induction hypothesis, $(A - \lambda I) x$ is a λ^{-1}-vector of X of grade r. Consequently,

$$(X - \lambda^{-1} I)^r (A - \lambda I) x = 0,$$

$$z = (X - \lambda^{-1} I)^{r-1} (A - \lambda I) x \neq 0,$$

$$Xz = \lambda^{-1} z.$$

Therefore,

$$(X - \lambda^{-1}I)^{r+1}x = -\lambda^{-1}X(X - \lambda^{-1}I)^r(A - \lambda I)x = 0,$$

$$(X - \lambda^{-1}I)^r x = -\lambda^{-1}Xz = -\lambda^{-2}z \neq 0.$$

This completes the induction. ■

The following theorem shows that for every matrix A of index 1, the group inverse is the only matrix in $A\{1\}$ or $A\{2\}$ having spectral properties comparable to those of the inverse of a nonsingular matrix. For convenience, let us call X an *S-inverse* of A (or A and X *S-inverses of each other*) if they share the property that, for every $\lambda \in C$ and every vector x, x is a λ-vector of A of grade p if and only if it is a $\lambda^\dagger$-vector of X of grade p.

THEOREM 5. Let $A \in C^{n \times n}$ have index 1. Then $A^\#$ is the unique *S-inverse* of A in $A\{1\} \cup A\{2\}$. If A is diagonable, $A^\#$ is the only *S-inverse* of A.

PROOF. First we shall show that $A^\#$ is an *S-inverse* of A. Since $X = A^\#$ satisfies (20) with $l = 1$, it follows from Lemma 3 that $A^\#$ satisfies the "if" part of the definition of *S-inverse* for $\lambda \neq 0$. Replacing A by $A^\#$ establishes the "only if" part for $\lambda \neq 0$, since $A^{\#\#} = A$ (see Ex. 9).

Since both A and $A^\#$ have index 1, all their 0-vectors are null vectors as pointed out in the second paragraph of this section. Thus, in order to prove that $A^\#$ satisfies the definition of *S-inverse* for $\lambda = 0$, we need only show that $N(A) = N(A^\#)$. But this follows from the commutativity of A and $A^\#$ and Ex. 1.11.

To show uniqueness, let $X \in A\{1\} \cup A\{2\}$ be an *S-inverse* of A. It follows that $N(X) = N(A)$. Thus A and X have the same nullity, and therefore the same rank. By Theorem 1.2, $X \in A\{1,2\}$.

Let $r = \text{rank } A$ and consider the equation

$$AP = PJ,$$

where P is nonsingular and J is a Jordan form of A. The columns of P are λ-vectors of A. Since A has index 1, those columns which are not null vectors are associated with nonzero eigenvalues, and are therefore in $R(A)$ by Lemma 2. Since there are r of them and they are linearly independent, they span $R(A)$. But, by hypothesis, these columns are also λ^{-1}-vectors of X and therefore in $R(X)$. Since rank $X = r$, these r vectors span $R(X)$, and so $R(X) = R(A)$. Thus X is a $\{1,2\}$-inverse of A such that $R(X) = R(A)$ and $N(X) = N(A)$. But $A^\#$ is the only such inverse, and so $X = A^\#$.

It was shown in Section 3 that if A is diagonable, an S-inverse of A must be a $\{1,2,5\}$-inverse. Since $A^{\#}$ is the only such inverse, this completes the proof. ∎

6. THE DRAZIN PSEUDOINVERSE.
INDEX OF A SQUARE MATRIX

We have seen that the group inverse does not exist for all square matrices, but only those of index 1. However, we shall show in this section that every square matrix has a unique $\{1^k, 2, 5\}$-inverse, where k is its index. This inverse is commonly known as the *Drazin pseudoinverse*, because it was first studied by Drazin [1] (though in the more general context of rings and semigroups without specific reference to matrices). The special properties of the Drazin pseudoinverse of a square matrix have been studied by Cline [3] and Greville [6]; not all of them will be mentioned here.

It is readily seen that the set of three equations (1^k), (2), (5) is equivalent to the set

$$AX = XA, \tag{5}$$

$$A^{k+1}X = A^k, \tag{21}$$

$$AX^2 = X. \tag{22}$$

It is evident also that if (21) holds for some positive integer k, then it holds for every integer $l > k$. It follows also from (21) that

$$\operatorname{rank} A^k = \operatorname{rank} A^{k+1}. \tag{23}$$

Therefore, a solution for X of (21) (and, consequently, of the set (5), (21), (22)) exists only if (23) holds. We shall show presently that if (23) does hold, there is a unique X (the Drazin pseudoinverse of A) satisfying (5), (21), and (22).

DEFINITION 1.   The smallest positive integer k for which (23) holds will be called the *index** of A.

The index has several properties, each of which might have been taken as the definition. The following lemma will play an important role in demonstrating these properties.

---

*"Index" is used in this sense by Wedderburn [1] and Drazin [1]. Some writers (e.g., MacDuffee [1]) define the index as the degree of the minimum polynomial.

LEMMA 4. Let $A \in C^{n \times n}$ have index k. Then all the matrices A^l for $l \geqslant k$ have the same rank, the same range, and the same null space. Similarly, all their transposes $(A^l)^T$ have the same range and the same null space, and their conjugate transposes $(A^l)^*$ have the same range and the same null space. Moreover, for no l less than k do A^l and a higher power of A (or their transposes or conjugate transposes) have the same range or the same null space.

PROOF. It may be well to point out first that (23) necessarily holds for *some* positive integer k (see Ex. 19).

It follows from (23) and Ex. 1.11 that

$$R(A^{k+1}) = R(A^k).$$

Therefore (21) holds for some X, and multiplication on the left by A^{l-k} gives

$$A^l = A^{l+1}X \qquad (l \geqslant k). \tag{24}$$

It follows from (24) that all the matrices A^l for $l \geqslant k$ have the same range and the same rank. From Ex. 1.11 and the fact that A^k and A^l have the same rank, we can deduce that they have the same null space.

The statements about the transposes and conjugate transposes are obtained by applying the preceding results to A^T and A^* and noting that $(A^l)^T = (A^T)^l$ and $(A^l)^* = (A^*)^l$.

If an equality of ranges of the kind ruled out by the last part of the theorem should occur, there must be some $l < k$ such that A^l or its transpose or conjugate transpose has the same range as the corresponding matrix with exponent $l+1$. But this would imply rank $A^l =$ rank A^{l+1}, and k would not be the index of A. Similarly, equality of null spaces would imply that A^l and A^{l+1} have the same nullity, and therefore the same rank. ∎

THEOREM 6. Let $A \in C^{n \times n}$. Then, the following statements are equivalent:

(a) The index of A is k.

(b) The smallest positive exponent for which (21) holds is k.

(c) If A is singular and $m(\lambda)$ is its minimum polynomial, k is the multiplicity of $\lambda = 0$ as a zero of $m(\lambda)$.

(d) If A is singular, k is the maximal grade of the 0-vectors of A.

PROOF. (a)$\Leftrightarrow$(b). Clearly (24) implies

$$\text{rank } A^{l+1} = \text{rank } A^l, \tag{25}$$

and by Ex. 1.11, (25) implies

$$R(A^{l+1}) = R(A^l),$$

so that (24) holds for some X. Thus (25) and (24) are equivalent. Consequently, (a)⟺(b).

(b)⟺(c). Let

$$m(\lambda) = \lambda^l p(\lambda),$$

where $p(0) \neq 0$. Let k be defined by (b), and we must now show that $k = l$. We have

$$p(A)A^l = \mathrm{O}.$$

If $l > k$, then

$$\mathrm{O} = p(A)A^l X = p(A)A^{l-1},$$

where $\lambda^{l-1} p(\lambda)$ is of lower degree than $m(\lambda)$, contrary to the definition of the minimum polynomial.

Since $p(0) \neq 0$, we can write*

$$m(\lambda) = c\lambda^l (1 - \lambda q(\lambda)), \tag{26}$$

where $c \neq 0$ and q is a polynomial. It follows that

$$A^{l+1} q(A) = A^l. \tag{27}$$

If $l < k$, this would contradict (b).

(a)⟺(d). Let A have index k and let h be the maximal grade of the 0-vectors of A. We must show that $h = k$. The definition of h implies that $N(A^l) = N(A^h)$ for all $l \geqslant h$, but $N(A^{h-1})$ is a proper subset of $N(A^h)$. It follows from Lemma 4 that $h = k$. ■

The following lemma will be used in proving the existence of a unique $\{1^k, 2, 5\}$-inverse of a square matrix of index k.

LEMMA 5. If Y is a $\{1^l, 5\}$-inverse of a square matrix A, then

$$X = A^l Y^{l+1}$$

is a $\{1^l, 2, 5\}$-inverse.

PROOF. We have

$$A^{l+1} Y = A^l, \qquad AY = YA.$$

---

*For this device we are indebted to M.R. Hestenes (see footnote 56, p. 687, of Ben-Israel and Charnes [1]).

Clearly X satisfies (5). We have then

$$A^{l}XA = A^{2l+1}Y^{l+1} = A^{2l}Y^{l} = A^{2l-1}Y^{l-1} = \cdots = A^{l},$$

and

$$XAX = A^{2l+1}Y^{2l+2} = A^{2l}Y^{2l+1} = \cdots = A^{l}Y^{l+1} = X. \quad \blacksquare$$

THEOREM 7. Let $A \in C^{n \times n}$ have index k. Then A has a unique $\{1^{k}, 2, 5\}$-inverse, which is expressible as a polynomial in A, and is also the unique $\{1^{l}, 2, 5\}$-inverse for every $l \geqslant k$.

PROOF. The matrix $q(A)$ of (27) is a $\{1^{k}, 5\}$-inverse of A. Therefore, by Lemma 5,

$$X = A^{k}(q(A))^{k+1} \tag{28}$$

is a $\{1^{k}, 2, 5\}$-inverse. This proves the existence of such an inverse.

A matrix X that satisfies (21) clearly satisfies (24) for all $l \geqslant k$. Therefore, a $\{1^{k}, 2, 5\}$-inverse of A is a $\{1^{l}, 2, 5\}$-inverse for all $l \geqslant k$.

Uniqueness will be proved by adapting the proof of uniqueness of the group inverse given in the remark following Theorem 1. Let $X, Y \in A\{1^{l}, 2, 5\}$, $E = AX = XA$, and $F = AY = YA$. Note that E and F are idempotent. Then $E = F$, since

$$E = AX = A^{l}X^{l} = AYA^{l}X^{l} = FAX = FE,$$

$$F = YA = Y^{l}A^{l} = Y^{l}A^{l}XA = YAE = FE.$$

The proof is then completed exactly as in the case of the group inverse. $\blacksquare$

This unique $\{1^{k}, 2, 5\}$-inverse is the Drazin pseudoinverse, and we shall denote it by $A^{(d)}$. The group inverse is the particular case of the Drazin pseudoinverse for matrices of index 1.

COROLLARY 1 (Englefield). Let $A \in C^{n \times n}$. Then there is a $\{1, 2\}$-inverse of A expressible as a polynomial in A if and only if A has index 1, in which case the only such inverse is the group inverse, which is given by

$$A^{\#} = A(q(A))^{2}, \tag{29}$$

where q is defined by (26).

PROOF. *Only if.* A $\{1, 2\}$-inverse of A that is a polynomial in A necessarily commutes with A, and is therefore a $\{1, 2, 5\}$-inverse. The group inverse $A^{\#}$ is the only such inverse, and A has a group inverse only if its index is 1.

If. If A has index 1, it has a group inverse, which is a $\{1, 2\}$-inverse, and in this case coincides with the Drazin pseudoinverse. It is therefore expressible as a polynomial in A by Theorem 7.

Formula (29) is merely the specialization of (28) for $k = 1$. ∎

COROLLARY 2   (Pearl [6]).   Let $A \in C^{n \times n}$. Then $A^{\dagger}$ is expressible as a polynomial in A if and only if A is range-Hermitian.

EXERCISES

19.   Let $A \in C^{n \times n}$. Show that (23) holds for some k between 1 and n, inclusive.

PROOF.   Since $n \geqslant \text{rank}\,(A^{k}) \geqslant \text{rank}\,(A^{k+1}) \geqslant 0$ for   all $k = 1, 2, \ldots, n$, eventually rank $A^{k} = \text{rank}\ A^{k+1}$ for some k between 1 and n.

20. $(A^{*})^{(d)} = (A^{(d)})^{*}$.

21. $(A^{T})^{(d)} = (A^{(d)})^{T}$.

22. $(A^{l})^{(d)} = (A^{(d)})^{l}$ for $l = 1, 2, \ldots$.

23.   If A has index k, A^{l} has index 1 and $(A^{l})^{\#} = (A^{(d)})^{l}$ for $l \geqslant k$.

24. $(A^{(d)})^{(d)} = A$ if and only if A has index 1 (Drazin).

25. $A^{(d)}$ has index 1 and $(A^{(d)})^{\#} = A^{2}A^{(d)}$.

26. $((A^{(d)})^{(d)})^{(d)} = A^{(d)}$ (Drazin).

27.   If A has index $k, R(A^{(d)}) = R(A^{l})$ and $N(A^{(d)}) = N(A^{l})$ for all $l \geqslant k$.

28. $R(A^{(d)})$ is the subspace spanned by all the λ-vectors of A for all nonzero eigenvalues λ, and $N(A^{(d)})$ is the subspace spanned by all the 0-vectors of A, and these are complementary subspaces.

29. $AA^{(d)} = A^{(d)}A$ is idempotent and is the projector on $R(A^{(d)})$ along $N(A^{(d)})$. Alternatively, if A has index k, it is the projector on $R(A^{l})$ along $N(A^{l})$ for all $l \geqslant k$.

30.   If A and X are S-inverses of each other, they have the same index.

31. $A^{(d)}(A^{(d)})^{\#} = AA^{(d)}$.

32.   Let $A \in C^{n \times n}$ have index k. Then, for all $l \geqslant k$,

$$A^{(d)} = A^{l}(q(A))^{l+1},$$

where q is defined by (26).

33.   If A is nilpotent, $A^{(d)} = O$.

34.   If $l > m > 0$, $A^{m}(A^{(d)})^{l} = (A^{(d)})^{l-m}$.

35.   If $m > 0$ and $l - m \geqslant k$, $A^{l}(A^{(d)})^{m} = A^{l-m}$.

36.   Let A have index k, and define as follows a set of matrices B_{j} where j ranges over all the integers. For $j \geqslant k$, $B_{j} = A^{j}$; for $0 \leqslant j < k$, $B_{j} = A^{k}(A^{(d)})^{k-j}$; for $j < 0$, $B_{j} = (A^{(d)})^{-j}$. Is the set of matrices $\{B_{j}\}$ an Abelian group under matrix multiplication with unit element B_{0} and multiplication rule $B_{l}B_{m} = B_{l+m}$? Is there an equivalent, but easier way of defining the matrices B_{j}?

37. If A has index k and $l \geqslant k$, show that

$$A^{(d)} = A^l (A^{2l+1})^{(1)} A^l, \tag{30}$$

where $(A^{2l+1})^{(1)}$ is an arbitrary element of $A^{2l+1}\{1\}$ (Greville [6]). Note that (17) is a particular case of (30).

38. Let $A \in C^{n \times n}$. Then A has index 1 if and only if the limit

$$\lim_{\lambda \to 0} (\lambda I_n + A)^{-1} A$$

exists, in which case

$$\lim_{\lambda \to 0} (\lambda I_n + A)^{-1} A = AA^{\#} \qquad \text{(Ben-Israel [10])}.$$

Remark. Here $\lambda \to 0$ means $\lambda \to 0$ through any neighborhood of 0 in C which excludes the nonzero eigenvalues of $-A$.

PROOF.   Let rank $A = r$ and let $A = FG$ be a full-rank factorization. Then the identity

$$(\lambda I_n + A)^{-1} A = F(\lambda I_r + GF)^{-1} G$$

holds whenever the inverses in question exist. Therefore the existence of $\lim_{\lambda \to 0} (\lambda I_n + A)^{-1}$ is equivalent to the existence of $\lim_{\lambda \to 0} (\lambda I_r + GF)^{-1}$ which, in turn, is equivalent to the nonsingularity of GF. The proof is completed by using Theorems 1 and 2. ∎

39. Let $A \in C^{n \times n}$. Then A is range-Hermitian if and only if

$$\lim_{\lambda \to 0} (\lambda I_n + A)^{-1} P_{R(A)} = A^{\dagger}.$$

PROOF.   Follows from Ex. 38 and Theorem 3.

40. Let $O \neq A \in C^{m \times n}$. Then

$$\lim_{\lambda \to 0} (\lambda I_n + A^*A)^{-1} A^* = A^{\dagger} \qquad \text{(den Broeder and Charnes [1])}. \tag{3.30}$$

PROOF.

$$\lim_{\lambda \to 0} (\lambda I_n + A^*A)^{-1} A^* = \lim_{\lambda \to 0} (\lambda I_n + A^*A)^{-1} P_{R(A^*A)} A^*$$

$$\text{(since } R(A^*) = R(A^*A))$$

$$= (A^*A)^{\dagger} A^* \qquad \text{(by Ex. 39 since } A^*A \text{ is range-Hermitian)}$$

$$= A^{\dagger} \qquad \text{(by Ex. 1.22).} \quad ∎$$

7. SPECTRAL PROPERTIES OF THE DRAZIN PSEUDOINVERSE

The spectral properties of the Drazin pseudoinverse are the same as those of the group inverse with regard to nonzero eigenvalues and the associated eigenvectors, but weaker for 0-vectors. The necessity for such weakening is apparent from the following theorem.

THEOREM 8. Let $A \in C^{n \times n}$ and let $X \in A\{1\} \cup A\{2\}$ be an S-inverse of A. Then both A and X have index 1.

PROOF. First, let $X \in A\{1\}$, and suppose x is a 0-vector of A of grade 2. Then, Ax is a null vector of A. Since X is an S-inverse of A, Ax is also a null vector of X. Thus,

$$0 = XAx = AXAx = Ax,$$

which contradicts the assumption that x is a 0-vector of A of grade 2. Hence, A has no 0-vectors of grade 2, and therefore has index 1, by Theorem 6(d). By Ex. 30, X also has index 1.

If $X \in A\{2\}$, we reverse the roles of A and X. ∎

Let us therefore call X an S'-inverse of A if, for all $\lambda \neq 0$, a vector x is a λ^{-1}-vector of X of grade p if and only if it is a λ-vector of A of grade p, and x is a 0-vector of X if and only if it is a 0-vector of A (without regard to grade).

THEOREM 9. For every square matrix A, A and $A^{(d)}$ are S'-inverses of each other.

PROOF. Since $A^{(d)}$ satisfies

$$A^{(d)} A^{k+1} = A^k, \qquad A(A^{(d)})^2 = A^{(d)},$$

the part of the definition of S'-inverses relating to nonzero eigenvalues follows from Lemma 3. Since $A^{(d)}$ has index 1 by Ex. 25, all its 0-vectors are null vectors. Thus the part of the definition of S'-inverses relating to 0-vectors follows from Ex. 28. ∎

8. INDEX 1-NILPOTENT DECOMPOSITION OF A SQUARE MATRIX

The following theorem plays an important role in the study of spectral generalized inverses of matrices of index greater than 1. It is implicit in Wedderburn's [1] results on idempotent and nilpotent parts, but is not stated by him in this form.

THEOREM 10. A square matrix A has a unique decomposition

$$A = B + N, \tag{31}$$

such that B has index 1, N is nilpotent, and

$$BN = NB = O. \tag{32}$$

Moreover,

$$B = \left(A^{(d)} \right)^{\#}. \tag{33}$$

PROOF. Suppose A has a decomposition (31) such that B has index 1, N is nilpotent, and (32) holds. We shall first show that this implies (33), and therefore the decomposition is unique if it exists.

Since

$$B^{\#} = B \left(B^{\#} \right)^2 = \left(B^{\#} \right)^2 B,$$

we have

$$B^{\#} N = N B^{\#} = O.$$

Consequently,

$$AB^{\#} = BB^{\#} = B^{\#} A. \tag{34}$$

Moreover,

$$A \left(B^{\#} \right)^2 = B \left(B^{\#} \right)^2 = B^{\#}. \tag{35}$$

Because of (32), we have

$$A^l = (B + N)^l = B^l + N^l \qquad (l = 1, 2, \ldots). \tag{36}$$

If l is sufficiently large so that $N^l = O$,

$$A^l = B^l,$$

and for such l,

$$A^{l+1} B^{\#} = B^{l+1} B^{\#} = B^l. \tag{37}$$

It follows from (34), (35), and (37) that $X = B^{\#}$ satisfies (5), (21), and (22), and therefore

$$B^{\#} = A^{(d)},$$

which is equivalent to (33).

It remains to show that this decomposition has the required properties. Clearly B has index 1. By taking

$$N = A - \left(A^{(d)} \right)^{\#} \tag{38}$$

and noting that

$$(A^{(d)})^{\#} = A^2 A^{(d)}$$

by Ex. 25, it is easily verified that (32) holds. Therefore (36) follows, and, if k is the index of A,

$$A^k = B^k + N^k = A^{2k}(A^{(d)})^k + N^k = A^k + N^k,$$

and therefore $N^k = \text{O}$. ∎

We shall call the matrix N given by (38) the *nilpotent part* of A, and shall denote it by $A^{(n)}$.

THEOREM 11.   Let $A \in C^{n \times n}$. Then A and X are S'-inverses of each other if

$$X^{(d)} = (A^{(d)})^{\#}. \tag{39}$$

Moreover, if $X \in A\{1\} \cup A\{2\}$, it is an S'-inverse of A only if (39) holds.

PROOF.   If (39) holds, $X^{(d)}$ and $A^{(d)}$ have the same range and the same null space, and consequently the projectors $XX^{(d)}$ and $AA^{(d)} = A^{(d)}(A^{(d)})^{\#}$ are equal. Thus, if l is the maximum of the indices of A and X,

$$XA^{l+1} = X(A^{(d)})^{\#}A^{(d)}A^{l+1} = XX^{(d)}A^l = A^l \tag{40}$$

by Ex. 29. By interchanging the roles of A and X we obtain also

$$AX^{l+1} = X^l. \tag{41}$$

From (40) and (41), Lemma 3, Ex. 28 and the fact that $A^{(d)}$ and $X^{(d)}$ have the same null space, we deduce that A and X are S'-inverses of each other.

On the other hand, let A and X be S'-inverses of each other, and let $X \in A\{1\}$. Then, by Ex. 28,

$$N(A^{(d)}) = N(X^{(d)}),$$

and so,

$$(A^{(d)})^{\#}X^{(n)} = (X^{(d)})^{\#}A^{(n)} = \text{O}.$$

Similarly, since

$$R(A^{(d)}) = R(X^{(d)}),$$

(2.44) gives

$$N(A^{(d)*}) = N(X^{(d)*}),$$

and therefore

$$X^{(n)}(A^{(d)})^{\#} = (X^{(d)})^{\#}A^{(n)} = \mathrm{O}.$$

Consequently,

$$A = AXA = (A^{(d)})^{\#}(X^{(d)})^{\#}(A^{(d)})^{\#} + A^{(n)}X^{(n)}A^{(n)},$$

and therefore

$$A^{(d)} = A^{(d)}AA^{(d)} = AA^{(d)}(X^{(d)})^{\#}AA^{(d)} = (X^{(d)})^{\#}, \qquad (42)$$

since $AA^{(d)}$ is the projector on the range of $(X^{(d)})^{\#}$ along its null space. But (42) is equivalent to (39).

If $X \in A\{2\}$, we reverse the roles of A and X. ∎

Referring back to the proof of Theorem 5, we note that if A has index 1, a matrix X that is an S-inverse of A and also either a $\{1\}$-inverse or a $\{2\}$-inverse, is automatically a $\{1,2\}$-inverse. However, a similar remark does not apply when the index of A is·greater than 1 and X is an S'-inverse of A. This is because $A^{(n)}$ is no longer a null matrix (as it is when A has index 1) and its properties must be taken into account. (For details see Ex. 48.)

9. QUASI-COMMUTING INVERSES

Erdelyi [3] calls A and X *quasi-commuting inverses* of each other if they are $\{1,2,5^k,6^k\}$-inverses of each other for some positive integer k. He noted that for such pairs of matrices the spectrum of X is obtained by replacing each eigenvalue λ of A by $\lambda^{\dagger}$. The following theorem shows that quasi-commuting inverses have much more extensive spectral properties.

THEOREM 12. If A and X are quasi-commuting inverses, they are S'-inverses.

PROOF. If A and X are $\{1,2,5^l,6^l\}$-inverses of each other, then

$$XA^{l+1} = A^lXA = A^l,$$

and similarly

$$AX^{l+1} = X^l. \qquad (43)$$

In view of Lemma 3 and Ex. 28, all that remains in order to prove that A and X are S'-inverses of each other is to show that $A^{(d)}$ and $X^{(d)}$ have the same null space. Now,

$$A^{(d)}x = 0$$

$$\Rightarrow 0 = A^{l+1}A^{(d)}x = A^l x$$

$$\Rightarrow 0 = X^{2l}A^l x = A^l X^{2l}x = X^l x \qquad [\text{by } (43)]$$

$$\Rightarrow 0 = (X^{(d)})^{l+1}X^l x = X^{(d)}x.$$

Since the roles of A and X are symmetrical, the reverse implication follows by interchanging them. ∎

COROLLARY 3. A and X are quasi-commuting inverses of each other if and only if (39) holds and $A^{(n)}$ and $X^{(n)}$ are $\{1,2\}$-inverses of each other.

PROOF. *If*: A and X are $\{1,2\}$-inverses of each other by Ex. 45. Choose l sufficiently large so that $(A^{(n)})^l = \text{O}$. Then

$$XA^l = \left((X^{(d)})^{\#} + X^{(n)} \right)\left((A^{(d)})^{\#} \right)^l$$

$$= \left((X^{(d)})^{\#} + X^{(n)} \right)(X^{(d)})^l = (X^{(d)})^{l-1} = A^l X.$$

By interchanging A and X, it follows also that A commutes with X^l.

Only if: By Theorem 12, A and X are S'-inverses of each other. Then, by Theorem 11, (39) holds, and by Ex. 48, $A^{(n)}$ and $X^{(n)}$ are $\{1,2\}$-inverses of each other. ∎

10. OTHER SPECTRAL GENERALIZED INVERSES

Greville [8] calls X a *strong spectral inverse* of A if Eqs. (9) are satisfied. Although this is not quite obvious, the relationship is a reciprocal one, and they can be called strong spectral inverses of each other. If A has index 1, Theorem 4 shows that $A^{\#}$ is the only strong spectral inverse. Greville has shown that strong spectral inverses are quasi-commuting, but, for a matrix A with index greater than 1, the set of strong spectral inverses is a proper subset of the set of quasi-commuting inverses. Strong spectral inverses have some remarkable and, in some respects, complicated properties, and there are a number of open questions concerning them. As these properties relate to matrices of index greater than 1, which are not for most purposes a very important class, they will not be discussed further here. The interested reader may consult Greville [6].

Cline [3] has pointed out that a square matrix A of index 1 has a $\{1,2,3\}$-inverse whose range is $R(A)$. This is, therefore, a "least-squares" inverse and also has some spectral properties (see Exs. 48 and 49). Greville [7] has extended this notion to square matrices of arbitrary index, but his extension raises some questions that have not been answered (see the conclusion of Greville [7]).

EXERCISES

41. If A has index 1, $A^{(n)} = \bigcirc$.

42. If A is nilpotent, rank $A^{l+1} <$ rank A^l unless $A^l = \bigcirc$.

43. If A is nilpotent, the smallest positive integer l such that $A^l = \bigcirc$ is called the *index of nilpotency* of A. Show that this is the same as the index of A as defined in Section 6.

44. A and $A^{(n)}$ have the same index.

45. rank $A =$ rank $A^{(d)} +$ rank $A^{(n)}$.

46. $A^{(d)}A^{(n)} = A^{(n)}A^{(d)} = \bigcirc$.

47. Every 0-vector of A of grade p is a 0-vector of $A^{(n)}$ of grade p.

48. Let A and X satisfy (39). Then $X \in A\{1\}$ if and only if $X^{(n)} \in A^{(n)}\{1\}$. Similar statements with $\{1\}$ replaced by $\{2\}$ and by $\{1,2\}$ are also true.

49. If A has index 1, show that $X = A^{\#}AA^{\dagger} \in A\{1,2,3\}$ (Cline). Show that this X has the properties of an S-inverse of A with respect to nonzero eigenvalues (but, in general, not with respect to 0-vectors). What is the condition on A that this X be an S-inverse of A?

50. For square A with arbitrary index, Greville has suggested as an extension of Cline's inverse

$$X = A^{(d)}AA^{\dagger} + A^{(1)}A^{(n)}A^{\dagger},$$

where $A^{(1)}$ is an arbitrary element of $A\{1\}$. Show that $X \in A\{1,2,3\}$ and has some spectral properties. Describe its spectral properties precisely.

51. Can a matrix A of index greater than 1 have an S-inverse? It can if we are willing to accept an "inverse" that is neither a $\{1\}$-inverse nor a $\{2\}$-inverse. Let

$$A^{(S)} = A^{(d)} + A^{(n)}.$$

Show that $A^{(S)}$ is an S-inverse of A and that $X = A^{(S)}$ is the unique solution of the four equations

$$AX = XA, \qquad A^{l+1}X = A^l,$$

$$AX^{l+1} = X^l, \qquad A - X = A^lX^l(A - X)$$

for every positive integer l not less than the index of A. Show also that $A^{(S)} = A^{\#}$ if A has index 1 and $(A^{(S)})^{(S)} = A$. In your opinion, can $A^{(S)}$ properly be called a generalized inverse of A?

52. Let F be a square matrix of index 1, and let G be such that $R(FG) \subset R(G)$. Then,

$$R(FG) = R(F) \cap R(G).$$

PROOF.   Evidently, $R(FG) \subset R(F)$, and therefore

$$R(FG) \subset R(F) \cap R(G).$$

Now let $x \in R(F) \cap R(G)$, and we must show that $x \in R(FG)$. Since F has index 1, it has a group inverse $F^{\#}$, which, by Corollary 1, can be expressed as a polynomial in F, say $p(F)$. We have

$$x = Fy = Gz$$

for some y, z, and therefore

$$x = FF^{\#}x = FF^{\#}Gz = Fp(F)Gz.$$

Since $R(FG) \subset R(G)$,

$$FG = GH$$

for some H, and, consequently,

$$F^l G = GH^l$$

for every non-negative integer l. Thus,

$$x = Fp(F)Gz = FGp(H)z \subset R(FG). \quad \blacksquare$$

(This is a slight extension of a result of Arghiriade [1].)

53.   *The "reverse-order" property for the Moore–Penrose inverse.*   For some pairs of matrices A, B the relation

$$(AB)^{\dagger} = B^{\dagger}A^{\dagger} \tag{44}$$

holds, and for others it does not. There does not seem to be a simple criterion for distinguishing the cases in which (44) holds. The following result is due to Greville [5].

For matrices A, B such that AB exists,

$$(AB)^{\dagger} = B^{\dagger}A^{\dagger} \tag{44}$$

if and only if $R(A^*AB) \subset R(B)$ and $R(BB^*A^*) \subset R(A^*)$.

PROOF. *If*: We have

$$BB^\dagger A^*AB = A^*AB \tag{45}$$

and

$$A^\dagger ABB^*A^* = BB^*A^*. \tag{46}$$

Taking conjugate transposes of both sides of (45) gives

$$B^*A^*ABB^\dagger = B^*A^*A, \tag{47}$$

and then multiplying on the right by $A^\dagger$ and on the left by $(AB)^{*\dagger}$ yields

$$ABB^\dagger A^\dagger = AB(AB)^\dagger. \tag{48}$$

Multiplying (46) on the left by $B^\dagger$ and on the right by $(AB)^{*\dagger}$ gives

$$B^\dagger A^\dagger AB = (AB)^\dagger AB. \tag{49}$$

It follows from (48) and (49) that $B^\dagger A^\dagger \in (AB)\{1,3,4\}$.

Finally, the equations,

$$B^*A^* = B^*BB^\dagger A^\dagger AA^*, \qquad B^\dagger A^\dagger = B^\dagger B^{*\dagger}B^*A^*A^{*\dagger}A^\dagger$$

show that

$$\text{rank } B^\dagger A^\dagger = \text{rank } B^*A^* = \text{rank } AB,$$

and therefore $B^\dagger A^\dagger \in (AB)\{2\}$ by Theorem 1.2, and so (44) holds.

Only if: We have

$$B^*A^* = B^\dagger A^\dagger ABB^*A^*,$$

and multiplying on the left by ABB^*B gives

$$ABB^*(I - A^\dagger A)BB^*A^* = \mathrm{O}.$$

Since the left member is Hermitian and $I - A^\dagger A$ is idempotent, it follows that

$$(I - A^\dagger A)BB^*A^* = \mathrm{O},$$

which is equivalent to (46). In an analogous manner, (45) is obtained. ∎
54. (Arghiriade [1]).   For matrices A, B such that AB exists, (44) holds if and only if A^*ABB^* is range-Hermitian.

PROOF. We shall ·show that the condition that A^*ABB^* be range-Hermitian is equivalent to the two conditions in Ex. 53, and the result will then follow from Ex. 53. Let C denote A^*ABB^*, and observe that

$$R(A^*AB) = R(C), \qquad R(BB^*A^*) = R(C^*)$$

because

$$CB^{*\dagger} = A^*AB, \qquad C^*A^\dagger = BB^*A^*.$$

Therefore it is sufficient to prove that $R(C) = R(C^*)$ if and only if $R(C) \subset R(B)$ and $R(C^*) \subset R(A^*)$.

If: A^*A and BB^* are Hermitian, and therefore of index 1 by Ex. 7. Since $R(BB^*) = R(B)$ by Corollary 1.2, it follows from Ex. 52 with $F = A^*A$, $G = BB^*$ that

$$R(C) = R(A^*) \cap R(B).$$

Reversing the assignments of F and G gives

$$R(C^*) = R(A^*) \cap R(B).$$

Thus $R(C) = R(C^*)$.

Only if: Obvious. ■

55. If l is any integer not less than the index of A,

$$(A^{(d)})^\dagger = (A^l)^\dagger A^{2l+1}(A^l)^\dagger \qquad \text{(Cline [3])}.$$

[*Hint*: Use Ex. 2.50, noting that $R(A^{(d)}) = R(A^l)$ and $N(A^{(d)}) = N(A^l)$.]

56. If the matrices A, E in $C^{m \times n}$ satisfy

$$R(E) \subset R(A), \tag{50}$$

$$R(E^*) \subset R(A^*), \tag{51}$$

and

$$\|A^\dagger E\| < 1 \tag{52}$$

for any multiplicative matrix norm, then

$$(A + E)^\dagger = (I + A^\dagger E)^{-1} A^\dagger. \tag{53}$$

PROOF.   The matrix $B = I + A^\dagger E$ is nonsingular by (52) and Exs. 1.42 and 1.48. Since

$$A + E = A + AA^\dagger E, \qquad \text{by (50)}$$

$$= A(I + A^\dagger E),$$

it suffices to show that the matrices A and $B = I + A^\dagger E$ have the "reverse order" property (44),

$$(A(I+A^\dagger E))^\dagger = (I+A^\dagger E)^{-1}A^\dagger,$$

which by Ex. 53 is equivalent to

$$R(A^*AB) \subset R(B) \tag{54}$$

and

$$R(BB^*A^*) \subset R(A^*). \tag{55}$$

Now (54) holds since B is nonsingular, and (55) follows from

$$R(BB^*A^*) = R((I+A^\dagger E)(I+A^\dagger E)^*A^*)$$

$$= R(A^* + E^*A^{\dagger *}A^* + A^\dagger E(I+A^\dagger E)^*A^*)$$

$$\subset R(A^*), \quad \text{by (51)}.$$

57. *Error bounds for generalized inverses.* Let A, E satisfy (50), (51), and (52). Then,

$$\|(A+E)^\dagger - A^\dagger\| \leqslant \frac{\|A^\dagger E\| \, \|A^\dagger\|}{1 - \|A^\dagger E\|}. \tag{56}$$

If (50) and (51) hold, but (52) is replaced by

$$\|A^\dagger\| \, \|E\| < 1, \tag{57}$$

then

$$\|(A+E)^\dagger - A^\dagger\| \leqslant \frac{\|A^\dagger\|^2 \, \|E\|}{1 - \|A^\dagger\| \, \|E\|} \quad \text{(Ben-Israel [5])}. \tag{58}$$

PROOF.   From Ex. 56 it follows that

$$(A+E)^\dagger - A^\dagger = (I+A^\dagger E)^{-1}A^\dagger - A^\dagger$$

$$= \sum_{k=0}^{\infty} (-1)^k (A^\dagger E)^k A^\dagger - A^\dagger, \quad \text{by (52) and Ex. 1.48,}$$

$$= \sum_{k=1}^{\infty} (-1)^k (A^\dagger E)^k A^\dagger$$

and hence,

$$\|(A+E)^\dagger - A^\dagger\| \leqslant \sum_{k=1}^{\infty} \|A^\dagger E\|^k \|A^\dagger\|$$

$$= \frac{\|A^\dagger E\|\,\|A^\dagger\|}{1 - \|A^\dagger E\|}, \qquad \text{by } (52).$$

The condition (57) [which is stronger than (52)] then implies (58).

For further results see Stewart [1], Wedin [1], [2], Pereyra [1], Golub and Pereyra [1] and Moore and Nashed [1].

SUGGESTED FURTHER READING

Section 4. For range-Hermitian matrices see Arghiriade [1], Katz [1], Katz and Pearl [1], and Pearl ([3], [4], [5], [6]). For matrices of index 1 see Ben-Israel [10]. For the group inverse see Robert [1].

Section 10. Poole and Boullion [1], Ward, Boullion, and Lewis [2], and Scroggs and Odell [1].

5

GENERALIZED INVERSES OF PARTITIONED MATRICES

1. INTRODUCTION

In this chapter we study linear equations and matrices in partitioned form. For example, in computing a (generalized or ordinary) inverse of a matrix $A \in C^{m \times n}$, the size or difficulty of the problem may be reduced if A is partitioned as

$$A = \begin{bmatrix} A_{11} & A_{12} \\ A_{21} & A_{22} \end{bmatrix}.$$

The typical result here is the sought inverse expressed in terms of the submatrices A_{ij}.

Partitioning by columns and by rows is used in Section 2 to solve linear equations, and to compute generalized inverses and related items.

Intersections of linear manifolds are studied in Section 3, and used in Section 4 to obtain common solutions of pairs of linear equations and to invert matrices partitioned by rows.

Greville's method for computing $A^\dagger$ for $A \in C^{m \times n}$, $n \geqslant 2$, is based on partitioning A as

$$A = [A_{n-1} \quad a_n]$$

where a_n is the nth column of A. $A^\dagger$ is then expressed in terms of a_n and

$A_{n-1}^\dagger$, which is computed in the same way, using the partition

$$A_{n-1} = [A_{n-2} \quad a_{n-1}], \qquad \text{etc.}$$

Greville's method and some of its consequences are studied in Section 5. Bordered matrices, the subject of Section 6, are matrices of the form

$$\begin{bmatrix} A & U \\ V^* & O \end{bmatrix}$$

where $A \in C^{m \times n}$ is given and U and V are chosen so that the resulting bordered matrix is nonsingular. Moreover

$$\begin{bmatrix} A & U \\ V^* & O \end{bmatrix}^{-1} = \begin{bmatrix} A^\dagger & V^{*\dagger} \\ U^\dagger & O \end{bmatrix}$$

expressing generalized inverses in terms of an ordinary inverse.

2. PARTITIONING MATRICES AND LINEAR EQUATIONS

Consider the linear equation

$$Ax = b \tag{1}$$

with given matrix A and vector b, in the following three cases.

Case 1.

$A \in C_r^{r \times n}$, i.e., A is of *full row rank*. Let the columns of A be rearranged, if necessary, so that the first r columns are linearly independent. A rearrangement of columns may be interpreted as postmultiplication by a suitable permutation matrix; thus,

$$AQ = [A_1 \quad A_2] \qquad \text{or} \qquad A = [A_1 \quad A_2]Q^T, \tag{2}$$

where Q is an $n \times n$ permutation matrix (hence $Q^{-1} = Q^T$) and A_1 consists of r linearly independent columns, so that $A_1 \in C_r^{r \times r}$, i.e., A_1 is nonsingular.

The matrix A_2 is in $C^{r \times (n-r)}$ and if $n = r$, this matrix and other items indexed by the subscript 2 are to be interpreted as absent.

Corresponding to (2), let the vector $x \in C^n$ be partitioned

$$x = \begin{bmatrix} x_1 \\ x_2 \end{bmatrix}, \qquad x_1 \in C^r. \tag{3}$$

Using (2) and (3) we rewrite (1) as

$$[A_1 \quad A_2]Q^T \begin{bmatrix} x_1 \\ x_2 \end{bmatrix} = b \tag{4}$$

easily shown to be satisfied by the vector

$$\begin{bmatrix} x_1 \\ x_2 \end{bmatrix} = Q \begin{bmatrix} A_1^{-1}b \\ 0 \end{bmatrix}, \tag{5}$$

which is thus a particular solution of (1).

The general solution of (1) is obtained by adding to (5) the general element of $N(A)$, i.e., the general solution of

$$Ax = 0. \tag{6}$$

In (2), the columns of A_2 are linear combinations of the columns of A_1, say,

$$A_2 = A_1 T \qquad \text{or} \qquad T = A_1^{-1}A_2 \in C^{r \times (n-r)}, \tag{7}$$

where the matrix T is called the *multiplier* corresponding to the partition (2), a name suggested by T being the "ratio" of the last $n-r$ columns of AQ to its first r columns.

Using (2), (3), and (7) permits rewriting (6) as

$$A_1[I_r \quad T]Q^T \begin{bmatrix} x_1 \\ x_2 \end{bmatrix} = 0, \tag{8}$$

whose general solution is clearly

$$\begin{bmatrix} x_1 \\ x_2 \end{bmatrix} = Q \begin{bmatrix} -T \\ I_{n-r} \end{bmatrix} y, \tag{9}$$

where $y \in C^{n-r}$ is arbitrary.

Adding (5) and (9) we obtain the general solution of (1):

$$\begin{bmatrix} x_1 \\ x_2 \end{bmatrix} = Q \begin{bmatrix} A_1^{-1}b \\ 0 \end{bmatrix} + Q \begin{bmatrix} -T \\ I_{n-r} \end{bmatrix} y, \qquad y \text{ arbitrary.} \tag{10}$$

Thus an advantage of partitioning A as in (2), is that it permits solving (1) by working with matrices smaller or more convenient than A. We also note that the null space of A is completely determined by the multiplier T and the permutation matrix Q, indeed (9) shows that the columns of the $n \times (n-r)$ matrix

$$Q \begin{bmatrix} -T \\ I_{n-r} \end{bmatrix} \tag{11}$$

form a basis for $N(A)$.

Case 2.

$A \in C_r^{m \times r}$, i.e., A is of *full column* rank. Unlike case 1, here the linear equation (1) may be inconsistent. If, however, (1) is consistent, then it has a unique solution. Partitioning the rows of A is useful for both checking the consistency of (1) and for computing its solution, if consistent.

Let the rows of A be rearranged, if necessary, so that the first r rows are linearly independent. This is written, analogously to (2), as

$$PA = \begin{bmatrix} A_1 \\ A_2 \end{bmatrix} \qquad \text{or} \qquad A = P^T \begin{bmatrix} A_1 \\ A_2 \end{bmatrix}, \tag{12}$$

where P is an $m \times m$ permutation matrix, and $A_1 \in C_r^{r \times r}$.

If $m = r$, the matrix A_2 and other items with the subscript 2 are to be interpreted as absent.

In (12) the rows of A_2 are linear combinations of the rows of A_1, say,

$$A_2 = SA_1 \quad \text{or} \quad S = A_2 A_1^{-1} \in C^{(m-r) \times r}, \tag{13}$$

where again S is called the *multiplier* corresponding to the partition (12), giving the "ratio" of the last $(m - r)$ rows of PA to its first r rows.

Corresponding to (12) let the permutation matrix P be partitioned as

$$P = \begin{bmatrix} P_1 \\ P_2 \end{bmatrix}, \qquad P_1 \in R^{r \times m}. \tag{14}$$

Equation (1) can now be rewritten, using (12), (13), and (14), as

$$\begin{bmatrix} I_r \\ S \end{bmatrix} A_1 x = \begin{bmatrix} P_1 \\ P_2 \end{bmatrix} b, \tag{15}$$

from which the conclusions below easily follow:

(a) Equation (1) is consistent if and only if

$$P_2 b = SP_1 b \tag{16}$$

i.e., the "ratio" of the last $m - r$ components of the vector Pb to its first r components is the multiplier S of (13).

(b) If (16) holds, then the unique solution of (1) is

$$x = A_1^{-1} P_1 b. \tag{17}$$

From (a) we note that the range of A is completely determined by the multiplier S and the permutation matrix P. Indeed, the columns of the $m \times r$ matrix

$$P^T \begin{bmatrix} I_r \\ S \end{bmatrix} \tag{18}$$

form a basis for $R(A)$.

Case 3.

$A \in C_r^{m \times n}$ with $r \leqslant m, n$. This *general case* has some of the characteristics of both cases 1 and 2, as here we partition both the columns and rows of A.

Since rank $A = r$, A has at least one nonsingular $r \times r$ submatrix A_{11}, which by a rearrangement of rows and columns can be brought to the top left corner of A, say

$$PAQ = \begin{bmatrix} A_{11} & A_{12} \\ A_{21} & A_{22} \end{bmatrix}, \tag{19}$$

where P and Q are permutation matrices, and $A_{11} \in C_r^{r \times r}$.

By analogy with (2) and (12), we may have to interpret some of these submatrices as absent, e.g., A_{12} and A_{22} are absent if $n = r$.

By analogy with (7) and (13) there are *multipliers* $T \in C^{r \times (n-r)}$ and $S \in C^{(m-r) \times r}$, satisfying

$$\begin{bmatrix} A_{12} \\ A_{22} \end{bmatrix} = \begin{bmatrix} A_{11} \\ A_{21} \end{bmatrix} T \quad \text{and} \quad [A_{21} \ A_{22}] = S[A_{11} \ A_{12}]. \tag{20}$$

These multipliers are given by

$$T = A_{11}^{-1} A_{12} \quad \text{and} \quad S = A_{21} A_{11}^{-1}. \tag{21}$$

Combining (19) and (20) results in the following *partition* of $A \in C_r^{m \times n}$

$$A = P^T \begin{bmatrix} A_{11} & A_{12} \\ A_{21} & A_{22} \end{bmatrix} Q^T$$

$$= P^T \begin{bmatrix} I_r \\ S \end{bmatrix} A_{11} [I_r \ T] Q^T, \tag{22}$$

where $A_{11} \in C_r^{r \times r}$. P and Q are permutation matrices, and S and T are given by (21).

As in cases 1 and 2 we conclude that the multipliers S and T, and the permutation matrices P and Q, carry all the information about the range and null space of A.

LEMMA 1.   Let $A \in C_r^{m \times n}$ be partitioned as in (22). Then
   (a) The columns of the $n \times (n - r)$ matrix

$$Q \begin{bmatrix} -T \\ I_{n-r} \end{bmatrix} \tag{11}$$

form a basis for $N(A)$.
   (b) The columns of the $m \times r$ matrix

$$P^T \begin{bmatrix} I_r \\ S \end{bmatrix} \tag{18}$$

form a basis for $R(A)$.   ∎

Returning to the linear equation (1), it may be partitioned by using (22) and (14), in analogy with (4) and (15), as follows:

$$\begin{bmatrix} I_r \\ S \end{bmatrix} A_{11}[I_r \quad T]Q^T \begin{bmatrix} x_1 \\ x_2 \end{bmatrix} = \begin{bmatrix} P_1 \\ P_2 \end{bmatrix} b. \tag{23}$$

The following theorem summarizes the situation, and includes the results of cases 1 and 2 as special cases.

THEOREM 1.   Let $A \in C_r^{m \times n}$, $b \in C^m$ be given, and let the linear equation

$$Ax = b \tag{1}$$

be partitioned as in (23). Then
   (a) Equation (1) is consistent if and only if*

$$P_2 b = SP_1 b \tag{16}$$

   (b) If (16) holds, the general solution of (1) is

$$\begin{bmatrix} x_1 \\ x_2 \end{bmatrix} = Q \begin{bmatrix} A_{11}^{-1} P_1 b \\ 0 \end{bmatrix} + Q \begin{bmatrix} -T \\ I_{n-r} \end{bmatrix} y, \tag{24}$$

*By convention, (16) is satisfied if $m = r$, in which case P_2, and S are interpreted as absent.

where $y \in C^{n-r}$ is arbitrary.

The partition (22) is useful also for computing generalized inverses. We collect some of these results in the following.

THEOREM 2. Let $A \in C_r^{m \times n}$ be partitioned as in (22). Then

(a) A $\{1,2\}$-inverse of A is

$$A^{(1,2)} = Q \begin{bmatrix} A_{11}^{-1} & O \\ O & O \end{bmatrix} P \qquad \text{(C.R. Rao [2]).} \qquad (25)$$

(b) A $\{1,2,3\}$-inverse of A is

$$A^{(1,2,3)} = Q \begin{bmatrix} A_{11}^{-1} \\ O \end{bmatrix} (I_r + S^*S)^{-1} [I_r \quad S^*] P \qquad (26)$$

(Meyer and Painter [1]).

(c) A $\{1,2,4\}$-inverse of A is

$$A^{(1,2,4)} = Q \begin{bmatrix} I_r \\ T^* \end{bmatrix} (I_r + TT^*)^{-1} [A_{11}^{-1} \quad O] P. \qquad (27)$$

(d) The Moore–Penrose inverse of A is

$$A^\dagger = Q \begin{bmatrix} I_r \\ T^* \end{bmatrix} (I_r + TT^*)^{-1} A_{11}^{-1} (I_r + S^*S)^{-1} [I_r \quad S^*] P \qquad (28)$$

(Noble [1]).

PROOF. The partition (22) is a full-rank factorization of A (see Lemma 1.4),

$$A = FG, \qquad F \in C_r^{m \times r}, \qquad G \in C_r^{r \times n} \qquad (29)$$

with

$$F = P^T \begin{bmatrix} I_r \\ S \end{bmatrix} A_{11}, \qquad G = [I_r \quad T] Q^T \qquad (30)$$

or alternatively

$$F = P^T \begin{bmatrix} I_r \\ S \end{bmatrix}, \qquad G = A_{11}[I_r \quad T]Q^T. \tag{31}$$

The theorem now follows from Ex. 1.26 and Ex. 1.16 by using (29) with either (30) or (31). ∎

EXERCISES

1. Let

$$A = \begin{bmatrix} A_{11} & A_{12} \\ A_{21} & A_{22} \end{bmatrix}, \qquad A_{11} \text{ nonsingular.} \tag{32}$$

Then

$$\text{rank } A = \text{rank } A_{11} \tag{33}$$

if and only if

$$A_{22} = A_{21}A_{11}^{-1}A_{12} \qquad (\text{Brand } [1]). \tag{34}$$

2. Let A, A_{11} satisfy (32) and (33). Then the general solution of

$$\begin{bmatrix} A_{11} & A_{12} \\ A_{21} & A_{22} \end{bmatrix} \begin{bmatrix} x_1 \\ x_2 \end{bmatrix} = \begin{bmatrix} 0 \\ 0 \end{bmatrix}$$

is given by

$$\begin{bmatrix} x_1 \\ x_2 \end{bmatrix} = \begin{bmatrix} -A_{11}^{-1}A_{12}x_2 \\ x_2 \end{bmatrix}, \qquad x_2 \text{ arbitrary.}$$

3. Let A, A_{11} satisfy (32) and (33). Then the linear equation

$$\begin{bmatrix} A_{11} & A_{12} \\ A_{21} & A_{22} \end{bmatrix} \begin{bmatrix} x_1 \\ x_2 \end{bmatrix} = \begin{bmatrix} b_1 \\ b_2 \end{bmatrix} \tag{35}$$

is consistent if and only if

$$A_{21}A_{11}^{-1}b_1 = b_2$$

in which case the general solution of (35) is given by

$$\begin{bmatrix} x_1 \\ x_2 \end{bmatrix} = \begin{bmatrix} A_{11}^{-1}b_1 - A_{11}^{-1}A_{12}x_2 \\ x_2 \end{bmatrix}, \qquad x_2 \text{ arbitrary.}$$

4. Let A, A_{11} satisfy (32) and (33). Then

$$A^\dagger = [A_{11} \quad A_{12}]^* T_{11}^* \begin{bmatrix} A_{11} \\ A_{21} \end{bmatrix}^*,$$

where

$$T_{11} = \left([A_{11} \quad A_{12}]A^* \begin{bmatrix} A_{11} \\ A_{21} \end{bmatrix} \right)^{-1} \qquad \text{(Zlobec [2])}.$$

5. Let $A \in C_r^{n \times n}$, $r < n$, be partitioned by

$$A = \begin{bmatrix} A_{11} & A_{12} \\ A_{21} & A_{22} \end{bmatrix} = \begin{bmatrix} I_r \\ S \end{bmatrix} A_{11}[I_r \quad T], \qquad A_{11} \in C_r^{r \times r}. \tag{36}$$

Then the group inverse $A^\#$ exists if and only if $I_r + ST$ is nonsingular, in which case

$$A^\# = \begin{bmatrix} I_r \\ S \end{bmatrix} ((I_r + TS)A_{11}(I_r + TS))^{-1}[I_r \quad T] \qquad \text{(Robert [1])}. \tag{37}$$

6. Let $A \in C_r^{n \times n}$ be partitioned as in (36). Then A is range-Hermitian if and only if $S = T^*$.

7. Let $A \in C_r^{m \times n}$ be partitioned as in (22). Then the following orthogonal projectors are given in terms of the multipliers S, T and the permutation matrices P, Q as:

(a) $\qquad P_{R(A)} = P^T \begin{bmatrix} I_r \\ S \end{bmatrix} (I_r + S^*S)^{-1} [I_r \quad S^*] P,$

(b) $\qquad P_{R(A^*)} = Q \begin{bmatrix} I_r \\ T^* \end{bmatrix} (I_r + TT^*)^{-1} [I_r \quad T] Q^T,$

(c) $\qquad P_{N(A)} = Q \begin{bmatrix} -T \\ I_{n-r} \end{bmatrix} (I_{n-r} + T^*T)^{-1} [-T^* \quad I_{n-r}] Q^T,$

(d) $\qquad P_{N(A^*)} = P^T \begin{bmatrix} -S^* \\ I_{m-r} \end{bmatrix} (I_{m-r} + SS^*)^{-1} [-S \quad I_{m-r}] P.$

Remark. (a) and (d) are alternative computations since

$$P_{R(A)} + P_{N(A^*)} = I_m.$$

The computation (a) requires inverting the $r \times r$ positive definite matrix $I_r + S^*S$, while in (d) the dimension of the positive definite matrix to be inverted is $(m-r) \times (m-r)$. Accordingly (a) may be preferred if $r < m-r$.

Similarly (b) and (c) are alternative computations since

$$P_{R(A^*)} + P_{N(A)} = I_n$$

with (b) preferred if $r < n-r$.

8. For a Hermitian matrix H we denote by

$H \geqslant O$ the fact that H is positive semidefinite,

$H > O$ that H is positive definite. Let

$$H = \begin{bmatrix} H_{11} & H_{12} \\ H_{12}^* & H_{22} \end{bmatrix},$$

where H_{11} and H_{22} are Hermitian. Then:

(a) $H \geqslant O$ if and only if

$$H_{11} \geqslant O, \qquad H_{11}H_{11}^{\dagger}H_{12} = H_{12} \qquad \text{and} \qquad H_{22} - H_{12}{}^*H_{11}^{\dagger}H_{12} \geqslant O$$

(b) $H > O$ if and only if

$$H_{11} > O, \qquad H_{11} - H_{12}H_{22}^{\dagger}H_{12}{}^* > O \qquad \text{and} \qquad H_{22} - H_{12}{}^*H_{11}^{-1}H_{12} > O$$

(Albert [2]).

9. Let

$$H = \begin{bmatrix} H_{11} & H_{12} \\ H_{12}{}^* & H_{22} \end{bmatrix}$$

be Hermitian positive semidefinite, and denote

$$H^{(\alpha)} = \begin{bmatrix} H_{11}^{(\alpha)} + H_{11}^{(\alpha)}H_{12}G^{(\alpha)}H_{12}{}^*H_{11}^{(\alpha)} & -H_{11}^{(\alpha)}H_{12}G^{(\alpha)} \\ -G^{(\alpha)}H_{12}{}^*H_{11}^{(\alpha)} & G^{(\alpha)} \end{bmatrix}, \quad (38)$$

where

$$G = H_{22} - H_{12}{}^*H_{11}^{(\alpha)}H_{12}$$

and α is an integer, or a set of integers, to be specified below. Then

(a) The relation (38) is an identity for $\alpha = 1$ and $\alpha = \{1,2\}$. This means that the right member of (38) is an $\{\alpha\}$-inverse of H if in it one substitutes the $\{\alpha\}$-inverses of H_{11} and G as indicated.

(b) If H_{22} is nonsingular and rank $H = $ rank $H_{11} + $ rank H_{22}, then (38) is an identity with $\alpha = \{1,2,3\}$ and $\alpha = \{1,2,3,4\}$ (Rohde [2]).

3. INTERSECTION OF MANIFOLDS

For any vector $f \in C^n$ and a subspace L of C^n, the set

$$f + L = \{ f + l : l \in L \} \tag{39}$$

is called a (linear) manifold of C^n. The vector f in (39) is not unique, indeed

$$f + L = (f + l) + L \qquad \text{for any } l \in L.$$

This nonuniqueness suggests singling out the representation

$$(f - P_L f) + L = P_{L^\perp} f + L \tag{40}$$

of the manifold (39) and calling it the *orthogonal representation* of the manifold $f + L$. We note that $P_{L^\perp} f$ is the unique vector of least Euclidean norm in the manifold (39).

In this section we study the intersection of two manifolds

$$\{f + L\} \cap \{g + M\} \tag{41}$$

for given vectors f and g, and given subspaces L and M in C^n. The results are needed in Section 4 below where the common solutions of pairs of linear equations are studied. Let such a pair be

$$Ax = a \tag{42}$$

and

$$Bx = b \tag{43}$$

where A and B are given matrices with n columns, and a and b are given vectors. Assuming (42) and (43) to be consistent, their solutions are the manifolds

$$A^\dagger a + N(A) \tag{44}$$

and

$$B^\dagger b + N(B), \tag{45}$$

respectively. If the intersection of these manifolds

$$\{A^\dagger a + N(A)\} \cap \{B^\dagger b + N(B)\} \tag{46}$$

is nonempty, then it is the set of common solutions of (42) and (43). This is the main explanation for our interest in intersections of manifolds, whose study here includes conditions for the intersection (41) to be nonempty, in which case its properties and representations are given.

Since linear subspaces are manifolds, this special case is considered first.

LEMMA 2. Let L and M be subspaces of C^n, with P_L and P_M the corresponding orthogonal projectors. Then

$$P_{L+M} = (P_L + P_M)(P_L + P_M)^\dagger$$

$$= (P_L + P_M)^\dagger (P_L + P_M). \tag{47}$$

PROOF.   Clearly $L + M = R([P_L \quad P_M])$. Therefore,

$$P_{L+M} = [P_L \quad P_M][P_L \quad P_M]^\dagger$$

$$= [P_L \quad P_M]\begin{bmatrix} P_L \\ P_M \end{bmatrix}(P_L + P_M)^\dagger \qquad \text{(by Ex. 10)}$$

$$= (P_L + P_M)(P_L + P_M)^\dagger, \qquad \text{since } P_L \text{ and } P_M \text{ are idempotent}$$

$$= (P_L + P_M)^\dagger(P_L + P_M), \qquad \text{since a Hermitian matrix}$$

commutes with its Moore–Penrose inverse.   ∎

The intersection of any two subspaces L and M in C^n is a subspace $L \cap M$ in C^n, nonempty since $0 \in L \cap M$. The orthogonal projector $P_{L \cap M}$ is given in terms of P_L and P_M, in the following.

THEOREM 3   (Anderson and Duffin [1]).   Let L, M, P_L, and P_M be as in Lemma 2. Then

$$P_{L \cap M} = 2P_L(P_L + P_M)^\dagger P_M$$

$$= 2P_M(P_L + P_M)^\dagger P_L. \tag{48}$$

PROOF.   Since $M \subset L + M$, it follows that

$$P_{L+M}P_M = P_M = P_M P_{L+M}, \tag{49}$$

and by using (47)

$$(P_L + P_M)(P_L + P_M)^\dagger P_M = P_M = P_M(P_L + P_M)^\dagger(P_L + P_M). \tag{50}$$

Subtracting $P_M(P_L + P_M)^\dagger P_M$ from the first and last expressions in (50) gives

$$P_L(P_L + P_M)^\dagger P_M = P_M(P_L + P_M)^\dagger P_L. \tag{51}$$

Now, let

$$H = 2P_L(P_L + P_M)^\dagger P_M = 2P_M(P_L + P_M)^\dagger P_L.$$

Evidently, $R(H) \subset L \cap M$, and therefore

$$H = P_{L \cap M} H = P_{L \cap M} \left(P_L (P_L + P_M)^\dagger P_M + P_M (P_L + P_M)^\dagger P_L \right)$$

$$= P_{L \cap M} (P_L + P_M)^\dagger (P_L + P_M)$$

$$= P_{L \cap M} P_{L+M} \qquad \text{(by Lemma 2)}$$

$$= P_{L \cap M},$$

since $L \cap M \subset L + M$. ∎

Other expressions for $L \cap M$ are given in the following theorem.

THEOREM 4   (Lent [1]).   Let L and M be subspaces of C^n. Then

(a) $L \cap M = [P_L \quad O]([P_L \quad -P_M]) = [O \quad P_M] N([P_L \quad -P_M])$

(b) $= N(P_{L^\perp} + P_{M^\perp})$

(c) $= N(I - P_L P_M) = N(I - P_M P_L)$.

PROOF.

(a) $x \in L \cap M$ if and only if

$$x = P_L y = P_M z \qquad \text{for some } y, z \in C^n,$$

which is equivalent to

$$x = [P_L \quad O] \begin{bmatrix} y \\ z \end{bmatrix} = [O \quad P_M] \begin{bmatrix} y \\ z \end{bmatrix}, \quad \text{where} \quad \begin{bmatrix} y \\ z \end{bmatrix} \in N([P_L \quad -P_M])$$

(b) Let $x \in L \cap M$. Then $P_{L^\perp} x = P_{M^\perp} x = 0$, proving that $x \in N(P_{L^\perp} + P_{M^\perp})$. Conversely, let $x \in N(P_{L^\perp} + P_{M^\perp})$, i.e.,

$$(I - P_L)x + (I - P_M)x = 0$$

or

$$2x = P_L x + P_M x$$

and therefore,

$$2\|x\| \leqslant \|P_L x\| + \|P_M x\|,$$

by the triangle inequality for norms. But by Ex. 2.41,

$$\|P_L x\| \leqslant \|x\|, \qquad \|P_M x\| \leqslant \|x\|.$$

Therefore,

$$\|P_L x\| = \|x\| = \|P_M x\|$$

and so, by Ex. 2.41,

$$P_L x = x = P_M x,$$

proving $x \in L \cap M$.

(c) Let $x \in L \cap M$. Then $x = P_L x = P_M x = P_L P_M x$, and therefore $x \in N(I - P_L P_M)$. Conversely, let $x \in N(I - P_L P_M)$ and therefore,

$$x = P_L P_M x \in L. \tag{52}$$

Also,

$$\|P_M x\|^2 + \|P_{M^\perp} x\|^2 = \|x\|^2$$

$$= \|P_L P_M x\|^2$$

$$\leqslant \|P_M x\|^2, \qquad \text{by Ex . 2.41.}$$

Therefore,

$$P_{M^\perp} x = 0, \qquad \text{i.e., } x \in M$$

and by (52),

$$x \in L \cap M.$$

The remaining equality in (c) is proved similarly. ∎

The intersection of manifolds, which if nonempty is itself a manifold, can now be determined.

THEOREM 5 (Ben-Israel [7], Lent [1]). Let f and g be vectors in C^n and let L and M be subspaces of C^n. Then the intersection of manifolds

$$\{f + L\} \cap \{g + M\} \tag{41}$$

is nonempty if and only if

$$g - f \in L + M, \tag{53}$$

in which case

(a) $\{f+L\}\cap\{g+M\}=f+P_L(P_L+P_M)^\dagger(g-f)+L\cap M$

(a') $=g-P_M(P_L+P_M)^\dagger(g-f)+L\cap M$

(b) $=f+(P_{L^\perp}+P_{M^\perp})^\dagger P_{M^\perp}(g-f)+L\cap M$

(b') $=g-(P_{L^\perp}+P_{M^\perp})^\dagger P_{L^\perp}(g-f)+L\cap M$

(c) $=f+(I-P_M P_L)^\dagger P_{M^\perp}(g-f)+L\cap M$

(c') $=g-(I-P_L P_M)^\dagger P_{L^\perp}(g-f)+L\cap M.$

PROOF. $\{f+L\}\cap\{g+M\}$ is nonempty if and only if

$$f+l=g+m, \quad \text{for some } l\in L, \quad m\in M,$$

which is equivalent to

$$g-f=l-m\in L+M.$$

We now prove (a), (b), and (c). The primed statements (a'), (b'), and (c')
are proved similarly to their unprimed counterparts.

(a) The points $x\in\{f+L\}\cap\{g+M\}$ are characterized by

$$x=f+P_L u=g+P_M v, \quad \text{for some } u,v\in C^n. \tag{54}$$

Thus

$$[P_L \quad -P_M]\begin{bmatrix} u \\ v \end{bmatrix}=g-f. \tag{55}$$

The linear equation (55) is consistent, since (41) is nonempty, and therefore
the general solution of (55) is

$$\begin{bmatrix} u \\ v \end{bmatrix}=[P_L \quad -P_M]^\dagger(g-f)+N([P_L \quad -P_M])$$

$$=\begin{bmatrix} P_L \\ -P_M \end{bmatrix}(P_L+P_M)^\dagger(g-f)+N([P_L \quad -P_M]) \tag{56}$$

by Ex. 10.

Substituting (56) in (54) gives

$$x = f + [P_L \quad O] \begin{bmatrix} u \\ v \end{bmatrix}$$

$$= f + P_L (P_L + P_M)^\dagger (g - f) + L \cap M$$

by Theorem 4(a).

(b) Writing (54) as

$$P_L u - P_M v = g - f$$

and multiplying by $P_{M^\perp}$ gives

$$P_{M^\perp} P_L u = P_{M^\perp} (g - f), \tag{57}$$

which implies

$$(P_{L^\perp} + P_{M^\perp}) P_L u = P_{M^\perp} (g - f). \tag{58}$$

The general solution of (58) is

$$P_L u = (P_{L^\perp} + P_{M^\perp})^\dagger P_{M^\perp} (g - f) + N(P_{L^\perp} + P_{M^\perp})$$

$$= (P_{L^\perp} + P_{M^\perp})^\dagger P_{M^\perp} (g - f) + L \cap M,$$

by Theorem 4(b), which when substituted in (54) proves (b).

(c) Equation (57) can be rewritten as

$$(I - P_M P_L) P_L u = P_{M^\perp} (g - f)$$

whose general solution is

$$P_L u = (I - P_M P_L)^\dagger P_{M^\perp} (g - f) + N(I - P_M P_L)$$

$$= (I - P_M P_L)^\dagger P_{M^\perp} (g - f) + L \cap M,$$

by Theorem 4(c), which when substituted in (54) proves (c). ∎

Theorem 5 verifies that the intersection (41), if nonempty, is itself a manifold. We note, in passing, that parts (a) and (a′) of Theorem 5 give the same representation of (41); i.e., if (53) holds, then

$$f + P_L (P_L + P_M)^\dagger (g - f) = g - P_M (P_L + P_M)^\dagger (g - f). \tag{59}$$

Indeed, (53) implies that

$$g - f = P_{L+M}(g - f)$$

$$= (P_L + P_M)(P_L + P_M)^\dagger(g - f),$$

which gives (59) by rearrangement of terms.

It will now be proved that parts (a), (a'), (b), and (b') of Theorem 5 give orthogonal representations of

$$\{f + L\} \cap \{g + M\} \tag{41}$$

if the representations $\{f + L\}$ and $\{g + M\}$ are orthogonal, i.e., if

$$f \in L^\perp, \qquad g \in M^\perp. \tag{60}$$

COROLLARY 1. Let L and M be subspaces of C^n, and let

$$f \in L^\perp, \qquad g \in M^\perp. \tag{60}$$

If (41) is nonempty, then each of its four representations given below is orthogonal.

(a) $\{f + L\} \cap \{g + M\} = f + P_L(P_L + P_M)^\dagger(g - f) + L \cap M$

(a') $\qquad\qquad\qquad = g - P_M(P_L + P_M)^\dagger(g - f) + L \cap M$

(b) $\qquad\qquad\qquad = f + (P_{L^\perp} + P_{M^\perp})^\dagger P_{M^\perp}(g - f) + L \cap M$

(b') $\qquad\qquad\qquad = g - (P_{L^\perp} + P_{M^\perp})^\dagger P_{L^\perp}(g - f) + L \cap M.$

PROOF.   Each of the above representations is of the form

$$\{f + L\} \cap \{g + M\} = v + L \cap M, \tag{61}$$

which is an orthogonal representation if and only if

$$P_{L \cap M} v = 0. \tag{62}$$

In the proof we use the facts

$$P_{L \cap M} = P_L P_{L \cap M} = P_{L \cap M} P_L = P_M P_{L \cap M} = P_{L \cap M} P_M, \tag{63}$$

which hold since $L \cap M$ is contained in both L and M.

(a) Here $v = f + P_L(P_L + P_M)^\dagger(g - f)$. The matrix $P_L + P_M$ is Hermitian, and therefore $(P_L + P_M)^\dagger$ is a polynomial in powers of $P_L + P_M$, by

Theorem 4.7. From (63) it follows therefore that

$$P_{L \cap M}(P_L + P_M)^\dagger = (P_L + P_M)^\dagger P_{L \cap M} \tag{64}$$

and (62) follows from

$$P_{L \cap M} v = P_{L \cap M} f + P_{L \cap M} P_L (P_L + P_M)^\dagger (g - f)$$

$$= P_{L \cap M} f + (P_L + P_M)^\dagger P_{L \cap M} (g - f) \quad \text{(by (63) and (64))}$$

$$= 0, \quad \text{by (60)}.$$

(a') Follows from (59) and (a).

(b) Here $v = f + (P_{L^\perp} + P_{M^\perp})^\dagger P_{M^\perp}(g - f)$. The matrix $P_{L^\perp} + P_{M^\perp}$ is Hermitian, and therefore $(P_{L^\perp} + P_{M^\perp})^\dagger$ is a polynomial in $P_{L^\perp} + P_{M^\perp}$, which implies that

$$P_{L \cap M}(P_{L^\perp} + P_{M^\perp})^\dagger = 0. \tag{65}$$

Finally, (62) follows from

$$P_{L \cap M} v = P_{L \cap M} f + P_{L \cap M}(P_{L^\perp} + P_{M^\perp})^\dagger P_{M^\perp}(g - f)$$

$$= 0, \quad \text{by (60) and (65)}.$$

(b') If (60) holds, then

$$g - f = P_{L^\perp + M^\perp}(g - f)$$

$$= (P_{L^\perp} + P_{M^\perp})^\dagger (P_{L^\perp} + P_{M^\perp})(g - f),$$

by Lemma 2, and therefore

$$f + (P_{L^\perp} + P_{M^\perp})^\dagger P_{M^\perp}(g - f) = g - (P_{L^\perp} + P_{M^\perp})^\dagger P_{L^\perp}(g - f),$$

which proves (b') identical to (b), if (60) is satisfied. ∎

Finally, we characterize subspaces L and M for which the intersection (41) is always nonempty.

COROLLARY 2. Let L and M be subspaces of C^n. Then the intersection

$$\{f + L\} \cap \{g + M\} \tag{41}$$

is nonempty for all $f, g \in C^n$ if and only if

$$L^\perp \cap M^\perp = \{0\}. \tag{66}$$

PROOF. The intersection (41) is by Theorem 5 nonempty for all $f, g \in C^n$, if and only if

$$L + M = C^n,$$

which is equivalent to

$$\{0\} = (L + M)^\perp$$
$$= L^\perp \cap M^\perp,$$

by Ex. 11(b). ∎

EXERCISES AND EXAMPLES

10. Let P_L and P_M be $n \times n$ orthogonal projectors. Then

$$[P_L \quad \pm P_M]^\dagger = \begin{bmatrix} P_L \\ \pm P_M \end{bmatrix} (P_L + P_M)^\dagger \tag{67}$$

PROOF. Use $A^\dagger = A^*(AA^*)^\dagger$ with $A = [P_L \quad \pm P_M]$, and the fact that P_L and P_M are Hermitian idempotents.

11. Let L and M be subspaces of C^n. Then:

(a) $(L \cap M)^\perp = L^\perp + M^\perp$

(b) $(L^\perp \cap M^\perp)^\perp = L + M$.

PROOF. (a) Evidently $L^\perp \subset (L \cap M)^\perp$ and $M^\perp \subset (L \cap M)^\perp$; hence

$$L^\perp + M^\perp \subset (L \cap M)^\perp.$$

Conversely, from $L^\perp \subset L^\perp + M^\perp$ it follows that

$$(L^\perp + M^\perp)^\perp \subset L^{\perp\perp} = L.$$

Similarly $(L^\perp + M^\perp)^\perp \subset M$, hence

$$(L^\perp + M^\perp)^\perp \subset L \cap M$$

and by taking orthogonal complements

$$(L \cap M)^{\perp} \subset L^{\perp} + M^{\perp}.$$

(b) Follows from (a) by replacing L and M by $L^{\perp}$ and $M^{\perp}$, respectively. ∎

12. Let $L_1, L_2, \ldots, L_k$ be any k linear subspaces of C^n, $k \geqslant 2$, and let

$$Q = P_{L_k} P_{L_{k-1}} \cdots P_{L_2} P_{L_1} P_{L_2} \cdots P_{L_{k-1}} P_{L_k}. \tag{68}$$

Then the orthogonal projector on $\bigcap\limits_{i=1}^{k} L_i$ is $\lim\limits_{m \to \infty} Q^m$ (von Neumann [4]).

13. The matrix Q of (68) is Hermitian, so let its spectral decomposition be given by

$$Q = \sum_{i=1}^{q} \lambda_i E_i$$

where

$$\lambda_1 \geqslant \lambda_2 \geqslant \cdots \geqslant \lambda_q$$

are the distinct eigenvalues of Q, and

$$E_1, E_2, \ldots, E_q$$

are the corresponding orthogonal projectors satisfying

$$E_1 + E_2 + \cdots + E_q = I$$

and

$$E_i E_j = O \qquad \text{if } i \neq j.$$

Then

$$1 \geqslant \lambda_1 \geqslant \lambda_2 \geqslant \cdots \geqslant \lambda_q \geqslant 0.$$

and

$$\bigcap_{i=1}^{k} L_i \neq \{0\} \qquad \text{if and only if } \lambda_1 = 1,$$

in which case the orthogonal projector on $\bigcap\limits_{i=1}^{k} L_i$ is E_1 (Pyle [2]).

14. *A closed-form expression.*   Using the notation of Ex. 13, the ortho-gonal projector on $\displaystyle\bigcap_{i=1}^{k} L_i$ is

$$Q^{\nu} + \left[\left(Q^{\nu+1} - Q^{\nu} \right)^{\dagger} - \left(Q^{\nu} - Q^{\nu-1} \right)^{\dagger} \right]^{\dagger}, \qquad \text{for } \nu = 2, 3, \dots \quad (69)$$

If λ_q, the smallest eigenvalue of Q, is positive then (69) holds also for $\nu = 1$, in which case Q^0 is taken as I (Pyle [2]).

4. COMMON SOLUTIONS OF LINEAR EQUATIONS AND GENERALIZED INVERSES OF PARTITIONED MATRICES

Consider the pair of linear equations

$$Ax = a, \qquad (42)$$

$$Bx = b, \qquad (43)$$

with given vectors a, b and matrices A, B having n columns.

Assuming (42) and (43) to be consistent, we study here their common solutions, if any, expressing them in terms of the solutions of (42) and (43).

The common solutions of (42) and (43) are the solutions of the partitioned linear equation

$$\begin{bmatrix} A \\ B \end{bmatrix} x = \begin{bmatrix} a \\ b \end{bmatrix}, \qquad (70)$$

which is often the starting point, the partitioning into (42) and (43) being used to reduce the size or difficulty of the problem.

The solutions of (42) and (43) constitute the manifolds

$$A^{\dagger} a + N(A) \qquad (44)$$

and

$$B^{\dagger} b + N(B), \qquad (45)$$

respectively. Thus, the intersection

$$\left\{ A^{\dagger} a + N(A) \right\} \cap \left\{ B^{\dagger} b + N(B) \right\} \qquad (46)$$

is the set of solutions of (70), and (70) is consistent if and only if (46) is nonempty.

The results of Section 3 are applicable to determining the intersection (46). In particular, Theorem 5 yields the following.

COROLLARY 3. Let A and B be matrices with n columns, and let a and b be vectors such that each of the equations (42) and (43) is consistent. Then (42) and (43) have common solutions if and only if

$$B^\dagger b - A^\dagger a \in N(A) + N(B), \tag{71}$$

in which case the set of common solutions is the manifold

(a) $\quad A^\dagger a + P_{N(A)}(P_{N(A)} + P_{N(B)})^\dagger (B^\dagger b - A^\dagger a) + N(A) \cap N(B)$

(a') $\quad = B^\dagger b - P_{N(B)}(P_{N(A)} + P_{N(B)})^\dagger (B^\dagger b - A^\dagger a) + N(A) \cap N(B)$

(b) $\quad = (A^\dagger A + B^\dagger B)^\dagger (A^\dagger a + B^\dagger b) + N(A) \cap N(B).$

PROOF. Follows from Theorem 5 by substituting

$$f = A^\dagger a, \qquad L = N(A), \qquad g = B^\dagger b, \qquad M = N(B). \tag{72}$$

Thus (71), (a), and (a') follow directly from (53), (a), and (a') of Theorem 5, respectively, by using (72).

That (b) follows from Theorem 5(b) or 5(b') is proved as follows. Substituting (72) in Theorem 5(b) gives

$$\{A^\dagger a + N(A)\} \cap \{B^\dagger b + N(B)\}$$

$$= A^\dagger a + (A^\dagger A + B^\dagger B)^\dagger B^\dagger B(B^\dagger b - A^\dagger a) + N(A) \cap N(B)$$

$$= (A^\dagger - (A^\dagger A + B^\dagger B)^\dagger B^\dagger B A^\dagger)a + (A^\dagger A + B^\dagger B)^\dagger B^\dagger b$$

$$+ N(A) \cap N(B), \tag{73}$$

since $P_{N(X)^\perp} = P_{R(X^*)} = X^\dagger X$ for $X = A, B$.

Now $R(A^\dagger) = R(A^*) \subset R(A^*) + R(B^*)$ and therefore

$$A^\dagger = (A^\dagger A + B^\dagger B)^\dagger (A^\dagger A + B^\dagger B) A^\dagger$$

by Lemma 2, from which it follows that

$$A^\dagger - (A^\dagger A + B^\dagger B)^\dagger B^\dagger B A^\dagger = (A^\dagger A + B^\dagger B)^\dagger A^\dagger,$$

which when substituted in (73) gives (b). ∎

Since each of the parts (a), (a'), and (b) of Corollary 3 gives the solutions of the partitioned equation (70), these expressions can be used to obtain the generalized inverses of partitioned matrices.

THEOREM 6 (Ben-Israel [7], Katz [2], Mihalyffy [1]). Let A and B be matrices with n columns. Then each of the following expressions is a

$\{1,2,4\}$-inverse of the partitioned matrix $\begin{bmatrix} A \\ B \end{bmatrix}$:

(a) $X = [A^\dagger \quad O] + P_{N(A)}(P_{N(A)} + P_{N(B)})^\dagger[-A^\dagger \quad B^\dagger],$ (74)

(a') $Y = [O \quad B^\dagger] - P_{N(B)}(P_{N(A)} + P_{N(B)})^\dagger[-A^\dagger \quad B^\dagger],$ (75)

(b) $Z = (A^\dagger A + B^\dagger B)^\dagger[A^\dagger \quad B^\dagger].$ (76)

Moreover, if

$$R(A^*) \cap R(B^*) = \{0\},$$ (77)

then each of the expressions (74), (75), (76) is the Moore–Penrose inverse

of $\begin{bmatrix} A \\ B \end{bmatrix}$.

PROOF.   From Corollary 3 it follows that whenever

$$\begin{bmatrix} A \\ B \end{bmatrix} x = \begin{bmatrix} a \\ b \end{bmatrix}$$ (70)

is consistent, then $X\begin{bmatrix} a \\ b \end{bmatrix}$, $Y\begin{bmatrix} a \\ b \end{bmatrix}$, and $Z\begin{bmatrix} a \\ b \end{bmatrix}$ are among its solutions. Also

the representations (44) and (45) are orthogonal, and therefore, by Corollary 1, the representations (a), (a') and (b) of Corollary 3 are also orthogonal. Thus $X\begin{bmatrix} a \\ b \end{bmatrix}$, $Y\begin{bmatrix} a \\ b \end{bmatrix}$, and $Z\begin{bmatrix} a \\ b \end{bmatrix}$ are all perpendicular to

$$N(A) \cap N(B) = N\begin{bmatrix} A \\ B \end{bmatrix}.$$

By Theorem 3.2, it follows therefore that X, Y, and Z are $\{1,4\}$-inverses of $\begin{bmatrix} A \\ B \end{bmatrix}$.

We show now that X, Y, and Z are $\{2\}$-inverses of $\begin{bmatrix} A \\ B \end{bmatrix}$.

(a) From (74) we get

$$X\begin{bmatrix} A \\ B \end{bmatrix} = A^\dagger A + P_{N(A)}(P_{N(A)} + P_{N(B)})^\dagger(-A^\dagger A + B^\dagger B).$$

But,

$$-A^\dagger A + B^\dagger B = P_{N(A)} - P_{N(B)} = (P_{N(A)} + P_{N(B)}) - 2P_{N(B)}.$$

Therefore, by Lemma 2 and Theorem 3,

$$X\begin{bmatrix} A \\ B \end{bmatrix} = A^\dagger A + P_{N(A)}P_{N(A)+N(B)} - P_{N(A) \cap N(B)}$$

$$= A^\dagger A + P_{N(A)} - P_{N(A) \cap N(B)} \qquad (\text{since } N(A) + N(B) \supset N(A))$$

$$= I_n - P_{N(A) \cap N(B)} \qquad (\text{since } P_{N(A)} = I - A^\dagger A). \tag{78}$$

Since $R(H^\dagger) = R(H^*) = N(H)^\perp$ for $H = A, B$,

$$P_{N(A) \cap N(B)}A^\dagger = O, \qquad P_{N(A) \cap N(B)}B^\dagger = O, \tag{79}$$

and therefore (78) gives

$$X\begin{bmatrix} A \\ B \end{bmatrix}X = X - P_{N(A)\cap N(B)}P_{N(A)}(P_{N(A)} + P_{N(B)})^{\dagger}[-A^{\dagger}\ \ B^{\dagger}].$$

Since

$$P_{N(A)\cap N(B)} = P_{N(A)\cap N(B)}P_{N(A)} = P_{N(A)\cap N(B)}P_{N(B)},$$

$$X\begin{bmatrix} A \\ B \end{bmatrix}X = X - \tfrac{1}{2}P_{N(A)\cap N(B)}(P_{N(A)} + P_{N(B)})(P_{N(A)} + P_{N(B)})^{\dagger}[-A^{\dagger}\ \ B^{\dagger}]$$

$$= X - \tfrac{1}{2}P_{N(A)\cap N(B)}P_{N(A)+N(B)}[-A^{\dagger}\ \ B^{\dagger}] \qquad \text{(by Lemma 2)}$$

$$= X - \tfrac{1}{2}P_{N(A)\cap N(B)}[-A^{\dagger}\ \ B^{\dagger}] \ \text{(since } N(A)+N(B) \supset N(A)\cap N(B))$$

$$= X \qquad \text{(by (79)).}$$

(a′) That Y given by (75) is a {2}-inverse of $\begin{bmatrix} A \\ B \end{bmatrix}$ is similarly proved.

(b) The proof that Z given by (76) is a {2}-inverse of $\begin{bmatrix} A \\ B \end{bmatrix}$ is easy since

$$Z\begin{bmatrix} A \\ B \end{bmatrix} = (A^{\dagger}A + B^{\dagger}B)^{\dagger}(A^{\dagger}A + B^{\dagger}B)$$

and therefore

$$Z\begin{bmatrix} A \\ B \end{bmatrix}Z = (A^{\dagger}A + B^{\dagger}B)^{\dagger}(A^{\dagger}A + B^{\dagger}B)(A^{\dagger}A + B^{\dagger}B)^{\dagger}[A^{\dagger}\ \ B^{\dagger}]$$

$$= (A^{\dagger}A + B^{\dagger}B)^{\dagger}[A^{\dagger}\ \ B^{\dagger}]$$

$$= Z.$$

Finally, we must show that (77) implies the X, Y, and Z given by (74), (75), and (76) respectively, are $\{3\}$-inverses of $\begin{bmatrix} A \\ B \end{bmatrix}$. Indeed (77) is equivalent to

$$N(A) + N(B) = C^n, \tag{80}$$

since $N(A) + N(B) = \{R(A^*) \cap R(B^*)\}^\perp$ by Ex. 11(b).

(a) From (74) it follows that

$$BX = [BA^\dagger \quad \text{O}] + BP_{N(A)}(P_{N(A)} + P_{N(B)})^\dagger[-A^\dagger \quad B^\dagger]. \tag{81}$$

But

$$(P_{N(A)} + P_{N(B)})(P_{N(A)} + P_{N(B)})^\dagger = I_n \tag{82}$$

by (80) and Lemma 2. Therefore,

$$BP_{N(A)}(P_{N(A)} + P_{N(B)})^\dagger = B(P_{N(A)} + P_{N(B)} - P_{N(B)})(P_{N(A)} + P_{N(B)})^\dagger = B,$$

and so (81) becomes

$$BX = [\text{O} \quad BB^\dagger].$$

Consequently,

$$\begin{bmatrix} A \\ B \end{bmatrix} X = \begin{bmatrix} AA^\dagger & \text{O} \\ \text{O} & BB^\dagger \end{bmatrix},$$

which proves that X is a $\{3\}$-inverse of $\begin{bmatrix} A \\ B \end{bmatrix}$.

(a') That Y given by (75) is a $\{3\}$-inverse of $\begin{bmatrix} A \\ B \end{bmatrix}$ whenever (77) holds is similarly proved, or, alternatively, (74) and (75) give

$$Y - X = [-A^\dagger \quad B^\dagger] - (P_{N(A)} + P_{N(B)})(P_{N(A)} + P_{N(B)})^\dagger[-A^\dagger \quad B^\dagger]$$

$$= \text{O}, \quad \text{by (82)}.$$

(b) Finally we show that Z is the Moore–Penrose inverse of $\begin{bmatrix} A \\ B \end{bmatrix}$ when (77) holds. By Ex. 2.32, the Moore–Penrose inverse of any matrix H is the only $\{1,2\}$-inverse U such that $R(U) = R(H^*)$ and $N(U) = N(H^*)$. Thus, $H^\dagger$ is also the unique matrix $U \in H\{1,2,4\}$ such that $N(U) \supset N(H^*)$.

Now, Z has already been shown to be a $\{1,2,4\}$-inverse of $\begin{bmatrix} A \\ B \end{bmatrix}$, and it is therefore sufficient to prove that

$$N(Z) \supset N([A^* \quad B^*]). \tag{83}$$

Let $\begin{bmatrix} u \\ v \end{bmatrix} \in N([A^* \quad B^*])$. Then

$$A^*u + B^*v = 0,$$

and therefore

$$A^*u = -B^*v = 0, \tag{84}$$

since, by (77), the only vector common to $R(A^*)$ and $R(B^*)$ is the zero vector. Since $N(H^\dagger) = N(H^*)$ for any H, (84) gives

$$A^\dagger u = B^\dagger v = 0,$$

and therefore by (76), $Z\begin{bmatrix} u \\ v \end{bmatrix} = 0$. Thus (83) is established, and the proof is complete. ∎

If a matrix is partitioned by columns instead of by rows, then Theorem 6 may still be used. Indeed,

$$[A \quad B] = \begin{bmatrix} A^* \\ B^* \end{bmatrix}^* \tag{85}$$

permits using Theorem 6 to obtain generalized inverses of $\begin{bmatrix} A^* \\ B^* \end{bmatrix}$, which is partitioned by rows, and then translating the results to the matrix $[A \quad B]$, partitioned by columns.

In working with the conjugate transpose of a matrix, we note that

$$X \in A\{i\} \Leftrightarrow X^* \in A^*\{i\} \qquad (i = 1, 2),$$

$$X \in A\{3\} \Leftrightarrow X^* \in A^*\{4\},$$

$$X \in A\{4\} \Leftrightarrow X^* \in A^*\{3\}. \tag{86}$$

Applying Theorem 6 to $\begin{bmatrix} A^* \\ B^* \end{bmatrix}$ as in (85), and using (86), we obtain the following.

COROLLARY 4. Let A and B be matrices with n rows. Then each of the following expressions is a $\{1, 2, 3\}$-inverse of the partitioned matrix $[A \quad B]$:

(a) $$X = \begin{bmatrix} A^\dagger \\ O \end{bmatrix} + \begin{bmatrix} -A^\dagger \\ B^\dagger \end{bmatrix} (P_{N(A^*)} + P_{N(B^*)})^\dagger P_{N(A^*)}, \tag{87}$$

(a') $$Y = \begin{bmatrix} O \\ B^\dagger \end{bmatrix} - \begin{bmatrix} -A^\dagger \\ B^\dagger \end{bmatrix} (P_{N(A^*)} + P_{N(B^*)})^\dagger P_{N(B^*)}, \tag{88}$$

(b) $$Z = \begin{bmatrix} A^\dagger \\ B^\dagger \end{bmatrix} (AA^\dagger + BB^\dagger)^\dagger. \tag{89}$$

Moreover, if

$$R(A) \cap R(B) = \{0\}, \tag{90}$$

then each of the expressions (87), (88), and (89) is the Moore–Penrose inverse of $[A \quad B]$.

Other and more general results on Moore–Penrose inverses of partitioned matrices were given in Cline [1]. However, these results are too formidable for reproduction here.

EXERCISE AND EXAMPLES

15. Let the partitioned matrix $\begin{bmatrix} A \\ B \end{bmatrix}$ be nonsingular. Then

(a) $\begin{bmatrix} A \\ B \end{bmatrix}^{-1} = [A^\dagger \quad O] + P_{N(A)}(P_{N(A)} + P_{N(B)})^{-1}[-A^\dagger \quad B^\dagger]$

(a') $= [O \quad B^\dagger] - P_{N(B)}(P_{N(A)} + P_{N(B)})^{-1}[-A^\dagger \quad B^\dagger]$

(b) $= (A^\dagger A + B^\dagger B)^{-1}[A^\dagger \quad B^\dagger]$

PROOF.   Follows from Theorem 6. Indeed the nonsingularity of $\begin{bmatrix} A \\ B \end{bmatrix}$ guarantees that (77) is satisfied, and also that the matrices $P_{N(A)} + P_{N(B)}$ and $A^\dagger A + B^\dagger B = P_{R(A^*)} + P_{R(B^*)}$ are nonsingular.

16. Let $A = [1 \quad 1]$, $B = [1 \quad 2]$. Then $\begin{bmatrix} A \\ B \end{bmatrix} = \begin{bmatrix} 1 & 1 \\ 1 & 2 \end{bmatrix}$ is nonsingular. We calculate now its inverse using Ex. 15(b).
   Here

$$A^\dagger = \tfrac{1}{2}\begin{bmatrix} 1 \\ 1 \end{bmatrix}, \quad A^\dagger A = \tfrac{1}{2}\begin{bmatrix} 1 & 1 \\ 1 & 1 \end{bmatrix},$$

$$B^\dagger = \tfrac{1}{5}\begin{bmatrix} 1 \\ 2 \end{bmatrix}, \quad B^\dagger B = \tfrac{1}{5}\begin{bmatrix} 1 & 2 \\ 2 & 4 \end{bmatrix},$$

$$A^\dagger A + B^\dagger B = \tfrac{1}{10}\begin{bmatrix} 7 & 9 \\ 9 & 13 \end{bmatrix}, \quad (A^\dagger A + B^\dagger B)^{-1} = \begin{bmatrix} 13 & -9 \\ -9 & 7 \end{bmatrix},$$

and finally,

$$\begin{bmatrix} A \\ B \end{bmatrix}^{-1} = (A^\dagger A + B^\dagger B)^{-1} [A^\dagger \quad B^\dagger]$$

$$= \begin{bmatrix} 13 & -9 \\ -9 & 7 \end{bmatrix} \tfrac{1}{10} \begin{bmatrix} 5 & 2 \\ 2 & 4 \end{bmatrix} = \begin{bmatrix} 2 & -1 \\ -1 & 1 \end{bmatrix}.$$

17. *Series expansions.* Let the partitioned matrix $\begin{bmatrix} A \\ B \end{bmatrix}$ be nonsingular.

Then

$$A^\dagger A + B^\dagger B = I + K, \tag{91}$$

where K is Hermitian and

$$\|K\| < 1. \tag{92}$$

From (91) and (92) it follows that

$$(A^\dagger A + B^\dagger B)^{-1} = \sum_{j=0}^{\infty} (-1)^j K^j. \tag{93}$$

Substituting (93) in Ex. 15(b) gives

$$\begin{bmatrix} A \\ B \end{bmatrix}^{-1} = \sum_{j=0}^{\infty} (-1)^j K^j [A^\dagger \quad B^\dagger]. \tag{94}$$

Similarly,

$$P_{N(A)} + P_{N(B)} = I - A^\dagger A + I - B^\dagger B$$

$$= I - K, \quad \text{with } K \text{ as in (91)}$$

and therefore

$$(P_{N(A)} + P_{N(B)})^{-1} = \sum_{j=0}^{\infty} K^j. \tag{95}$$

Substituting (95) in Ex. 15(a) gives

$$\left[\begin{array}{c} A \\ B \end{array}\right]^{-1} = [A^\dagger \quad O] + (I - A^\dagger A) \sum_{j=0}^{\infty} K^j [-A^\dagger \quad B^\dagger]. \tag{96}$$

18. Let the partitioned matrix $\left[\begin{array}{c} A \\ B \end{array}\right]$ be nonsingular. Then the solution of

$$\left[\begin{array}{c} A \\ B \end{array}\right] x = \left[\begin{array}{c} a \\ b \end{array}\right], \tag{70}$$

for any given a and b is

$$x = \sum_{j=0}^{\infty} (-1)^j K^j (A^\dagger a + B^\dagger b) \tag{97}$$

$$= A^\dagger a + (I - A^\dagger A) \sum_{j=0}^{\infty} K^j (B^\dagger b - A^\dagger a), \tag{98}$$

with K given by (91).

PROOF.   Use (94) and (96).

Remark.   If the nonsingular matrix $\left[\begin{array}{c} A \\ B \end{array}\right]$ is ill-conditioned, then slow convergence may be expected in (93) and (95), and hence in (94) and (96). Even then the convergence of (97) or (98) may be reasonable for certain vectors $\left[\begin{array}{c} a \\ b \end{array}\right]$. Thus for example, if $\| B^\dagger b - A^\dagger a \|$ is sufficiently small, then (98) may be reasonably approximated by its first few terms.

19. *Common solutions for n matrix equations.*   For $(i = 1, \ldots, n)$ let the matrices $A_i \in C^{p \times q}$, $B_i \in C^{p \times r}$ be given. For $(k = 1, \ldots, n)$ define recursively

$$C_k = A_k F_{k-1}, \qquad D_k = B_k - A_k E_{k-1},$$

$$E_k = E_{k-1} + F_{k-1} C_k^\dagger D_k \quad \text{and} \quad F_k = F_{k-1}(I - C_k^\dagger C_k), \tag{99}$$

where

$$E_0 = O_{q \times r}, \qquad F_0 = I_q.$$

Then the n matrix equations

$$A_i X = B_i \qquad (i = 1, \ldots, n) \qquad (100)$$

have a common solution if and only if

$$C_i C_i^\dagger D_i = D_i \qquad (i = 1, \ldots, n), \qquad (101)$$

in which case the general common solution of (100) is

$$X = E_n + F_n Z, \qquad (102)$$

where $Z \in C^{q \times r}$ is arbitrary (Morris and Odell [1]).

20. For $i = 1, \ldots, n$ let $A_i \in C^{1 \times q}$, and let C_i $(i = 1, \ldots, n)$ be defined by (99). Let the vectors $\{A_1, A_2, \ldots, A_k\}$ be linearly independent. Then the vectors $\{A_1, A_2, \ldots, A_{k+1}\}$ are linearly independent if and only if $C_{k+1} \neq O$ (Morris and Odell [1]).

21. For $i = 1, \ldots, n$ let A_i, C_i be as in Ex. 20. For any $k \leqslant n$, the vectors $\{C_1, C_2, \ldots, C_k\}$ are orthogonal and span the subspace spanned by $\{A_1, A_2, \ldots, A_k\}$ (Morris and Odell [1]).

5. GREVILLE'S METHOD AND RELATED RESULTS

Greville's method for computing the Moore–Penrose inverse $A^\dagger$ of a matrix $A \in C^{m \times n}$ is a finite iterative method. The main variant of this method, described in Theorem 7 below, uses n iterations. At the kth iteration $(k = 1, \ldots, n)$ it computes $A_k^\dagger$, where A_k is the submatrix of A consisting of its first k columns.

First we need some notation. For $k = 2, \ldots, n$ the matrix A_k is partitioned as

$$A_k = [A_{k-1} \quad a_k] \qquad (103)$$

where a_k is the kth column of A. For $k = 2, \ldots, n$ let the vectors d_k and c_k be defined by

$$d_k = A_{k-1}^\dagger a_k \qquad (104)$$

$$c_k = a_k - A_{k-1} d_k \qquad (105)$$

$$= a_k - A_{k-1} A_{k-1}^\dagger a_k$$

$$= a_k - P_{R(A_{k-1})} a_k$$

$$= P_{N(A_{k-1}^*)} a_k.$$

THEOREM 7   (Greville [3]).   Let $A \in C^{m \times n}$. Using the above notation, the Moore–Penrose inverse of A_k $(k = 2, \ldots, n)$ is

$$[A_{k-1} \quad a_k]^\dagger = \begin{bmatrix} A_{k-1}^\dagger - d_k b_k^* \\ b_k^* \end{bmatrix}, \tag{106}$$

where

$$b_k^* = c_k^\dagger \qquad \text{if } c_k \neq 0, \tag{107}$$

$$b_k^* = (1 + d_k^* d_k)^{-1} d_k^* A_{k-1}^\dagger \qquad \text{if } c_k = 0. \tag{108}$$

Remark.   If both $c_k = 0$ and $d_k = 0$, then it follows from (104) and (105) that $a_k = 0$, and as expected $b_k = 0$ by (108).

PROOF.   Let $A_k^\dagger = [A_{k-1} \quad a_k]^\dagger$ be partitioned as

$$A_k^\dagger = \begin{bmatrix} B_k \\ b_k^* \end{bmatrix}, \tag{109}$$

where b_k^* is the kth row of $A_k^\dagger$. Multiplying (103) and (109) gives

$$A_k A_k^\dagger = A_{k-1} B_k + a_k b_k^*. \tag{110}$$

Now by (103), Ex. 2.32 and Corollary 2.7,

$$N(A_{k-1}^\dagger) = N(A_{k-1}^*) \supset N(A_k^*) = N(A_k^\dagger) = N(A_k A_k^\dagger),$$

and it follows from Ex. ·2.23 that

$$A_{k-1}^\dagger A_k A_k^\dagger = A_{k-1}^\dagger. \tag{111}$$

Moreover, since

$$R(A_k^\dagger) = R(A_k^*)$$

by Ex. 2.32, it follows from (103), (109), and Corollary 2.7 that

$$R(B_k) \subset R(A_{k-1}^*) = R(A_{k-1}^\dagger) = R(A_{k-1}^\dagger A_{k-1}),$$

and therefore

$$A_{k-1}^\dagger A_{k-1} B_k = B_k \tag{112}$$

by Ex. 2.23. It follows from (111) and (112) that premultiplication of (110) by $A_{k-1}^\dagger$ gives

$$A_{k-1}^\dagger = B_k + A_{k-1}^\dagger a_k b_k^*$$

$$= B_k + d_k b_k^*, \tag{113}$$

by (104). Thus, we may write

$$[A_{k-1} \quad a_k]^\dagger = \begin{bmatrix} A_{k-1}^\dagger - d_k b_k^* \\ b_k^* \end{bmatrix}, \tag{106}$$

with b_k^* still to be determined. We distinguish two cases according as d_k is or is not in $R(A_{k-1})$, i.e., according as c_k is or is not 0.

Case I $(c_k \neq 0)$.

By using (113), (110) becomes

$$A_k A_k^\dagger = A_{k-1} A_{k-1}^\dagger + (a_k - A_{k-1} d_k) b_k^*$$

$$= A_{k-1} A_{k-1}^\dagger + c_k b_k^* \tag{114}$$

by (105). Since $A_k A_k^\dagger$ is Hermitian, it follows from (114) that $c_k b_k^*$ is Hermitian, and therefore

$$b_k^* = \delta c_k^*, \tag{115}$$

where δ is some real scalar. From (103) and (105) we obtain

$$A_k = A_k A_k^\dagger A_k = [A_{k-1} + c_k b_k^* A_{k-1} \quad a_k - c_k + (b_k^* a_k) c_k],$$

and comparison with (103) shows that

$$b_k^* a_k = 1, \tag{116}$$

since $c_k \neq 0$. Now, by (105),

$$c_k = P a_k,$$

where P denotes the orthogonal projector on $N(A_{k-1}^*)$. Therefore, (115) and (116) give

$$1 = b_k^* a_k = \delta c_k^* a_k = \delta a_k^* P a_k$$

$$= \delta a_k^* P^2 a_k = \delta c_k^* c_k, \tag{117}$$

since P is idempotent. By (115), (117), and Ex. 1.18(a)

$$b_k^* = \delta c_k^* = c_k^\dagger.$$

Case II $(c_k \neq 0)$.

Here $R(A_k) = R(A_{k-1})$, and so, by (109) and (2.50),

$$N(b^*) \supset N(A_k^\dagger) = N(A_k^*) = N(A_{k-1}^*) = N(A_{k-1}^\dagger)$$

$$= N(A_{k-1}A_{k-1}^\dagger).$$

Therefore, by Ex. 2.23,

$$b_k^* A_{k-1} A_{k-1}^\dagger = b_k^*. \tag{118}$$

Now, (103) and (106) give

$$A_k^\dagger A_k = \begin{bmatrix} A_{k-1}^\dagger A_{k-1} - d_k b_k^* A_{k-1} & (1-\alpha)d_k \\ b_k^* A_{k-1} & \alpha \end{bmatrix}, \tag{119}$$

where

$$\alpha = b_k^* a_k \tag{120}$$

is a scalar (real, in fact, since it is a diagonal element of a Hermitian matrix). Since (119) is Hermitian we have

$$b_k^* A_{k-1} = (1-\alpha)d_k^*.$$

Thus, by (118),

$$b_k^* = b_k^* A_{k-1} A_{k-1}^\dagger = (1-\alpha)d_k^* A_{k-1}^\dagger. \tag{121}$$

Substitution of (121) in (120) gives

$$\alpha = (1-\alpha)d_k^* d_k, \tag{122}$$

by (104). Adding $1-\alpha$ to both sides of (122) gives

$$(1-\alpha)(1+d_k^* d_k) = 1$$

and substitution for $1-\alpha$ in (121) gives (108). ∎

Greville's method as described above, thus computes $A^\dagger$ recursively in terms of $A_k^\dagger$ $(k=1,2,\ldots,n)$. This method was adapted by Greville [3] for the computation of $A^\dagger y$, for any $y \in C^m$, without computing $A^\dagger$. This is done as follows:

Let

$$\tilde{A} = [A \quad y].$$ (123)

Then (106) gives

$$A_k^\dagger \tilde{A} = \begin{bmatrix} A_{k-1}^\dagger \tilde{A} - d_k b_k^* \tilde{A} \\ b_k^* \tilde{A} \end{bmatrix}.$$ (124)

By (104) it follows that d_k is the kth column of $A_{k-1}^\dagger \tilde{A}$, for $k=2,\ldots,n$. Therefore only the vector $b_k^* \tilde{A}$ is needed to get $A_k^\dagger \tilde{A}$ from $A_{k-1}^\dagger \tilde{A}$ by (124). If $c_k = 0$, then (108) gives $b_k^* \tilde{A}$ as

$$b_k^* \tilde{A} = (1 + d_k^* d_k)^{-1} d_k^* A_{k-1}^\dagger \tilde{A} \qquad (c_k = 0).$$ (125)

If $c_k \neq 0$, then from (107)

$$b_k^* \tilde{A} = (c_k^* c_k)^{-1} c_k^* \tilde{A} \qquad (c_k \neq 0).$$ (126)

The computation of (126) is simplified by noting that the kth element of the vector $c_k^* \tilde{A}$ is $c_k^* a_k$ $(k = 1, 2, \ldots, n)$. Premultiplying (105) by c_k^* we obtain

$$c_k^* c_k = c_k^* a_k,$$ (127)

since $c_k^* A_{k-1} = 0$ by (105). Thus the vector (126) may be computed by computing $c_k^* \tilde{A}$ and normalizing it by dividing by its kth element. In the Greville method as described above, the matrix to be inverted is modified at each iteration by adjoining an additional column. This is the natural approach to some applications. Consider, for example, the *least-squares polynomial approximation problem* where a given real function $y(t)$ is to be approximated by the polynomials $\sum_{j=0}^k x_j t^j$. In the discrete version of this problem, the function $h(t)$ is represented by the m-dimensional vector

$$y = [y_i] = [y(t_i)] \qquad (i = 1, \ldots, m),$$ (128)

whose ith component is the function y evaluated at $t = t_i$, where the points $t_1, t_2, \ldots, t_m$ are given. Similarly, the polynomial t^j $(j = 0, 1, \ldots)$ is represented by the m-dimensional vector

$$a_{j+1} = [a_{i,j+1}] = \left[(t_i)^j \right] \qquad (i = 1, \ldots, m).$$ (129)

The problem is, therefore, for a given approximation error $\epsilon > 0$ to find an integer $k = k(\epsilon)$ and a vector $x \in R^{k-1}$ such that

$$\|A_{k-1}x - y\| \leq \epsilon, \tag{130}$$

where y is given by (128) and $A_{k-1} \in R^{m \times (k-1)}$ is the matrix

$$A_{k-1} = [a_1 \ a_2 \cdots a_{k-1}] \tag{131}$$

for a_j given by (129).

For any k, the Euclidean norm of $\|A_{k-1}x - y\|$ is minimized by

$$x = A_{k-1}^\dagger y. \tag{132}$$

If for a given k, the vector (132) does not satisfy (130), i.e., if

$$\|A_{k-1}A_{k-1}^\dagger y - y\| > \epsilon, \tag{133}$$

then we try achieving (130) with the matrix

$$A_k = [A_{k-1} \ a_k], \tag{103}$$

where, in effect, the degree of the approximating polynomial has been increased from $k-2$ to $k-1$. Greville's method described above computes $A_k^\dagger y$ in terms of $A_{k-1}^\dagger y$, and is thus the natural method for solving the above polynomial approximation problem and similar problems in approximation and regression.

There are applications on the other hand which call for modifying the matrix to be inverted by adjoining additional rows. Consider, for example, the problem of solving (or approximating the solution of) the following linear equation:

$$\sum_{j=1}^{n} a_{ij}x_j = y_i \qquad (i = 1, \dots, k-1), \tag{134}$$

where n is fixed and the data $\{a_{ij}, y_i : i = 1, \dots, k-1; j = 1, \dots, n\}$ are the result of some experiment or observation repeated $k-1$ times, with the row

$$[a_{i1} \ a_{i2} \cdots a_{in} \ y_i] \qquad (i = 1, \dots, k-1)$$

the result of the ith experiment.

Let $\hat{x}_{k-1}$ be the least-norm least-squares solution of (134), i.e.,

$$\hat{x}_{k-1} = A_{(k-1)}^\dagger y_{(k-1)}, \tag{135}$$

where $A_{(k-1)} = [a_{ij}]$ and $y_{(k-1)} = [y_i]$, $i = 1, \dots, k-1$; $j = 1, \dots, n$. If the results of an additional experiment or observation become available after (134) is solved, then it is necessary to update the solution (135) in light of the additional information. This explains the need for the variant of Greville's method described in Corollaries 5 and 6 below, for which some notation is needed.

Let n be fixed and let $A_{(k)} \in C^{k \times n}$ be partitioned as

$$A_{(k)} = \begin{bmatrix} A_{(k-1)} \\ a_k^* \end{bmatrix}, \qquad a_k^* \in C^{1 \times n} \tag{136}$$

Also, in analogy with (104) and (105), let

$$d_k^* = a_k^* A_{(k-1)}^\dagger, \tag{137}$$

$$c_k^* = a_k^* - d_k^* A_{(k-1)}. \tag{138}$$

COROLLARY 5 (Kishi [1]).   Using the above notation

$$A_{(k)}^\dagger = [A_{(k-1)}^\dagger - b_k d_k^* \quad b_k], \tag{139}$$

where

$$b_k = c_k^{*\dagger}, \qquad \text{if } c_k^* \neq 0 \tag{140}$$

$$b_k = (1 + d_k^* d_k)^{-1} A_{(k-1)}^\dagger d_k, \qquad \text{if } c_k^* = 0. \tag{141}$$

PROOF.   Follows by applying Theorem 7 to the conjugate transpose of the matrix (136). ■

In some applications it is necessary to compute

$$\hat{x}_k = A_{(k)}^\dagger y_{(k)} \qquad \text{for given } y_{(k)} \in C^k$$

but $A_{(k)}^\dagger$ is not needed. Then $\hat{x}_k$ may be obtained from $\hat{x}_{k-1}$ very simply, as follows.

COROLLARY 6   (Albert and Sittler [1]).   Let the vector $y_{(k)} \in C^k$ be partitioned as

$$y_{(k)} = \begin{bmatrix} y_{(k-1)} \\ y_k \end{bmatrix}, \qquad y_k \in C, \tag{142}$$

and let

$$\hat{x}_k = A^{\dagger}_{(k)} y_{(k)}, \qquad \hat{x}_{k-1} = A^{\dagger}_{(k-1)} y_{(k-1)} \tag{143}$$

using the notation (136). Then

$$\hat{x}_k = \hat{x}_{k-1} + (y_k - a^*_k \hat{x}_{k-1}) b_k, \tag{144}$$

with b_k given by (140) or (141).

PROOF.   Follows directly from Corollary 5.   ∎

Further references on these and related results are Albert and Sittler [1] and Kishi [1], where Greville's method is studied in relation to Kalman's filtering and prediction methods (Kalman [2] and [3]).

EXERCISES AND EXAMPLES

22. *A converse of Theorem 7.* Let the matrix $A_{k-1} \in C^{m \times (k-1)}$ be obtained from $A_k \in C^{m \times k}$ by deleting its kth column a_k. If A_k is of full column rank

$$\begin{bmatrix} A^{\dagger}_{k-1} \\ 0^T \end{bmatrix} = A^{\dagger}_k - \frac{A^{\dagger}_k b_k b^*_k}{b^*_k b_k}, \tag{145}$$

where b^*_k is the last row of $A^{\dagger}_k$ (Fletcher [2]).

23. Let

$$A_2 = [a_1 \quad a_2] = \begin{bmatrix} 1 & 0 \\ 2 & 1 \\ 0 & -1 \end{bmatrix}.$$

Then

$$A^{\dagger}_1 = a^{\dagger}_1 = \begin{bmatrix} 1 \\ 2 \\ 0 \end{bmatrix}^{\dagger} = \tfrac{1}{5}[1 \quad 2 \quad 0],$$

$$d_2 = A_1^\dagger a_2 = \tfrac{1}{5}\begin{bmatrix} 1 & 2 & 0 \end{bmatrix}\begin{bmatrix} 0 \\ 1 \\ -1 \end{bmatrix} = \tfrac{2}{5},$$

$$c_2 = a_2 - A_1 d_2 = \begin{bmatrix} 0 \\ 1 \\ -1 \end{bmatrix} - \begin{bmatrix} 1 \\ 2 \\ 0 \end{bmatrix}\tfrac{2}{5} = \begin{bmatrix} -\tfrac{2}{5} \\ \tfrac{1}{5} \\ -1 \end{bmatrix},$$

and by (107)

$$b_2^* = c_2^\dagger = \tfrac{1}{6}\begin{bmatrix} -2 & 1 & -5 \end{bmatrix}.$$

$A_2^\dagger$ is now computed by (106) as

$$A_2^\dagger = \tfrac{1}{5}\begin{bmatrix} 1 & 2 & 0 \\ 0 & 0 & 0 \end{bmatrix} + \begin{bmatrix} -\tfrac{2}{5} \\ 1 \end{bmatrix}\begin{bmatrix} -\tfrac{2}{6} & \tfrac{1}{6} & -\tfrac{5}{6} \end{bmatrix}$$

$$= \begin{bmatrix} \tfrac{1}{3} & \tfrac{1}{3} & \tfrac{1}{3} \\ -\tfrac{1}{3} & \tfrac{1}{6} & -\tfrac{5}{6} \end{bmatrix}.$$

Let now $a_2^\dagger$ be computed by (145), i.e., by deleting a_1 from A_2. Interchanging columns of A_2 and rows of $A_2^\dagger$ we obtain

$$A_2^\dagger b_2 = \begin{bmatrix} -\tfrac{1}{3} & \tfrac{1}{6} & -\tfrac{5}{6} \\ \tfrac{1}{3} & \tfrac{1}{3} & \tfrac{1}{3} \end{bmatrix}\begin{bmatrix} \tfrac{1}{3} \\ \tfrac{1}{3} \\ \tfrac{1}{3} \end{bmatrix} = \begin{bmatrix} -\tfrac{1}{3} \\ \tfrac{1}{3} \end{bmatrix}$$

and

$$b_2^* b_2 = \begin{bmatrix} \tfrac{1}{3} & \tfrac{1}{3} & \tfrac{1}{3} \end{bmatrix}\begin{bmatrix} \tfrac{1}{3} \\ \tfrac{1}{3} \\ \tfrac{1}{3} \end{bmatrix} = \tfrac{1}{3},$$

and finally from (145)

$$
\begin{bmatrix} A_1^\dagger \\ 0^T \end{bmatrix} = \begin{bmatrix} -\frac{1}{3} & \frac{1}{6} & -\frac{5}{6} \\ \frac{1}{3} & \frac{1}{3} & \frac{1}{3} \end{bmatrix} - \begin{bmatrix} -1 \\ 1 \end{bmatrix} \begin{bmatrix} \frac{1}{3} & \frac{1}{3} & \frac{1}{3} \end{bmatrix}
$$

$$
= \begin{bmatrix} 0 & \frac{1}{2} & -\frac{1}{2} \\ 0 & 0 & 0 \end{bmatrix},
$$

or

$$
a_2^\dagger = \begin{bmatrix} 0 \\ 1 \\ -1 \end{bmatrix}^\dagger = \tfrac{1}{2}[0 \quad 1 \quad -1] \qquad \text{(Fletcher [2])}.
$$

6. GENERALIZED INVERSES OF BORDERED MATRICES

Partitioning was shown above to permit working with submatrices smaller in size and better behaved (e.g., nonsingular) than the original matrix. In this section a nonsingular matrix is obtained from the original matrix by adjoining to it certain matrices. Thus from a given $A \in C^{m \times n}$ we obtain the matrix

$$
\begin{bmatrix} A & U \\ V^* & O \end{bmatrix}, \tag{146}
$$

which, under certain conditions on U and V^*, is nonsingular, and from its inverse $A^\dagger$ can be read off. These ideas find applications in differential equations (Reid [2]) and eigenvalue computations (Blattner [1]).

The following theorem is based on the results of Blattner [1].

THEOREM 8   Let $A \in C_r^{m \times n}$ and let the matrices U and V satisfy:

(a) $U \in C_{m-r}^{m \times (m-r)}$ and the columns of U are a basis for $N(A^*)$.

(b) $V \in C_{n-r}^{n \times (n-r)}$ and the columns of V are a basis for $N(A)$.

Then the matrix

$$\begin{bmatrix} A & U \\ V^* & O \end{bmatrix}, \tag{146}$$

is nonsingular and its inverse is

$$\begin{bmatrix} A^\dagger & V^{*\dagger} \\ U^\dagger & O \end{bmatrix}. \tag{147}$$

PROOF.   Premultiplying (147) by (146) gives

$$\begin{bmatrix} AA^\dagger + UU^\dagger & AV^{*\dagger} \\ V^*A^\dagger & V^*V^{*\dagger} \end{bmatrix}. \tag{148}$$

Now, $R(U) = N(A^*) = R(A)^\perp$ by assumption (a) and (2.49), and therefore

$$AA^\dagger + UU^\dagger = I_n \tag{149}$$

by Ex. 2.45. Moreover,

$$V^*A^\dagger = V^*A^\dagger AA^\dagger = V^*A^*A^{\dagger *}A^\dagger = (AV)^*A^{\dagger *}A^\dagger = O, \tag{150}$$

by (1.2), (1.4), and assumption (b), while

$$AV^{*\dagger} = AV^{\dagger *} = A(V^\dagger VV^\dagger)^* = A(V^\dagger V^{\dagger *}V^*)^*$$

$$= AVV^\dagger V^{\dagger *} = O, \tag{151}$$

by (1.2), Ex. 1.17(b), (1.3), and assumption (b). Finally, V^* is of full row rank by assumption (b), and therefore

$$V^*V^{*\dagger} = I_{n-r}, \tag{152}$$

by Lemma 1.2(b). By (149)–(152), (148) reduces to I_{m+n-r}, and therefore (146) is nonsingular and (147) is its inverse. ∎

Further references on bordered matrices are Blattner [1], Reid [2], Hearon [1], and Germain-Bonne [1].

EXERCISES

24.   Let $A \in C_r^{m \times n}$ and let $U \in C^{m \times (m-r)}$ and $V \in C^{n \times (n-r)}$ satisfy

$$AV = O, \qquad V^*V = I_{n-r}, \qquad A^*U = O, \qquad \text{and} \qquad U^*U = I_{m-r}. \quad (153)$$

Then the matrix

$$\begin{bmatrix} A & U \\ V^* & O \end{bmatrix} \qquad\qquad\qquad (146)$$

is nonsingular and its inverse is

$$\begin{bmatrix} A^\dagger & V \\ U^* & O \end{bmatrix} \qquad \text{(Reid [2])}. \qquad\qquad (154)$$

25.   Let A, U, and V be as in Ex. 24, and let

$$\alpha = \min\{\|Ax\| : x \in R(A^*), \|x\| = 1\}, \qquad (155)$$

$$\beta = \max\{\|A^\dagger y\| : y \in C^n, \|y\| = 1\}. \qquad (156)$$

Then

$$\alpha\beta = 1 \qquad \text{(Reid [2])}. \qquad\qquad (157)$$

PROOF.   If $y \in C^n$, $\|y\| = 1$, then

$$z = A^\dagger y$$

is the solution of

$$Az = (I_m - UU^*)y, \qquad V^*z = 0.$$

Therefore,

$$\alpha\|A^\dagger y\| = \alpha\|z\| \leqslant \|Az\| \qquad \text{(by (155))}$$

$$= \|(I_m - UU^*)y\|$$

$$\leqslant \|y\| \qquad \text{(by Ex. 2.41 since } I_m - UU^* \text{ is an orthogonal projector)}$$

$$= 1. \qquad\qquad\qquad\qquad (158)$$

From (158) it follows that $\alpha\beta \leqslant 1$. On the other hand, let

$$x \in R(A^*), \qquad \|x\| = 1;$$

then

$$x = A^\dagger A x,$$

so that

$$1 = \|x\| = \|A^\dagger A x\| \leqslant \beta \|Ax\|,$$

proving that $\alpha\beta \geqslant 1$, and completing the proof of (157). See also Exs. 6.13 and 6.15.

26. *A generalization of Theorem 8.* Let

$$A = \begin{bmatrix} B & C \\ D & O \end{bmatrix}$$

be nonsingular of order n, where B is $m \times p$, $0 < m < n$ and $0 < p < n$. Then A^{-1} is of the form

$$A^{-1} = \begin{bmatrix} E & F \\ G & O \end{bmatrix}, \qquad (159)$$

where E is $p \times m$, if and only if B is of rank $m + p - n$, in which case

$$E = B^{(1,2)}_{N(D), R(C)}, \qquad F = D^{(1,2)}_{N(B), \{0\}}, \qquad G = C^{(1,2)}_{R(I_{n-p}), R(B)} \qquad (160)$$

PROOF. We first observe that since A is nonsingular, C is of full column rank $n - p$, for otherwise the columns of A would not be linearly independent. Similarly, D is of full row rank $n - m$. Since C is $m \times (n - p)$, it follows that $n - p \leqslant m$, or, in other words,

$$m + p \geqslant n.$$

If. Since A is nonsingular, the $m \times n$ matrix $[B \quad C]$ is of full row rank m, and therefore of column rank m. Therefore, a basis for C^m can be chosen from among its columns. Moreover, this basis can be chosen so that it includes all $n - p$ columns of C, and the remaining $m + p - n$ basis elements are columns of B. Since B is of rank $m + p - n$, the latter columns span $R(B)$. Therefore $R(B) \cap R(C) = \{0\}$, and consequently $R(B)$ and $R(C)$ are complementary subspaces. Similarly, we can show that $R(B^*)$ and $R(D^*)$ are complementary subspaces of C^p, and therefore their orthogonal complements $N(B)$ and $N(D)$ are complementary spaces.

The results of the preceding paragraph guarantee the existence of all three $\{1,2\}$-inverses in the right members of Eqs. (160). If X now denotes the right member of (159) with E, F, G given by (160), an easy computation shows that $AX = I_n$.

Only if. It was shown in the "if" part of the proof that rank B is at least $m + p - n$. If A^{-1} is of the form (162), we must have

$$BF = O. \tag{161}$$

Since A^{-1} is nonsingular, it follows from (159) that F is of full column rank $n - m$. Thus, (161) exhibits $n - m$ independent linear relations among the columns of B. Therefore, the rank of B is at most $p - (n - m) = m + p - n$. This completes the proof. ∎

SUGGESTED FURTHER READING

Section 2.   Ben-Israel [13], Burns, Carlson, Haynsworth and Markham [1], and Carlson, Haynsworth, and Markham [1].

Section 3.   Afriat [1].

Section 4.   Hartwig [4] and Harwood, Lovass-Nagy, and Powers [1].

Section 5.   Meyer [1].

6

A SPECTRAL THEORY FOR RECTANGULAR MATRICES

1. INTRODUCTION

Linear transformations in $\mathcal{L}(C^n, C^m)$ and their matrix representations in $C^{m \times n}$ are studied in this chapter, resulting in the simplest (diagonal) representations of linear transformations.

Given a linear transformation $A : C^n \rightarrow C^m$ and two bases $U = \{u_1, u_2, \ldots, u_m\}$ and $V = \{v_1, v_2, \ldots, v_n\}$ of C^m and C^n, respectively, the *matrix representation of A relative to the bases U and V* is the matrix $A_{\{U,V\}} = [a_{ij}] \in C^{m \times n}$ determined (uniquely) by

$$Av_j = \sum_{i=1}^{m} a_{ij} u_i, \qquad j = 1, \ldots, n. \tag{1}$$

For any such pair of bases $\{U, V\}$, (1) is a one-to-one correspondence between the linear transformations $\mathcal{L}(C^n, C^m)$ and the matrices $C^{m \times n}$, allowing the customary practice of using the same symbol A to denote both the linear transformation $A : C^n \rightarrow C^m$ and its matrix representation $A_{\{U,V\}}$.

The main result, Theorem 2 (a restatement of the Autonne–Eckart–Young theorem) states that for any $A \in C_r^{m \times n}$ with *singular values** $\alpha(A) = \{\alpha_1, \alpha_2, \ldots, \alpha_r\}$ ordered by

$$\alpha_1 \geqslant \alpha_2 \geqslant \cdots \geqslant \alpha_r > 0, \tag{2}$$

*See, for example, Exs. 9–19 below.

and for any r scalars $d(A) = \{d_1, d_2, \ldots, d_r\}$ satisfying

$$|d_i| = \alpha_i, \qquad i = 1, \ldots, r \tag{3}$$

there exist two unitary matrices $U \in U^{m \times m}$ (the set of $m \times m$ unitary matrices) and $V \in U^{n \times n}$ such that the $m \times n$ matrix

$$D = U^*AV = \begin{bmatrix} d_1 & & & & \\ & \searrow & & & O \\ & & \searrow & & \\ & & & d_r & \\ \hline & O & & & O \end{bmatrix} \tag{4}$$

is diagonal. Thus any $m \times n$ complex matrix A is unitarily equivalent[†] to a diagonal matrix

$$A = UDV^*. \tag{5}$$

The corresponding statement for linear transformations is that for any linear transformation $A : C^n \to C^m$ with $\dim R(A) = r$, and for any set of scalars $d(A) = \{d_1, \ldots, d_r\}$ satisfying (3), there exist two orthonormal bases $V = \{v_1, v_2, \ldots, v_n\}$ and $U = \{u_1, u_2, \ldots, u_m\}$ of C^n and C^m, respectively, such that the corresponding matrix representation $A_{\{U,V\}}$ is diagonal,

$$A_{\{U,V\}} = \begin{bmatrix} d_1 & & & & \\ & \searrow & & & O \\ & & \searrow & & \\ & & & d_r & \\ \hline & O & & & O \end{bmatrix},$$

i.e.,

$$\begin{cases} Av_j = d_j u_j, & j = 1, \ldots, r \\[2mm] Av_j = 0, & j = r+1, \ldots, n. \end{cases} \tag{6}$$

If $d(A) = \alpha(A)$, i.e., if the scalars $\{d_1, \ldots, d_r\}$ in (3) are chosen as the

[†]See, e.g., Ex. 8.

singular values of A,

$$d_i = \alpha_i, \qquad i = 1, \ldots, r, \tag{7}$$

then (5) is called the *singular-value decomposition* of A. In the general case, we will call (5) a UDV^*-*decomposition* of A.

While (7) is the most common choice, there are cases where other choices seem more natural. Thus, if $A \in C_r^{n \times n}$ is normal, then the choice $d(A) = \sigma(A)/\{0\}$, i.e., choosing the scalars $\{d_1, \ldots, d_r\}$ to be the nonzero eigenvalues of A, guarantees that $U = V$ in (4) and (5), giving the spectral theorem for normal matrices* as a special case of (5).

The UDV^*-decomposition studied in Section 2 is the basis for a generalized spectral theory for rectangular matrices; this theory generalizes and extends the classical spectral theory for normal matrices (Theorem 2.13), replacing orthogonal projectors and eigenvalues by partial isometries and scalars $d(A)$ satisfying (3), respectively. This generalized spectral theory, essentially due to Penrose [1], Lanczos [1], Hestenes [2], [3], [4], [5] and Hawkins and Ben-Israel [1], is developed in Section 4, following the discussion of partial isometries in Section 3.

EXERCISES AND EXAMPLES

1. *Similar matrices.* We recall that two square matrices A, B are called *similar* if

$$B = S^{-1}AS \tag{8}$$

for some nonsingular matrix S. If S in (8) is *unitary* [*orthogonal*] then A, B are called *unitarily similar* [*orthogonally similar*].

Given a linear transformation $A : C^n \to C^n$ and a basis $V = \{v_1, v_2, \ldots, v_n\}$ of C^n, the *matrix representation of A relative to the basis V* is the matrix $A_{\{V\}} = [a_{ij}] \in C^{n \times n}$ determined (uniquely) by

$$Av_j = \sum_{i=1}^{n} a_{ij} v_i, \qquad j = 1, \ldots, n. \tag{9}$$

Show that two $n \times n$ complex matrices are similar if and only if each is a matrix representation of the same linear transformation relative to a basis of C^n.

PROOF. *If*: Let $V = \{v_1, v_2, \ldots, v_n\}$ and $W = \{w_1, w_2, \ldots, w_n\}$ be two bases of C^n and let $A_{\{V\}}$ and $A_{\{W\}}$ be the corresponding matrix representation

*Theorem 2.13; see also Exs. 4 and 33 below.

of a given linear transformation $A : C^n \to C^n$. The bases V and W determine a (unique) nonsingular matrix $S = [s_{ij}]$ such that

$$w_j = \sum_{i=1}^{n} s_{ij} v_i, \qquad j = 1, \ldots, n,$$

or using the rules of matrix multiplication

$$[w_1, w_2, \ldots, w_n] = [v_1, v_2, \ldots, v_n] S,$$

i.e.,

$$[v_1, v_2, \ldots, v_n] = [w_1, w_2, \ldots, w_n] S^{-1}. \tag{10}$$

Rewriting (9) as

$$A[v_1, v_2, \ldots, v_n] = [v_1, v_2, \ldots, v_n] A_{\{V\}}, \tag{11}$$

we conclude, by substituting (10) in (11), that

$$A[w_1, w_2, \ldots, w_n] = [w_1, w_2, \ldots, w_n] S^{-1} A_{\{V\}} S,$$

and by the uniqueness of the matrix representation,

$$A_{\{W\}} = S^{-1} A_{\{V\}} S.$$

Only if. Similarly proved.

2. *Schur triangularization.* Any $A \in C^{n \times n}$ is unitarily similar to a triangular matrix.

PROOF. See, e.g., Marcus and Minc [1], p. 67.

3. *Perron's approximate diagonalization.* Let $A \in C^{n \times n}$. Then for any $\epsilon > 0$ there is a nonsingular matrix S such that $S^{-1} A S$ is a triangular matrix

$$S^{-1}AS = \begin{bmatrix} \lambda_1 & b_{12} & - & - & - & - & b_{1n} \\ 0 & \lambda_2 & & & & & \\ & & \ddots & & & & \\ & & & \ddots & & & \\ 0 & - & - & - & - & 0 & \lambda_n \end{bmatrix}$$

with the off-diagonal elements satisfying

$$\sum_{i,j} |b_{ij}| \leq \epsilon \qquad \text{(Bellman [1], p. 205)}.$$

4. A matrix in $C^{n \times n}$ is normal if and only if it is unitarily similar to a diagonal matrix.

5. A matrix in $C^{n \times n}$ is Hermitian if and only if it is unitarily similar to a real diagonal matrix.

6. A matrix in $C^{n \times n}$ is range-Hermitian if and only if it is similar to a matrix of the form

$$\begin{bmatrix} A & O \\ O & O \end{bmatrix},$$

where A is nonsingular (Katz and Pearl [1]).

PROOF. See Lemma 4.1.

7. For any $n \geqslant 2$ there is an $n \times n$ real matrix which is not similar to a triangular matrix in $R^{n \times n}$.

Hint. The diagonal elements of a triangular matrix are its eigenvalues.

8. *Equivalent matrices.* Two matrices A, B in $C^{m \times n}$ are called *equivalent* if there are nonsingular matrices $S \in C^{m \times m}$, $T \in C^{n \times n}$ such that

$$B = S^{-1}AT. \tag{12}$$

Note that this definition agrees with the definition of equivalence given in Section 2.11. If S and T in (12) are unitary matrices, then A, B are called *unitarily equivalent*.

Show that two matrices in $C^{m \times n}$ are equivalent if, and only if, each is a matrix representation of the same linear transformation relative to a pair of bases of C^m and C^n.

PROOF. Similar to the proof of Ex. 1.

9. *Singular values.* Let $A \in C_r^{m \times n}$ and let $\lambda_j(A^*A)$, $j = 1, \dots, n$, denote the eigenvalues of A^*A ordered by

$$\lambda_1(A^*A) \geqslant \lambda_2(A^*A) \geqslant \cdots \geqslant \lambda_r(A^*A) > \lambda_{r+1}(A^*A)$$

$$= \cdots = \lambda_n(A^*A) = 0. \tag{13}$$

The *singular values* of A, denoted by $\alpha_j(A)$, $j = 1, \dots, r$, are defined as

$$\alpha_j(A) = +\sqrt{\lambda_j(A^*A)}, \qquad j = 1, \dots, r. \tag{14}$$

The *set of singular values* of A is denoted by $\alpha(A)$. Ordering the eigenvalues of AA^* as in (13), it follows from

$$\lambda_j(AA^*) = \lambda_j(A^*A), \qquad j = 1, \dots, \min\{m, n\}$$

that the singular values can be defined equivalently by

$$\alpha_j(A) = +\sqrt{\lambda_j(AA^*)}\ , \qquad j = 1,\ldots,r. \tag{15}$$

10. A and A^* have the same singular values.
11. Unitarily equivalent matrices have the same singular values.

PROOF. Let $A \in C^{m \times n}$, and let $U \in U^{m \times m}$ and $V \in U^{n \times n}$ be any two unitary matrices. Then the matrix

$$(UAV)(UAV)^* = UAVV^*A^*U^* = UAA^*U^*$$

is similar to AA^*, and thus has the same eigenvalues. Therefore the matrices UAV and A have the same singular values. ∎

12. Let $A \in C_r^{m \times n}$. Then the matrix $\begin{bmatrix} O & A \\ A^* & O \end{bmatrix}$ has $2r$ nonzero eigenvalues given by $\pm \alpha_j(A), j = 1,\ldots,r$ (Lanczos [1]).

13. *An extremal characterization of singular values.* Let $A \in C_r^{m \times n}$. Then

$$\alpha_k(A) = \max\{\,\|Ax\| : \|x\| = 1,\, x \perp x_1,\ldots,x_{k-1}\,\}, \qquad k = 1,\ldots,r, \tag{16}$$

where
    $\|\ \|$ denotes the Euclidean norm,
    $\{x_1, x_2,\ldots,x_{k-1}\}$ is an orthonormal set of vectors in C^n, defined recursively by

$$\|Ax_1\| = \max\{\,\|Ax\| : \|x\| = 1\,\}$$

$$\|Ax_j\| = \max\{\,\|Ax\| : \|x\| = 1, x \perp x_1,\ldots,x_{j-1}\,\}, \qquad j = 2,\ldots,k-1,$$

and the right member of (16) is the (attained) supremum of $\|Ax\|$ over all vectors $x \in C^n$ with norm one, which are perpendicular to $x_1, x_2,\ldots,x_{k-1}$.

PROOF. Follows from the corresponding extremal characterization of the eigenvalues of A^*A (see, e.g., Marcus and Minc [1], p. 114),

$$\lambda_k(A^*A) = \max\{\,(x, A^*Ax) : \|x\| = 1, x \perp x_1,\ldots,x_{k-1}\,\}$$

$$= (x_k, A^*Ax_k), \qquad k = 1,\ldots,n$$

since $(x, A^*Ax) = (Ax, Ax) = \|Ax\|^2$. Here the vectors $\{x_1,\ldots,x_n\}$ are an orthonormal set of eigenvectors of A^*A,

$$A^*Ax_k = \lambda_k(A^*A)x_k, \qquad k = 1,\ldots,n. \quad ∎$$

The singular values can be characterized equivalently as

$$\alpha_k(A) = \max\{\|A^*y\| : \|y\| = 1, y \perp y_1, \ldots, y_{k-1}\}$$

$$= \|A^*y_k\|, \qquad k = 1, \ldots, r$$

where the vectors $\{y_1, \ldots, y_r\}$ are an orthonormal set of eigenvectors of AA^*, corresponding to its positive eigenvalues

$$AA^*y_k = \lambda_k(AA^*)y_k, \qquad k = 1, \ldots, r.$$

We can interpret this extremal characterization as follows: Let the columns of A be $a_j, j = 1, \ldots, n$. Then

$$\|A^*y\|^2 = \sum_{j=1}^{n} |(a_j, y)|^2.$$

Thus y_1 is a normalized vector maximizing the sum of squares of moduli of its inner products with the columns of A, the maximum value being $\alpha_1^2(A)$, etc.

14. If $A \in C_r^{n \times n}$ is normal and its eigenvalues are ordered by

$$|\lambda_1(A)| \geqslant |\lambda_2(A)| \geqslant \cdots \geqslant |\lambda_r(A)| > |\lambda_{r+1}(A)| = \cdots = |\lambda_n(A)| = 0$$

then the singular values of A are

$$\alpha_j(A) = |\lambda_j(A)|, \qquad j = 1, \ldots, r.$$

Hint. Use Ex. 13 and the spectral theorem for normal matrices, Theorem 2.13.

15. Let $A \in C_r^{m \times n}$, and let the singular values of $A^\dagger$ be ordered by

$$\alpha_1(A^\dagger) \geqslant \alpha_2(A^\dagger) \geqslant \cdots \geqslant \alpha_r(A^\dagger).$$

Then

$$\alpha_j(A^\dagger) = \frac{1}{\alpha_{r-j+1}(A)}, \qquad j = 1, \ldots, r. \tag{17}$$

PROOF.

$$\alpha_j^2(A^\dagger) = \lambda_j(A^{\dagger *}A^\dagger), \qquad \text{by definition (14)}$$

$$= \lambda_j\big((AA^*)^\dagger\big), \qquad \text{since } A^{\dagger *}A^\dagger = A^{*\dagger}A^\dagger = (AA^*)^\dagger$$

$$= \frac{1}{\lambda_{r-j+1}(AA^*)}$$

$$= \frac{1}{\alpha_{r-j+1}^2(A)}, \qquad \text{by definition (15).} \quad \blacksquare$$

16. Let $\|\ \ \|$ be the matrix norm

$$\|A\| = (\operatorname{trace} A^*A)^{1/2} = \left(\sum_{i=1}^m \sum_{j=1}^n |a_{ij}|^2\right)^{1/2} \tag{18}$$

defined on $C^{m \times n}$; see, e.g., Ex. 1.33.
Then for any $A \in C_r^{m \times n}$

$$\|A\|^2 = \sum_{j=1}^r \alpha_j^2(A). \tag{19}$$

PROOF.   Follows from $\operatorname{trace} A^*A = \sum_{j=1}^n \lambda_j(A^*A)$.

See also Ex. 59 below.

17. Let $\|\ \ \|_2$ be the *spectral norm*, defined on $C^{m \times n}$ by

$$\|A\|_2 = \max\big\{\sqrt{\lambda} : \lambda \text{ an eigenvalue of } A^*A\big\}$$

$$= \alpha_1(A); \tag{20}$$

see, e.g., Ex. 1.39. Then for any $A \in C_r^{m \times n}$, $r \geq 1$,

$$\|A\|_2 \|A^\dagger\|_2 = \frac{\alpha_1(A)}{\alpha_r(A)}. \tag{21}$$

PROOF.   Follows from Ex. 15 and definition (20).

18. *A condition number.* Let $A \in C_n^{n \times n}$, i.e., an $n \times n$ nonsingular matrix, and consider the equation

$$Ax = b \tag{22}$$

for $b \in C^n$. The sensitivity of the solution of (22) to changes in the right-hand side b, is indicated by the *condition number* of A, defined for any multiplicative matrix norm $\| \quad \|$ by

$$\text{cond}(A) = \|A\| \|A^{-1}\|. \tag{23}$$

Indeed, changing b to $(b + \delta b)$ results in a change of the solution $x = A^{-1}b$ to $x + \delta x$, with

$$\delta x = A^{-1} \delta b. \tag{24}$$

For a consistent pair of vector and matrix norms (see Exs. 1.34–1.36), it follows from (22) that

$$\|b\| \leqslant \|A\| \|x\|. \tag{25}$$

Similarly, from (24)

$$\|\delta x\| \leqslant \|A^{-1}\| \|\delta b\|. \tag{26}$$

From (25) and (26) we get the following bound:

$$\frac{\|\delta x\|}{\|x\|} \leqslant \|A\| \|A^{-1}\| \frac{\|\delta b\|}{\|b\|} = \text{cond}(A) \frac{\|\delta b\|}{\|b\|} \tag{27}$$

relating the change of the solution to the change in data and the condition number (23).

The condition number (23) corresponding to the matrix norm (20) is, by (21)

$$\text{cond}(A) = \frac{\alpha_1(A)}{\alpha_n(A)}. \tag{28}$$

Show that for the condition number (28)

$$\text{cond}(A^*A) = (\text{cond}(A))^2,$$

proving that A^*A is worse conditioned than A, if $\text{cond}(A) > 1$ (Taussky [1]).

19. *Weyl's inequalities.* Let $A \in C_r^{n \times n}$ have eigenvalues $\lambda_1, \ldots, \lambda_n$ ordered by

$$|\lambda_1| \geqslant |\lambda_2| \geqslant \cdots \geqslant |\lambda_n|$$

and singular values

$$\alpha_1 \geqslant \alpha_2 \geqslant \cdots \geqslant \alpha_r.$$

Then

$$\sum_{j=1}^{k} |\lambda_j| \leqslant \sum_{j=1}^{k} \alpha_j, \tag{29}$$

$$\prod_{j=1}^{k} |\lambda_j| \leqslant \prod_{j=1}^{k} \alpha_j, \tag{30}$$

for $k = 1, \ldots, r$ (Weyl [2], Marcus and Minc [1], pp. 115–116).

A Historical Note

Singular values were originally introduced for integral equations as early as 1907 (Schmidt [1], [2]; see also Smithies [1] and [2], Chapter VIII). In 1937 von Neumann [3] studied gauge functions of singular values in relation to matrix norms and inequalities; see, e.g., Exs. 59–60 below. At about the same time, Eckart and Young [1], [2] extended the singular-value decomposition (see next section) and utilized it in problems of matrix approximations; see, e.g., Exs. 21–27 below. In 1949–1950 Weyl [1] and Fan [1], [2] used singular values to establish bounds on eigenvalues; see, e.g., Ex. 19. For recent results see, e.g., Thompson [1].

2. THE UDV^* DECOMPOSITION

The UDV^* decomposition studied here is a variation of the singular value decomposition proved by Beltrami, Jordan, and Sylvester for square real matrices (see, e.g., MacDuffee [1], p. 78), by Autonne [1] for square complex matrices, and by Eckart and Young [2] for rectangular matrices. Our approach follows that of Eckart and Young [2]. First we require the following theorem.

THEOREM 1. Let $O \neq A \in C_r^{m \times n}$, let $\alpha(A)$, the singular values of A, be

$$\alpha_1 \geqslant \alpha_2 \geqslant \ldots \geqslant \alpha_r > 0, \tag{2}$$

and let $d(A) = \{d_1, \ldots, d_r\}$ be any complex scalars satisfying

$$|d_i| = \alpha_i, \qquad i = 1, \ldots, r. \tag{3}$$

Let $\{u_1, u_2, \ldots, u_r\}$ be an orthonormal set of eigenvectors of AA^* corresponding to its nonzero eigenvalues:

$$AA^* u_i = \alpha_i^2 u_i, \qquad i = 1, \ldots, r \tag{31}$$

$$(u_i, u_j) = \delta_{ij}, \qquad i, j = 1, \ldots, r. \tag{32}$$

Let $\{v_1, v_2, \ldots, v_r\}$ be defined by

$$v_i = \frac{1}{\bar{d}_i} A^* u_i, \qquad i = 1, \ldots, r. \tag{33}$$

Then $\{v_1, v_2, \ldots, v_r\}$ is an orthonormal set of eigenvectors of A^*A corresponding to its nonzero eigenvalues

$$A^*Av_i = \alpha_i^2 v_i, \qquad i = 1, \ldots, r \tag{34}$$

$$(v_i, v_j) = \delta_{ij}, \qquad i, j = 1, \ldots, r. \tag{35}$$

Furthermore

$$u_i = \frac{1}{d_i} Av_i, \qquad i = 1, \ldots, r. \tag{36}$$

Dually, let the vectors $\{v_1, v_2, \ldots, v_r\}$ satisfy (34) and (35) and let the vectors $\{u_1, u_2, \ldots, u_r\}$ be defined by (36). Then $\{u_1, \ldots, u_r\}$ satisfy (31), (32), and (33).

PROOF. Let v_i be given by (33), $i = 1, \ldots, r$. Then

$$A^*Av_i = \frac{1}{\bar{d}_i} A^*AA^* u_i$$

$$= d_i A^* u_i, \qquad \text{by (31) and (3)}$$

$$= \alpha_i^2 v_i, \qquad \text{by (33) and (3)}$$

and

$$(v_i, v_j) = \frac{1}{\bar{d}_i d_j} (A^* u_i, A^* u_j)$$

$$= \frac{1}{\bar{d}_i d_j} (AA^* u_i, u_j)$$

$$= \frac{d_i}{d_j} (u_i, u_j), \qquad \text{by (31) and (3)}$$

$$= \delta_{ij}, \qquad \text{by (32).} \quad \blacksquare$$

Equations (36) follow from (33) and (31). The dual statement follows by interchanging A and A^*.

An easy consequence of Theorem 1 is the following.

THEOREM 2   (Autonne [1], Eckart and Young [2]).   Let $O \neq A \in C_r^{m \times n}$, and let $d(A) = \{d_1, \ldots, d_r\}$ be complex scalars satisfying

$$|d_i| = \alpha_i, \qquad i = 1, \ldots, r \tag{3}$$

where

$$\alpha_1 \geqslant \alpha_2 \geqslant \cdots \geqslant \alpha_r > 0 \tag{2}$$

are the singular values of A.

Then there exist unitary matrices $U \in U^{m \times m}$ and $V \in U^{n \times n}$ such that the matrix

$$D = U^*AV = \begin{bmatrix} d_1 & & & | & \\ & \ddots & & | & O \\ & & d_r & | & \\ \text{---} & \text{---} & \text{---} & | & \text{---} \\ & O & & | & O \end{bmatrix} \tag{4}$$

is diagonal.

PROOF.   For the given $A \in C_r^{m \times n}$ we construct two such matrices U and V as follows.

Let the vectors $\{u_1, \ldots, u_r\}$ in C^m satisfy (31) and (32), and thus form an orthonormal basis of $R(AA^*) = R(A)$; see, e.g., Corollary 1.2. Let $\{u_{r+1}, \ldots, u_m\}$ be an orthonormal basis of $R(A)^\perp = N(A^*)$. Then the set $\{u_1, \ldots, u_r, u_{r+1}, \ldots, u_m\}$ is an orthonormal basis of C^m satisfying (31) and

$$A^*u_i = 0, \qquad i = r+1, \ldots, m. \tag{37}$$

The matrix U defined by

$$U = [u_1, \ldots, u_r, u_{r+1}, \ldots, u_m] \tag{38}$$

is thus an $m \times m$ unitary matrix.

Let now the vectors $\{v_1, \ldots, v_r\}$ in C^n be defined by (33). Then these vectors satisfy (34) and (35), and thus form an orthonormal basis of $R(A^*A) = R(A^*)$. Let $\{v_{r+1}, \ldots, v_n\}$ be an orthonormal basis of $R(A^*)^\perp = N(A)$. Then $\{v_1, \ldots, v_r, v_{r+1}, \ldots, v_n\}$ is an orthonormal basis of C^n satisfying (34) and

$$Av_j = 0, \qquad j = r+1, \ldots, n. \tag{39}$$

The matrix V defined by

$$V = [v_1, \ldots, v_r, v_{r+1}, \ldots, v_n] \tag{40}$$

is thus an $n \times n$ unitary matrix.

With U and V as given above, the matrix

$$D = U^*AV = [d_{ij}], \qquad i = 1, \ldots, m, j = 1, \ldots, n$$

satisfies

$$d_{ij} = u_i^*Av_j = 0 \qquad \text{if } i > r \text{ or } j > r, \qquad \text{by (37) and (39)},$$

and for $i, j = 1, \ldots, r$

$$d_{ij} = u_i^*Av_j$$

$$= \frac{1}{\bar{d}_j} u_i^* AA^* u_j, \qquad \text{by (33)}$$

$$= d_j u_i^* u_j, \qquad \text{by (31) and (3)}$$

$$= d_j \delta_{ij}, \qquad \text{by (32)},$$

completing the proof. ∎

A corresponding decomposition of $A^\dagger$ is given in

COROLLARY 1 (Penrose [1]. Let A, D, U, and V be as in Theorem 2. Then

$$A^\dagger = VD^\dagger U^* \tag{41}$$

where

$$D^\dagger = \begin{bmatrix} \dfrac{1}{d_1} & & & & \\ & \ddots & & & O \\ & & \dfrac{1}{d_r} & & \\ \hline & O & & & O \end{bmatrix}. \tag{42}$$

PROOF. Equation (41) follows from (5) and Ex. 1.22. The form (42) for

$D^{\dagger}$ is obvious since

$$D = \begin{bmatrix} d_1 & & & \vdots & \\ & \ddots & & \vdots & O \\ & & d_r & \vdots & \\ \hline & O & & \vdots & O \\ & & & \vdots & \end{bmatrix}. \quad \blacksquare$$

EXERCISES AND EXAMPLES

20. Let $A \in C_r^{m \times n}$, let $\{u_1, \ldots, u_r\}$ satisfy (31) and (32), and let $\{v_1, \ldots, v_r\}$ be given by (33). Then

$$A = \sum_{i=1}^{r} d_i u_i v_i^*. \qquad (43)$$

PROOF. The vectors $\{v_1, \ldots, v_r\}$ form an orthonormal basis of $R(A^*)$. Therefore,

$$\sum_{i=1}^{r} d_i u_i v_i^* x = 0 \qquad \text{for all } x \in R(A^*)^{\perp} = N(A),$$

and for any $j = 1, \ldots, r$

$$\sum_{i=1}^{r} d_i u_i v_i^* v_j = d_j u_j, \qquad \text{by (35)}$$

$$= A v_j, \qquad \text{by (36)}$$

proving that for all $x \in C^n$

$$\sum_{i=1}^{r} d_i u_i v_i^* x = A x. \quad \blacksquare$$

21. *Best matrix approximations of given rank.* For a given $A \in C_r^{m \times n}$ and an integer k, $1 \leqslant k \leqslant r$, a *best rank-k approximation of A* is a matrix $A_{(k)} \in C_k^{m \times n}$ satisfying

$$\|A - A_{(k)}\| = \inf_{X \in C_k^{m \times n}} \|A - X\|, \qquad (44)$$

where $\| \quad \|$ is the matrix norm (18).

For the matrices D, U, and V of Theorem 2, let $D_{(k)}$, $U_{(k)}$, and $V_{(k)}$ denote their submatrices defined by

$$D_{(k)} = \begin{bmatrix} d_1 & & \\ & \ddots & \\ & & d_k \end{bmatrix} \in C^{k \times k}, \qquad U_{(k)} = [u_1, \ldots, u_k] \in C^{m \times k},$$

$$V_{(k)} = [v_1, \ldots, v_k] \in C^{n \times k}. \tag{45}$$

Then a best rank-k approximation of A is

$$A_{(k)} = U_{(k)} D_{(k)} V_{(k)}^*, \tag{46}$$

which is unique if, and only if, the kth and the $(k+1)$st singular values of A are distinct:

$$\alpha_k \neq \alpha_{k+1}. \tag{47}$$

The approximation error of $A_{(k)}$ is

$$\|A - A_{(k)}\| = \left(\sum_{i=k+1}^{r} \alpha_i^2 \right)^{1/2} \qquad \text{(Eckart and Young [2])}. \tag{48}$$

PROOF. The matrix norm (18) is unitarily invariant (see, e.g., Ex. 1.41); thus for any $X \in C^{m \times n}$

$$\|A - X\| = \|U^*(A - X)V\|$$

$$= \|D - U^*XV\|, \qquad \text{by (4).} \tag{49}$$

Denoting

$$U^*XV = Y = [y_{ij}], \tag{50}$$

we rewrite (49) as

$$\|A - X\|^2 = \|D - Y\|^2$$

$$= \sum_{i=1}^{r} |d_i - y_{ii}|^2 + \sum_{\substack{i \neq j \\ \text{or} \\ i > r}} |y_{ij}|^2. \tag{51}$$

The (unique) $Y \in C_k^{m \times n}$ which minimizes (51) is therefore

$$y_{ij} = \begin{cases} d_i & \text{if } 1 \leqslant i = j \leqslant k \\ 0 & \text{otherwise} \end{cases} \tag{52}$$

and (46) follows from (52) and (50).

Reasoning as in Ex. 20, we can rewrite $A_{(k)}$ as

$$A_{(k)} = \sum_{i=1}^{k} d_i u_i v_i^*$$

$$= \left(\sum_{i=1}^{k} u_i u_i^* \right) A, \qquad \text{by (33)} \tag{53}$$

which can be used to prove the statement on the uniqueness of $A_{(k)}$.

Finally (48) follows from

$$\|A - A_{(k)}\|^2 = \sum_{i=k+1}^{r} \alpha_i^2 \qquad \text{by (51), (52), and (3).} \quad \blacksquare$$

This proof is due to Golub and Kahan [1]; see also Householder and Young [1], Gaches, Rigal, and Rousset de Pina [1], and Franck [1].

22. Let $A \in C_r^{m \times n}$. Then, using the notation of Ex. 21,

$$A = U_{(r)} D_{(r)} V_{(r)}^* = A_{(r)}, \tag{54}$$

$$A^\dagger = V_{(r)} D_{(r)}^{-1} U_{(r)}^* = A_{(r)}^\dagger. \tag{55}$$

23. Let $O \neq A \in C_r^{m \times n}$ have singular values

$$\alpha_1 \geqslant \alpha_2 \geqslant \cdots \geqslant \alpha_r > 0$$

and let $M_{r-1} = \cup_{k=0}^{r-1} C_k^{m \times n}$ be the set of $m \times n$ matrices of ranks $\leqslant r - 1$. Then the distance, using either norm (18) or the spectral norm (20), of A from M_{r-1} is

$$\inf_{X \in M_{r-1}} \|A - X\| = \alpha_r. \tag{56}$$

Two easy consequences of (56) are:

(a) Let A be as above, and let $B \in C^{m \times n}$ satisfy

$$\|B\| < \alpha_r;$$

then

$$\text{rank } (A+B) \geqslant \text{rank } A.$$

(b) For any $0 \leqslant k \leqslant \min\{m,n\}$, the $m \times n$ matrices of ranks $\leqslant k$, form a closed set in $C^{m \times n}$.

In particular, the $n \times n$ singular matrices form a closed set in $C^{n \times n}$. For any nonsingular $A \in C^{n \times n}$ with singular values

$$\alpha_1 \geqslant \alpha_2 \geqslant \cdots \geqslant \alpha_n > 0,$$

the smallest singular value α_n is a measure of the nonsingularity of A.

24. *A minimal rank matrix approximation.* Let $A \in C^{m \times n}$ and let $\epsilon > 0$. Find a matrix $B \in C^{m \times n}$ of minimal rank, satisfying for the norm (18)

$$\|A - B\| \leqslant \epsilon.$$

Solution. Using the notation of Ex. 21,

$$B = A_{(k)},$$

where k is determined by

$$\left(\sum_{i=k+1}^{r} \alpha_i^2 (A) \right)^{1/2} \leqslant \epsilon$$

$$\left(\sum_{i=k}^{r} \alpha_i^2 (A) \right)^{1/2} > \epsilon \qquad \text{(Golub [2])}.$$

25. *A unitary matrix approximation.* Let $U^{n \times n}$ denote the class of $n \times n$ unitary matrices. Let $A \in C_r^{n \times n}$ with a singular-value decomposition

$$A = UDV^*, \qquad D = \begin{bmatrix} \alpha_1 & & & \vdots & \\ & \ddots & & \vdots & O \\ & & \alpha_r & \vdots & \\ \text{-----} & & \text{-----} & + & \text{--} \\ & O & & \vdots & O \end{bmatrix}.$$

Then

$$\inf_{W \in U^{n \times n}} \|A - W\| = \|D - I\| = \sqrt{\sum_{i=1}^{r} (1 - \alpha_i)^2 + n - r}$$

is attained for

$$W = UV^* \qquad \text{(Fan and Hoffman [1], Mirsky [1], Golub [2]).}$$

26. The following generalization of Ex. 25 arises in factor analysis; see, e.g., Green [1] and Schönemann [1].

For given $A, B \in C^{m \times n}$, find a $W \in U^{n \times n}$ such that

$$\|A - BW\| \leqslant \|A - BX\| \qquad \text{for any } X \in U^{n \times n}.$$

Solution. $W = UV^*$ where $B^*A = UDV^*$ is a singular-value decomposition of B^*A.

27. Let $A_{(k)}$ be a best rank-k approximation of $A \in C_r^{m \times n}$ (as given by Ex. 21). Then $A_{(k)}^*$, $A_{(k)}A_{(k)}^*$, and $A_{(k)}^*A_{(k)}$ are best rank-k approximations of A^*, AA^*, and A^*A, respectively. If A is square, then $A_{(k)}^j$ is a best rank-k approximation of A^j for all $j = 1, 2, \ldots$ (Householder and Young [1]).

28. *Real matrices.* If $A \in R_r^{m \times n}$, then the unitary matrices U and V in its singular-value decomposition

$$A = UDV^*, \qquad D = \begin{bmatrix} \alpha_1 & & & & \\ & \diagdown & & \bigcirc & \\ & & \diagdown & & \\ \hline & & & \alpha_r & \\ & \bigcirc & & & \bigcirc \end{bmatrix},$$

can also be taken to be real, hence orthogonal.

29. *Simultaneous diagonalization.* Let $A_1, A_2 \in C^{m \times n}$. Then the following are equivalent:

(a) There exist two unitary matrices U, V such that both

$$D_1 = U^* A_1 V,$$

$$D_2 = U^* A_2 V$$

are diagonal real matrices (in which case one of them, say D_1, can be assumed to be non-negative).

(b) $A_1 A_2^*$ and $A_2^* A_1$ are both Hermitian (Eckart and Young [2]).

30. Let $A_1, A_2 \in C^{n \times n}$ be Hermitian matrices. Then the following are equivalent:

(a) There is a unitary matrix U such that both

$$D_1 = U^* A_1 U$$

and

$$D_2 = U^* A_2 U$$

are diagonal real matrices.

   (b) $A_1 A_2$ and $A_2 A_1$ are both Hermitian.

   (c) $A_1 A_2 = A_2 A_1$.

31. Let $A_1, A_2 \in C^{m \times n}$. Then the following are equivalent:

   (a) There are two unitary matrices U, V such that both

$$D_1 = U^* A_1 V$$

and

$$D_2 = U^* A_2 V$$

are diagonal matrices.

   (b) There is a polynomial f such that

$$A_1 A_2^* = f(A_2 A_1^*),$$

$$A_2^* A_1 = f(A_1^* A_2) \qquad \text{(Williamson [1])}.$$

32. *Normal matrices.* If $O \neq A \in C_r^{n \times n}$ is normal and its nonzero eigenvalues are ordered by

$$|\lambda_1| \geqslant |\lambda_2| \geqslant \cdots \geqslant |\lambda_r| > 0,$$

then the scalars $d(A) = \{d_1, \ldots, d_r\}$ in (3) can be chosen as the corresponding eigenvalues

$$d_i = \lambda_i, \qquad i = 1, \ldots, r. \tag{57}$$

This choice reduces both (33) and (36) to

$$u_i = v_i, \qquad i = 1, \ldots, r. \tag{58}$$

PROOF. The first claim follows from Ex. 14.

   Using Exs. 4 and 5 it can be shown that all four matrices A, A^*, AA^*, and A^*A have common eigenvectors. Therefore, the vectors $\{u_1, \ldots, u_r\}$ of (31) and (32) are also eigenvectors of A^*, and (33) reduces to (58).

33. If $O \neq A \in C_r^{n \times n}$ is normal, and the scalars $d(A)$ are chosen by (57), then the UDV^*-decomposition (5) of A reduces to the statement that A is unitarily similar to a diagonal matrix

$$A = UDU^*; \qquad \text{see Ex. 4.}$$

3. PARTIAL ISOMETRIES AND THE POLAR DECOMPOSITION THEOREM

A linear transformation $U: C^n \to C^m$ is called a *partial isometry* (sometimes also a *subunitary transformation*) if it is norm preserving on the orthogonal complement of its null space, i.e., if

$$\| Ux \| = \| x \| \quad \text{for all } x \in N(U)^{\perp} = R(U^*), \tag{59}$$

or equivalently, if it is distance preserving

$$\| Ux - Uy \| = \| x - y \| \quad \text{for all } x, y \in N(U)^{\perp}.$$

Except where otherwise indicated, the norms used here are the Euclidean vector norm and the corresponding spectral norm for matrices; see Ex. 1.39.

Partial isometries in Hilbert spaces were studied extensively by von Neumann [2], Halmos [2], Halmos and McLaughlin [1], Erdelyi [5], and others. Most of the results given here are special cases, for the finite-dimensional space C^n.

A nonsingular partial isometry is called an *isometry* (or a *unitary transformation*). Thus a linear transformation $U: C^n \to C^n$ is an isometry if (59) holds for all $x \in C^n$.

We recall that $U \in C^{n \times n}$ is a unitary matrix if and only if $U^* = U^{-1}$. Analogous characterizations of partial isometries are collected in the following theorem, drawn from Halmos [2], Hestenes [2], and Erdelyi [1].

THEOREM 3. Let $U \in C^{m \times n}$. Then the following eight statements are equivalent.

    (a) U is a partial isometry.
    (a*) U^* is a partial isometry.
    (b) $U^* U$ is an orthogonal projector.
    (b*) UU^* is an orthogonal projector.
    (c) $UU^* U = U$
    (c*) $U^* UU^* = U^*$
    (d) $U^* = U^{\dagger}$
    (e) $U^{\dagger}$ is a partial isometry.

PROOF. We prove (a)⇔(b), (a)⇔(e), and (b)⇔(c)⇔(d). The obvious equivalence (c)⇔(c*) then takes care of the dual statements (a*) and (b*).

(a)⇒(b). Since $R(U^* U) = R(U^*)$, (b) can be rewritten as

$$U^* U = P_{R(U^*)}. \tag{60}$$

From Ex. 5 it follows for any Hermitian $H \in C^{n \times n}$ that

$$(Hx, x) = 0, \qquad \text{for all } x \in C^n, \tag{61}$$

implies $H = O$. Consider now the matrix

$$H = P_{R(U^*)} - U^* U.$$

Clearly,

$$(Hx, x) = 0 \qquad \text{for all } x \in R(U^*)^\perp = N(U),$$

while for $x \in R(U^*)$

$$(P_{R(U^*)} x, x) = (x, x)$$

$$= (Ux, Ux) \qquad \text{by (a)}$$

$$= (U^* Ux, x).$$

Thus, (a) implies that the Hermitian matrix $H = P_{R(U^*)} - U^* U$ satisfies (61), which in turn implies (60).

(b)$\Rightarrow$(a). This follows from

$$(Ux, Ux) = (U^* Ux, x)$$

$$= (P_{R(U^*)} x, x), \qquad \text{by (60)}$$

$$= (x, x) \qquad \text{if } x \in R(U^*).$$

(a)$\Leftrightarrow$(e). Since

$$y = Ux, \qquad x \in R(U^*)$$

is equivalent to

$$x = U^\dagger y, \qquad y \in R(U),$$

it follows that

$$(Ux, Ux) = (x, x) \qquad \text{for all } x \in R(U^*)$$

is equivalent to

$$(y, y) = (U^\dagger y, U^\dagger y) \qquad \text{for all } y \in R(U) = N(U^\dagger)^\perp.$$

(b)$\Leftrightarrow$(c)$\Leftrightarrow$(d). The obvious equivalence (c)$\Leftrightarrow$(c*) states that $U^* \in U\{1\}$ if, and only if, $U^* \in U\{2\}$. Since U^* is (always) a $\{3,4\}$-inverse of U, it follows that U^* is a $\{1\}$-inverse of U if, and only if, $U^* = U^\dagger$. ■

Returning to the UDV^*-decomposition of Section 2, we identify some useful partial isometries in the following theorem.

THEOREM 4   (Hestenes [2]).   Let $O \neq A \in C_r^{m \times n}$, and let

$$A = UDV^*, \tag{5}$$

where the unitary matrices $U \in U^{m \times m}, V \in U^{n \times n}$ and the diagonal matrix $D \in C^{m \times n}$ are given as in Theorem 2. Let $U_{(r)}, D_{(r)},$ and $V_{(r)}$ be defined by (45).

Then

(a) The matrices $U_{(r)}, V_{(r)}$ are partial isometries with

$$U_{(r)} U_{(r)}^* = P_{R(A)}, \qquad U_{(r)}^* U_{(r)} = I_r, \tag{62}$$

$$V_{(r)} V_{(r)}^* = P_{R(A^*)}, \qquad V_{(r)}^* V_{(r)} = I_r. \tag{63}$$

(b) The matrix

$$E = U_{(r)} V_{(r)}^* \tag{64}$$

is a partial isometry with

$$EE^* = P_{R(A)}, \qquad E^* E = P_{R(A^*)}. \tag{65}$$

PROOF.   (a) That $U_{(r)}, V_{(r)}$ are partial isometries is obvious from their definitions and the unitarity of U and V (see, e.g., Ex. 36). Now

$$U_{(r)}^* U_{(r)} = I_r,$$

by definition (45), since U is unitary, and

$$P_{R(A^*)} = A^\dagger A = A_{(r)}^\dagger A_{(r)}, \qquad \text{by Ex. 21}$$

$$= V_{(r)} D_{(r)}^{-1} U_{(r)}^* U_{(r)} D_{(r)} V_{(r)}^*, \qquad \text{by (54) and (55)}$$

$$= V_{(r)} V_{(r)}^*,$$

with the remaining statements in (a) similarly proved.

(b) Using (62) and (63), it can be verified that

$$E^\dagger = V_{(r)} U_{(r)}^* = E^*,$$

from which (65) follows easily.   ■

The partial isometry E thus maps $R(A^*)$ isometrically onto $R(A)$. Since A also maps $R(A^*)$ onto $R(A)$, we should expect A to be a "multiple" of E. This is the essence of the following Theorem, proved by Autonne [1] and Williamson [1] for nonsingular matrices, by Penrose [1] for rectangular matrices and by Murray and von Neumann [1] for linear operators in Hilbert spaces.

THEOREM 5   (**The polar decomposition theorem**).   Let $O \neq A \in C_r^{m \times n}$. Then A can be written as

$$A = GE = EH, \qquad (66)$$

where $E \in C^{m \times n}$ is a partial isometry and $G \in C^{m \times m}$, $H \in C^{n \times n}$ are Hermitian and positive semi-definite.

The matrices E, G, and H are uniquely determined by

$$R(E) = R(G), \qquad (67)$$

$$R(E^*) = R(H), \qquad (68)$$

in which case

$$G^2 = AA^*, \qquad (69)$$

$$H^2 = A^*A, \qquad (70)$$

and E is given by

$$E = U_{(r)}V_{(r)}^*. \qquad (64)$$

PROOF.   Let

$$A = UDV^* \qquad (5)$$

be the singular-value decomposition of A, i.e., let the r nonzero diagonal elements of D be chosen by (7). For any k, $r \leqslant k \leqslant \min\{m,n\}$, we use (45) to define the three matrices

$$D_{(k)} = \begin{bmatrix} \alpha_1 & & & & \\ & \diagdown & & O & \\ & & \diagdown & & \\ & & & \alpha_r & \\ \hline & O & & & O \end{bmatrix} \in C^{k \times k},$$

$$U_{(k)} = [u_1,\dots,u_k] \in C^{m \times k}, \qquad V_{(k)} = [v_1,\dots,v_k] \in C^{n \times k}.$$

Then (5) can be rewritten as

$$A = U_{(k)}D_{(k)}V_{(k)}^*$$

$$= (U_{(k)}D_{(k)}U_{(k)}^*)(U_{(k)}V_{(k)}^*), \qquad \text{since } U_{(k)}^*U_{(k)} = I_k$$

$$= (U_{(k)}V_{(k)}^*)(V_{(k)}D_{(k)}V_{(k)}^*), \qquad \text{since } V_{(k)}^*V_{(k)} = I_k,$$

which proves (66) with the partial isometry

$$E = U_{(k)}V_{(k)}^*$$ (71)

and the positive semi-definite matrices

$$G = U_{(k)}D_{(k)}U_{(k)}^*, \qquad H = V_{(k)}D_{(k)}V_{(k)}^*.$$ (72)

This also shows E to be nonunique if $r < \min\{m,n\}$, in which case G and H also are nonunique, for (72) can then be replaced by

$$G = U_{(k)}D_{(k)}U_{(k)}^* + u_{k+1}u_{k+1}^*,$$

$$H = V_{(k)}D_{(k)}V_{(k)}^* + v_{k+1}v_{k+1}^*,$$

which satisfy (66) for the E given in (71).

Let now E and G satisfy (67). Then from (66)

$$AA^* = GEE^*G = GEE^\dagger G = GP_{R(E)}G = G^2,$$

which proves (69) and the uniqueness of G; see also Ex. 34 below. The uniqueness of E follows from

$$E = EE^\dagger E = GG^\dagger E = G^\dagger GE = G^\dagger A.$$ (73)

Similarly (68) implies (70) and the uniqueness of H, E.

Finally from

$$G^2 = AA^*$$

$$= U_{(r)}D_{(r)}V_{(r)}^*V_{(r)}D_{(r)}U_{(r)}^*, \qquad \text{by (54)}$$

$$= U_{(r)}D_{(r)}^2U_{(r)}^*,$$

we conclude that

$$G = U_{(r)}D_{(r)}U_{(r)}^*$$

and consequently

$$G^\dagger = U_{(r)}D_{(r)}^{-1}U_{(r)}^*.$$ (74)

Therefore,

$$E = G^\dagger A, \qquad \text{by (73)}$$

$$= U_{(r)}D_{(r)}^{-1}U_{(r)}^*U_{(r)}D_{(r)}V_{(r)}^*, \qquad \text{by (74) and (54)}$$

$$= U_{(r)}V_{(r)}^*, \qquad \text{proving (64).} \quad \blacksquare$$

If, in the proof of Theorem 5, one uses a general UDV^*-decomposition of A instead of the singular-value decomposition, then the matrices G and H defined by (72) are merely normal matrices, and need not be Hermitian. Hence, the following corollary.

COROLLARY 2.   Let $O \neq A \in C^{m \times n}$. Then, for any choice of the scalars $d(A)$ in (3), there exist a partial isometry $E \in C^{m \times n}$ and two normal matrices $G \in C^{m \times m}$, $H \in C^{n \times n}$, satisfying (66). The matrices E, G, and H are uniquely determined by (67) and (68), in which case

$$GG^* = AA^*, \tag{75}$$

$$H^*H = A^*A, \tag{76}$$

and E is given by (64). ∎

Theorem 5 is the matrix analog of the polar decomposition of a complex number

$$z = x + iy, \qquad x,y \text{ real}$$

as

$$z = |z|e^{i\theta}, \tag{77}$$

where

$$|z| = (z\bar{z})^{1/2} = (x^2 + y^2)^{1/2}$$

and

$$\theta = \arctan\frac{y}{x}.$$

Indeed, the complex scalar z in (77) corresponds to the matrix A in (66), while $\bar{z}$, $|z|$, and $e^{i\theta}$ correspond to A^*, G (or H) and E, respectively. This analogy is natural since $|z| = (z\bar{z})^{1/2}$ corresponds to the square roots $G = (AA^*)^{1/2}$ or $H = (A^*A)^{1/2}$, while the scalar $e^{i\theta}$ satisfies

$$|ze^{i\theta}| = |z| \qquad \text{for all } z \in C,$$

which justifies its comparison to the partial isometry E; see also Exs. 52 and 56.

EXERCISES AND EXAMPLES

34.   *Square roots.* Let $A \in C_r^{n \times n}$ be Hermitian positive semi-definite. Then there exists a unique Hermitian positive semi-definite matrix $B \in C_r^{n \times n}$

satisfying

$$B^2 = A; \tag{78}$$

B is called the *square root* of A, denoted by $A^{1/2}$.

PROOF. Writing A as

$$A = UDU^*, \qquad U \text{ unitary}, \qquad D = \begin{bmatrix} \lambda_1 & & & \vdots & \\ & \ddots & & \vdots & O \\ & & \lambda_r & \vdots & \\ \hline & O & & \vdots & O \end{bmatrix},$$

we see that

$$B = UD^{1/2}U^*, \qquad D^{1/2} = \begin{bmatrix} \lambda_1^{1/2} & & & \vdots & \\ & \ddots & & \vdots & O \\ & & \lambda_r^{1/2} & \vdots & \\ \hline & O & & \vdots & O \end{bmatrix}$$

is a Hermitian positive semi-definite matrix satisfying (78). To prove uniqueness, assume that B is a Hermitian matrix satisfying (78). Then, since B and $A = B^2$ commute, it follows from Ex. 30 that

$$B = U\tilde{D}U^*$$

where $\tilde{D}$ is diagonal and real, by Ex. 5, hence

$$\tilde{D} = D^{1/2}, \qquad \text{by (78)}.$$

35. *Linearity of isometries.* Let X, Y be real normed linear spaces and let $f: X \to Y$ be isometric, i.e.,

$$\|f(x_1) - f(x_2)\|_Y = \|x_1 - x_2\|_X \qquad \text{for all } x_1, x_2 \in X$$

where $\| \quad \|_X$ and $\| \quad \|_Y$ are the norms in X and Y, respectively. If $f(0) = 0$ then f is a linear transformation (Mazur and Ulam). For extensions and references see Dunford and Schwartz [1], p. 91 and Vogt [1].

36. *Partial isometries.* If the $n \times n$ matrix is unitary and $U_{(k)}$ is any $n \times k$ submatrix of U, then $U_{(k)}$ is a partial isometry. Conversely, if $W \in C_k^{n \times k}$ is a partial isometry, then there is an $n \times (n-k)$ partial isometry V such that the matrix $U = [W \; V]$ is unitary.

37. Any matrix unitarily equivalent to a partial isometry is a partial isometry.

PROOF.   Let $A = UBV^*$, $U \in U^{m \times m}$, $V \in U^{n \times n}$. Then

$$A^\dagger = VB^\dagger U^*, \qquad \text{by Ex. 1.22}$$

$$= VB^* U^*, \qquad \text{if } B \text{ is a partial isometry}$$

$$= A^*. \quad \blacksquare$$

38. Let $A \in C_r^{m \times n}$ be a partial isometry with singular values $\alpha(A) = \{\alpha_i : i = 1, \ldots, r\}$. Then

$$\alpha_i = 1, \qquad i = 1, \ldots, r.$$

Consequently, in any UDV^*-decomposition of a partial isometry, the diagonal factor

$$D = \begin{bmatrix} d_1 & & & & \\ & \ddots & & & O \\ & & d_r & & \\ & O & & & O \end{bmatrix}$$

has $|d_i| = 1$, $i = 1, \ldots, r$.

39. A linear transformation $E : C^n \to C^m$ with $\dim R(E) = r$ is a partial isometry if, and only if, there are two orthonormal bases $\{v_1, \ldots, v_r\}$ and $\{u_1, \ldots, u_r\}$ of $R(E^*)$ and $R(E)$, respectively, such that

$$u_i = Ev_i, \qquad i = 1, \ldots, r.$$

40. *Contractions.* A matrix $A \in C^{m \times n}$ is called a *contraction* if

$$\|Ax\| \leqslant \|x\| \qquad \text{for all } x \in C^n. \tag{79}$$

For any $A \in C^{m \times n}$ the following statements are equivalent:

(a) A is a contraction.

(b) A^* is a contraction.

(c) For any subspace L of C^m containing $R(A)$, the matrix $P_L - AA^*$ is positive semi-definite.

PROOF. (a)$\Leftrightarrow$(b). By Exs. 1.34 and 1.39,(a) is equivalent to

$$\|A\|_2 \leqslant 1,$$

but

$$\|A\|_2 = \|A^*\|_2 \quad \text{by (20) and Ex. 10}.$$

(b)$\Leftrightarrow$(c). By definition (79), the statement (b) is equivalent to

$$0 \leqslant (x,x) - (A^*x, A^*x)$$
$$= ((I - AA^*)x, x) \quad \text{for all } x \in C^m,$$

which in turn is equivalent to (c). ∎

41. Let $A \in C^{m \times n}$ be a contraction and let L be any subspace of C^m containing $R(A)$. Then the $(m+n) \times (m+n)$ matrix $M(A)$ defined by

$$M(A) = \left[\begin{array}{c|c} A & \sqrt{P_L - AA^*} \\ \hline O & O \end{array} \right]$$

is a partial isometry (Halmos and McLaughlin [1], Halmos [2]).

PROOF. The square root $\sqrt{P_L - AA^*}$ exists and is unique by Exs. 40(c) and 34. The proof then follows by verifying that

$$M(A)(M(A))^* M(A) = M(A). \quad ∎$$

42. *Eigenvalues of partial isometries.* Let U be an $n \times n$ partial isometry and let λ be an eigenvalue of U corresponding to the eigenvector x. Then

$$|\lambda| = \frac{\|P_{R(U^*)}x\|}{\|x\|};$$

hence

$$|\lambda| \leqslant 1 \quad (\text{Erdelyi } [1]).$$

PROOF. From $Ux = \lambda x$ we conclude

$$|\lambda| \|x\| = \|Ux\| = \|UP_{R(U^*)}x\| = \|P_{R(U^*)}x\|. \quad ∎$$

43. The partial isometry

$$U = \begin{bmatrix} 1 & 0 & 0 \\ 0 & \sqrt{3}/2 & 0 \\ 0 & \frac{1}{2} & 0 \end{bmatrix}$$

has the following eigensystem:

$$\lambda = 0, \quad x = \begin{bmatrix} 0 \\ 0 \\ 1 \end{bmatrix} \in N(U),$$

$$\lambda = 1, \quad x = \begin{bmatrix} 1 \\ 0 \\ 0 \end{bmatrix} \in R(U^*),$$

$$\lambda = \sqrt{3}/2, \quad x = \begin{bmatrix} 0 \\ \sqrt{3}/2 \\ \frac{1}{2} \end{bmatrix} = \begin{bmatrix} 0 \\ \sqrt{3}/2 \\ 0 \end{bmatrix} + \begin{bmatrix} 0 \\ 0 \\ \frac{1}{2} \end{bmatrix},$$

$$\begin{bmatrix} 0 \\ \sqrt{3}/2 \\ 0 \end{bmatrix} \in R(U^*), \qquad \begin{bmatrix} 0 \\ 0 \\ \frac{1}{2} \end{bmatrix} \in N(U).$$

44. *Normal partial isometries.* Let U be an $n \times n$ partial isometry. Then U is normal if and only if it is range-Hermitian.

PROOF. Since any normal matrix is range-Hermitian, only the "if" part needs proof. Let U be range-Hermitian, i.e., let $R(U) = R(U^*)$. Then $UU^* = U^*U$, by Theorem 3. ∎

45. Let U be an $n \times n$ partial isometry. If U is normal, then its eigenvalues have absolute values 0 or 1.

PROOF. For any nonzero eigenvalue λ of a normal partial isometry U, it follows from

$$Ux = \lambda x$$

that $x \in R(U) = R(U^*)$, and therefore

$$|\lambda| \, \|x\| = \|Ux\| = \|x\|. \quad \blacksquare$$

46. The converse of Ex. 45 is false. Consider, for example, the partial isometry

$$U = \begin{bmatrix} 0 & 1 \\ 0 & 0 \end{bmatrix}.$$

47. Let $E \in C^{n \times n}$ be a contraction. Then E is a normal partial isometry if, and only if, the eigenvalues of E have absolute values 0 or 1 and rank $E = $ rank E^2 (Erdelyi [6], Lemma 2).

48. A matrix $E \in C^{n \times n}$ is a normal partial isometry if, and only if,

$$E = U \begin{bmatrix} W & O \\ O & O \end{bmatrix} U^*,$$

where U and W are unitary matrices (Erdelyi [6]).

49. *Polar decompositions.* Let $A \in C^{n \times n}$ and let

$$A = GE, \tag{66}$$

where G is positive semi-definite and E is a partial isometry satisfying

$$R(E) = R(G). \tag{67}$$

Then A is normal if, and only if

$$GE = EG,$$

in which case E is a normal partial isometry (Hearon [2], Theorem 1; Halmos [2], Problem 108).

50. Let $A \in C^{n \times n}$ have the polar decomposition (66) and (67). Then A is a partial isometry if and only if G is an orthogonal projector.

PROOF.   *If.* Let

$$G = G^* = G^2. \tag{80}$$

Then

$AA^* = GEE^*G,$ by (66)

$= G^2,$ since $EE^* = P_{R(G)}$ by Theorem 3(b*) and (67)

$= G,$ by (80),

proving that A is a partial isometry by Theorem 3(b*).

Only if. Let A be a partial isometry and let $A = GE$ be its unique polar decomposition determined by (67). Then

$$AA^* = G^2$$

is a Hermitian idempotent, by Theorem 3(b*), and hence its square root G is also idempotent. ∎

51.   Let $A \in C^{m \times n}$ have the polar decomposition (66) satisfying (67) and (68). Then α is a singular value of A if, and only if,

$$Ax = \alpha Ex, \quad \text{for some } 0 \neq x \in R(E^*) \tag{81}$$

or equivalently, if and only if

$$A^*y = \alpha E^*y, \quad \text{for some } 0 \neq y \in R(E) \quad \text{(Hestenes [2]).} \tag{82}$$

PROOF.   From (66) it follows that (81) is equivalent to

$$G(Ex) = \alpha(Ex),$$

which, by (69), is equivalent to

$$AA^*(Ex) = \alpha^2(Ex).$$

The equivalence of (82) is similarly proved. ∎

52.   Let z be any complex scalar with the polar decomposition

$$z = |z|e^{i\theta}. \tag{77}$$

Then, for any real α, the following inequalities are obvious:

$$|z - e^{i\theta}| \leqslant |z - e^{i\alpha}| \leqslant |z + e^{i\theta}|.$$

Fan and Hoffman [1] established the following analogous matrix inequalities:

Let $A \in C^{n \times n}$ be decomposed as

$$A = UH,$$

where U is unitary and H is positive semi-definite. Then for any unitary $W \in U^{n \times n}$, the inequalities

$$\|A - U\| \leqslant \|A - W\| \leqslant \|A + U\|$$

hold for every unitarily invariant norm.

Give the analogous inequalities for the polar decomposition of rectangular matrices given in Theorem 5.

53. *Generalized Cayley transforms.* Let L be a subspace of C^n. Then the equations

$$U = (P_L + iH)(P_L - iH)^\dagger, \tag{83}$$

$$H = i(P_L - U)(P_L + U)^\dagger \tag{84}$$

establish a one-to-one correspondence between all Hermitian matrices H with

$$R(H) \subset L \tag{85}$$

and all normal partial isometries U with

$$R(U) = L \tag{86}$$

whose spectrum excludes -1 (Ben-Israel [6]).

PROOF. Note that

$$(P_L \pm iH) \quad \text{and} \quad (P_L + U)$$

map L onto itself for Hermitian H satisfying (85) and normal partial isometry U satisfying (86), whose spectrum excludes -1. Since on L, $(P_L \pm iH)$ and $(P_L \pm U)$ reduce to $(I \pm iH)$ and $(I \pm U)$, respectively, the proof follows from the classical theorem; see, e.g., Gantmacher [1], Vol. I, p. 279. ∎ See also Pearl [1], [2] and Nanda [1].

54. Let H be a given Hermitian matrix. Let L_1 and L_2 be two subspaces containing $R(H)$, and let U_1 and U_2 be the normal partial isometries defined, respectively, by (83). If $L_1 \subset L_2$ then $U_1 = U_2 P_{L_1}$, i.e., U_1 is the restriction of U_2 to L_1. Thus, the "minimal" normal partial isometry corresponding to a given Hermitian matrix H is

$$U = (P_{R(H)} + iH)(P_{R(H)} - iH)^\dagger.$$

55. A well-known inequality of Fan and Hoffman ([1], Theorem 3) is extended to the singular case as follows.

If H_1, H_2 are Hermitian with $R(H_1) = R(H_2)$ and if

$$U_k = (P_{R(H_k)} + iH_k)(P_{R(H_k)} - iH_k)^\dagger, \qquad k = 1, 2,$$

then

$$\|U_1 - U_2\| \leqslant 2\|H_1 - H_2\|$$

for every unitarily invariant norm (Ben-Israel [6]).

Trace Inequalities

56. Let z be a complex scalar. Then, for any real α, the following inequality is obvious:

$$|z| \geqslant \text{Re}\{ze^{i\alpha}\}.$$

An analogous matrix inequality can be stated as follows:
Let $H \in C^{n \times n}$ be Hermitian positive semi-definite. Then

$$\text{trace } H \geqslant \text{Re}\{\text{trace}(HW)\}, \qquad \text{for all } W \in U^{n \times n},$$

where $U^{n \times n}$ is the class of $n \times n$ unitary matrices.

PROOF. Suppose there is a $W_0 \in U^{n \times n}$ with

$$\text{trace } H < \text{Re}\{\text{trace}(HW_0)\}. \tag{87}$$

Let

$$H = UDU^* \qquad \text{with } U \in U^{n \times n}$$

and

$$D = \begin{bmatrix} \alpha_1 & & & \\ & \ddots & & \\ & & \ddots & \\ & & & \alpha_n \end{bmatrix},$$

where $\{\alpha_1, \ldots, \alpha_n\}$ are the eigenvalues of H. Then

$$\sum \alpha_i = \text{trace } H < \text{Re}\{\text{trace}(UDU^*W_0)\}, \qquad \text{by (87)}$$

$$= \text{Re}\{\text{trace } A\}, \qquad \text{where } A = UDV^*, \ V^* = U^*W_0 \tag{88}$$

$$= \text{Re}\{\sum \lambda_i\}, \qquad \text{where } \{\lambda_i, \ldots, \lambda_n\} \text{ are the eigenvalues of } A.$$

But $AA^* = UDV^*VDU^* = UD^2U^*$, proving that the nonzero $\{\alpha_i\}$ are the singular values of A. Thus (88) implies that

$$\sum \alpha_i < \sum |\lambda_i|,$$

a contradiction of Weyl's inequality (29). ∎

57. Let $A \in C_r^{m \times n}$ be given, and let $W_l^{n \times m}$ denote the class of all partial isometries in $C_l^{n \times m}$, where $l = \min\{m, n\}$. Then

$$\sup_{W \in W_l^{n \times m}} \mathrm{Re}\{\mathrm{trace}(AW)\}$$

is attained for some $W_0 \in W_l^{n \times m}$. Moreover, AW_0 is Hermitian positive semi-definite, and

$$\sup_{W \in W_l^{n \times m}} \mathrm{Re}\{\mathrm{trace}(AW)\} = \mathrm{trace}(AW_0) = \sum_{i=1}^{r} \alpha_i, \qquad (89)$$

where $\{\alpha_1, \ldots, \alpha_r\}$ are the singular values of A. (For $m = n$, and unitary W, this result is due to von Neumann [3].)

PROOF. Without a loss of generality, assume that $m \leqslant n$. Let

$$A = GE \qquad (66)$$

be a polar decomposition, where the partial isometry E is taken to be of full rank (using (71) with $k = m$), so $E \in W_m^{m \times n}$. Then, for any $W \in W_m^{n \times m}$,

$$\mathrm{trace}(AW) = \mathrm{trace}(GEW)$$

$$= \mathrm{trace}\left(\begin{bmatrix} G & O \\ O & O \end{bmatrix}\begin{bmatrix} E \\ E^\perp \end{bmatrix}\begin{bmatrix} W & W^\perp \end{bmatrix}\right), \qquad (90)$$

where the submatrices $E^\perp$ and $W^\perp$ are chosen so as to make

$$\begin{bmatrix} E \\ E^\perp \end{bmatrix} \quad \text{and} \quad \begin{bmatrix} W & W^\perp \end{bmatrix}$$

unitary matrices; see, e.g., Ex. 36. Since

$$\begin{bmatrix} G & O \\ O & O \end{bmatrix}$$

is positive semi-definite, and

$$\left[\begin{array}{c} E \\ \hline E^{\perp} \end{array} \right] \left[\; W \; \vdots \; W^{\perp} \; \right]$$

is unitary, it follows from Ex. 56 and (90), that

$$\sup_{W \in W_m^{n \times m}} \text{Re} \{ \text{trace}(AW) \}$$

is attained for $W_0 \in W_m^{n \times m}$ satisfying

$$A W_0 = G,$$

and (89) follows from (72). ∎

58. Let $A \in C_r^{m \times n}$ and $B \in C_s^{n \times m}$ have singular values

$$\alpha_1 \geqslant \alpha_2 \geqslant \cdots \geqslant \alpha_r > 0$$

and

$$\beta_1 \geqslant \beta_2 \geqslant \cdots \geqslant \beta_s > 0,$$

respectively. Then

$$\sup_{\substack{X \in U^{n \times n} \\ W \in U^{m \times m}}} \text{Re} \{ \text{trace}(AXBW) \}$$

is attained for some $X_0 \in U^{n \times n}$, $W_0 \in U^{m \times m}$, and is given by

$$\text{trace}(AX_0BW_0) = \sum_{i=1}^{\min\{r,s\}} \alpha_i \beta_i \qquad (\text{von Neumann [3]}).$$

This result was proved by von Neumann ([3], Theorem 1) for the case $m = n$. The general case is proved by "squaring" the matrices A and B, i.e., adjoining zero rows and columns to make them square.

Gauge Functions and Singular Values

The following two exercises relate gauge functions (Ex. 3.50) to matrix norms and inequalities. The unitarily invariant matrix norms are characterized in Ex. 59 as symmetric gauge functions of the singular values. For square matrices these results were proved by von Neumann [3]; see also Mirsky [1].

59. *Unitarily invariant matrix norms.* We use here the notation of Ex. 3.50.

Let the functions $\| \ \|_\varphi : C^{m \times n} \to R$ and $\hat{\varphi} : C^{mn} \to R$ be defined, for any function $\varphi : R^l \to R, l = \min\{m,n\}$, as follows: For any $A = [a_{ij}] \in C^{m \times n}$ with singular values

$$\alpha_1 \geqslant \alpha_2 \geqslant \cdots \geqslant \alpha_r > 0,$$

$\|A\|_\varphi$ and $\hat{\varphi}(a_{11}, \ldots, a_{mn})$ are defined as

$$\|A\|_\varphi = \hat{\varphi}(a_{11}, \ldots, a_{mn}) = \varphi(\alpha_1, \ldots, \alpha_r, 0, \ldots, 0). \tag{91}$$

Then:

(a) If $\varphi : R^l \to R$ satisfies conditions $(G1)$–$(G3)$ of Ex. 3.50, so does $\hat{\varphi} : C^{mn} \to R$.

(b) $\|UAV\|_\varphi = \|A\|_\varphi$ for all $A \in C^{m \times n}$, $U \in U^{m \times m}$, $V \in U^{n \times n}$.

(c) Let $\varphi : R^l \to R$ satisfy conditions $(G1)$–$(G3)$ of Ex. 3.50, and let $\varphi_D : R^l \to R$ be its dual, defined by (3.85). Then, for any $A \in C^{m \times n}$, the following supremum is attained, and

$$\sup_{\substack{X \in C^{n \times m} \\ \|X\|_\varphi = 1}} \mathrm{Re}\{ \mathrm{trace}(AX) \} = \|A\|_{\varphi_D}. \tag{92}$$

(d) If $\varphi : R^l \to R$ is a symmetric gauge function, then $\hat{\varphi} : C^{mn} \to R$ is a gauge function, and $\| \ \|_\varphi : C^{m \times n} \to R$ is a unitarily invariant norm.

(e) If $\| \ \| : C^{m \times n} \to R$ is a unitarily invariant norm, then there is a symmetric gauge function $\varphi : R^l \to R$ such that $\| \ \| = \| \ \|_\varphi$.

PROOF. (a) Follows from definition (91).

(b) Obvious by Ex. 11.

(c) For the given $A \in C^{m \times n}$

$$\sup_{\substack{X \in C^{m \times n} \\ \|X\|_\varphi = 1}} \mathrm{Re}\{ \mathrm{trace}(AX) \} = \sup_{\substack{X \in C^{n \times m} \\ \|X\|_\varphi = 1 \\ U \in U^{n \times n}, V \in U^{m \times m}}} \mathrm{Re}\{ \mathrm{trace}(AUXV) \}, \qquad \text{by (b)}$$

$$= \sup_{\varphi(\xi_1, \ldots, \xi_l) = 1} \sum_i \alpha_i \xi_i, \qquad \text{by Ex. 58}$$

$$= \varphi_D(\alpha_1, \ldots, \alpha_r), \qquad \text{by (3.86) and (3.88)}$$

$$= \|A\|_{\varphi_D}, \qquad \text{by (91),}$$

where

$$\alpha_1 \geqslant \alpha_2 \geqslant \cdots \geqslant \alpha_r > 0$$

and

$$\xi_1 \geqslant \cdots \qquad > 0$$

are the singular values of A and X, respectively.

(d) Let φ_D be the dual of φ, and let $\hat{\varphi}_D : C^{mn} \to R$ be defined by (91) as

$$\hat{\varphi}_D(a_{11}, \ldots, a_{mn}) = \|A\|_{\varphi_D}, \qquad \text{for } A = [a_{ij}].$$

Then

$$\hat{\varphi}_D(a_{11}, \ldots, a_{mn}) = \|A^*\|_{\varphi_D}, \qquad \text{by Ex. 10}$$

$$= \sup_{\substack{X = [x_{ij}] \in C^{m \times n} \\ \hat{\varphi}(x_{11}, \ldots, x_{mn}) = 1}} \operatorname{Re}\{\operatorname{trace}(A^*X)\}, \quad \text{by (92)}$$

$$= \sup_{\hat{\varphi}(x_{11}, \ldots, x_{mn}) = 1} \sum_{i,j} \bar{a}_{ij} x_{ij}$$

proving that $\hat{\varphi}_D : C^{mn} \to R$ is the dual of $\hat{\varphi} : C^{mn} \to R$, by using (3.86) and (3.88). Since φ is the dual of φ_D (by Ex. 3.50(d)), it follows that $\hat{\varphi}$ is the dual of $\hat{\varphi}_D$ and, by Ex. 3.50(b), $\hat{\varphi} : C^{mn} \to R$ is a gauge function. That $\| \ \|_\varphi$ is a unitarily invariant matrix norm follows then from (b) and Ex. 3.54.

(e) Let $\| \ \| : C^{m \times n} \to R$ be a unitarily invariant matrix norm, and define $\varphi : R^l \to R$ by

$$\varphi(x) = \varphi(x_1, x_2, \ldots, x_l) = \|[\operatorname{diag}|x_i|]\|,$$

where

$$[\operatorname{diag}|x_i|] = \begin{bmatrix} |x_1| & & \\ & \ddots & \\ & & |x_l| \end{bmatrix} \in C^{m \times n}.$$

Then φ is a symmetric gauge function and $\| \ \| = \| \ \|_\varphi$. ∎

60. *Inequalities for singular values.* Let $A, B \in C^{m \times n}$ and let

$$\alpha_1 \geqslant \cdots \geqslant \alpha_r > 0$$

and

$$\beta_1 \geqslant \cdots \geqslant \beta_s > 0$$

be the singular values of A and B, respectively. Then for any symmetric gauge function $\varphi : R^l \to R$, $l = \min\{m, n\}$, the singular values

$$\gamma_1 \geqslant \cdots \geqslant \gamma_t > 0$$

of $A + B$ satisfy

$$\varphi(\gamma_1, \ldots, \gamma_t, 0, \ldots, 0) \leqslant \varphi(\alpha_1, \ldots, \alpha_r, 0, \ldots, 0) + \varphi(\beta_1, \ldots, \beta_s, 0, \ldots, 0) \quad (93)$$

(von Neumann [3]).

PROOF.   The inequality (93) follows from (91) and Ex. 59(d), since

$$\|A + B\|_\varphi \leqslant \|A\|_\varphi + \|B\|_\varphi. \quad \blacksquare$$

4.   A SPECTRAL THEORY FOR RECTANGULAR MATRICES

The following theorem, due to Penrose [1], is a generalization to rectangular matrices of the classical spectral theorem for normal matrices (Theorem 2.13).

THEOREM 6   **(Spectral theorem for rectangular matrices).** Let $O \neq A \in C_r^{m \times n}$, and let $d(A) = \{d_1, \ldots, d_r\}$ be complex scalars satisfying

$$|d_i| = \alpha_i, \qquad i = 1, \ldots, r \quad (3)$$

where

$$\alpha_1 \geqslant \alpha_2 \geqslant \cdots \geqslant \alpha_r > 0 \quad (2)$$

are the singular values, $\alpha(A)$, of A.

Then there exist r partial isometries $\{E_i : i = 1, \ldots, r\}$ in $C_1^{m \times n}$ satisfying

$$E_i E_j^* = O, \qquad E_i^* E_j = O, \qquad 1 \leqslant i \neq j \leqslant r \quad (94)$$

$$E_i E^* A = A E^* E_i, \qquad i = 1, \ldots, r \quad (95)$$

where

$$E = \sum_{i=1}^r E_i \quad (96)$$

is the partial isometry given by (64), and

$$A = \sum_{i=1}^r d_i E_i \quad (97)$$

Furthermore, for each $i = 1, \ldots, r$, the partial isometry $(\bar{d}_i / |d_i|) E_i$ is unique if the corresponding singular value is simple, i.e., if $\alpha_i < \alpha_{i-1}$ and $\alpha_i > \alpha_{i+1}$ for $2 \leqslant i \leqslant r$ and $1 \leqslant i \leqslant r-1$, respectively.

PROOF. Let the vectors $\{u_1, u_2, \ldots, u_r\}$ satisfy (31) and (32), let vectors $\{v_1, v_2, \ldots, v_r\}$ be defined by (33), and let

$$E_i = u_i v_i^*, \qquad i = 1, \ldots, r. \tag{98}$$

Then E_i is a partial isometry by Theorem 3(c), since $E_i E_i^* E_i = E_i$ by (32) and (35), from which (94) also follows. The statement on uniqueness follows from (98), (3), (31), (32), and (33). The result (97) was proved in Ex. 20, which also shows the matrix E of (64) to be given by (96). Finally, (95) follows from (96), (97), and (94). ∎

As shown by the proof of Theorem 6, the spectral representation (97) of A is just a way of rewriting its UDV^*-decomposition. The following spectral representation of $A^\dagger$ similarly follows from Corollary 1.

COROLLARY 3. Let A, d_i, and E_i, $i = 1, \ldots, r$, be as in Theorem 6. Then

$$A^\dagger = \sum_{i=1}^{r} \frac{1}{d_i} E_i^*. \quad \blacksquare \tag{99}$$

If $A \in C_r^{n \times n}$ is a normal matrix with nonzero eigenvalues $\{\lambda_i : i = 1, \ldots, r\}$ ordered by

$$|\lambda_1| \geqslant |\lambda_2| \geqslant \cdots \geqslant |\lambda_r|,$$

then, by Ex. 32, the choice

$$d_i = \lambda_i, \qquad i = 1, \ldots, r \tag{57}$$

guarantees that

$$u_i = v_i, \qquad i = 1, \ldots, r \tag{58}$$

and consequently, the partial isometries E_i of (98) are orthogonal projectors

$$P_i = u_i u_i^*, \qquad i = 1, \ldots, r \tag{100}$$

and (97) reduces to

$$A = \sum_{i=1}^{r} \lambda_i P_i, \tag{101}$$

giving the spectral theorem for normal matrices as a special case of Theorem 5.

The classical spectral theory for square matrices (see, e.g., Dunford and Schwartz [1], pp. 556–565) makes extensive use of matrix functions $f: C^{n \times n} \to C^{n \times n}$, induced by scalar functions $f: C \to C$, according to the definition given in Ex. 61. Similarly, the spectral theory for rectangular matrices given here uses matrix functions $f: C^{m \times n} \to C^{m \times n}$ which correspond to scalar functions $f: C \to C$, according to the following.

DEFINITION 1.   Let $f: C \to C$ be any scalar function. Let $A \in C_r^{m \times n}$ have a spectral representation

$$A = \sum_{i=1}^{r} d_i E_i \qquad (97)$$

as in Theorem 6. Then the *matrix function* $f: C^{m \times n} \to C^{m \times n}$ corresponding to $f: C \to C$ is defined at A by

$$f(A) = \sum_{i=1}^{r} f(d_i) E_i. \qquad (102)$$

Note that the value of $f(A)$ defined by (102) depends on the particular choice of the scalars $d(A)$ in (3). In particular, for a normal matrix $A \in C^{n \times n}$, the choice of $d(A)$ by (57) reduces (102) to the classical definition—see (119) below—in the case that $f(0) = 0$ or that A is nonsingular.

Let

$$A = U_{(r)} D_{(r)} V_{(r)}^*, \qquad D_{(r)} = \begin{bmatrix} d_1 & & \\ & \ddots & \\ & & d_r \end{bmatrix} \qquad (54)$$

be a UDV^*-decomposition of a given $A \in C_r^{m \times n}$. Then Definition 1 gives $f(A)$ as

$$f(A) = U_{(r)} f(D_{(r)}) V_{(r)}^*, \qquad f(D_{(r)}) = \begin{bmatrix} f(d_1) & & \\ & \ddots & \\ & & f(d_r) \end{bmatrix}. \qquad (103)$$

An easy consequence of Theorem 6 and Definition 1 is the following:

THEOREM 7.   Let $f, g, h: C \to C$ be scalar functions and let f, g, h:

$C^{m\times n}\to C^{m\times n}$ be the corresponding matrix functions defined by Definition 1.

Let $A\in C_r^{m\times n}$ have a UDV^*-decomposition

$$A = U_{(r)}D_{(r)}V_{(r)}^*$$ (54)

and let the partial isometry E be given by

$$E = U_{(r)}V_{(r)}^*.$$ (64)

Then

(a) If $f(z)=g(z)+h(z)$, then $f(A)=g(A)+h(A)$.
(b) If $f(z)=g(z)h(z)$, then $f(A)=g(A)E^*h(A)$.
(c) If $f(z)=g(h(z))$, then $f(A)=g(h(A))$.

PROOF.   Parts (a) and (c) are obvious by Definition (102).
   (b) If $f(z)=g(z)h(z)$, then

$$g(A)E^*h(A)=\left(\sum_{i=1}^r g(d_i)E_i\right)\left(\sum_{j=1}^r E_j^*\right)\left(\sum_{k=1}^r h(d_k)E_k\right),$$

by (102) and (96),

$$=\sum_{i=1}^r g(d_i)h(d_i)E_i,\qquad \text{by (94) and Theorem 3(c),}$$

$$=\sum_{i=1}^r f(d_i)E_i=f(A). \quad\blacksquare$$

For matrix functions defined as above, an analog of Cauchy's integral theorem is given in Corollary 4 below. First we require

LEMMA 1.   Let $A\in C_r^{m\times n}$ be represented by

$$A = \sum_{i=1}^r d_iE_i.$$ (97)

Let $\{\hat{d}_j:j=1,\dots,q\}$ be the set of distinct $\{d_i:i=1,\dots,r\}$ and let

$$\hat{E}_j = \sum_{\substack{i\\ \{d_i=\hat{d}_j\}}} E_i,\qquad j=1,\dots,q$$ (104)

For each $j=1,\dots,q$ let Γ_j be a contour (i.e., a closed rectifiable Jordan curve, positively oriented in the customary way) surrounding $\hat{d}_j$ but no other $\hat{d}_k$.

Then

(a) For each $j = 1, \ldots, q$, $\hat{E}_j$ is a partial isometry and

$$\hat{E}_j^* = \frac{1}{2\pi i} \int_{\Gamma_j} (zE - A)^\dagger dz. \tag{105}$$

(b) If $f : C \to C$ is analytic in a domain containing the set surrounded by

$$\Gamma = \bigcup_{j=1}^{q} \Gamma_j,$$

then

$$\sum_{j=1}^{r} f(d_j) E_j^* = \frac{1}{2\pi i} \int_{\Gamma} f(z)(zE - A)^\dagger dz; \tag{106}$$

in particular,

$$A^\dagger = \frac{1}{2\pi i} \int_{\Gamma} \frac{1}{z}(zE - A)^\dagger dz. \tag{107}$$

PROOF.   (a) From (94) and Theorem 3 it follows that $\hat{E}_j$ and $\hat{E}_j^*$ are partial isometries for each $j = 1, \ldots, q$. Also, from (96), (97), and Corollary 3,

$$(zE - A)^\dagger = \sum_{k=1}^{r} \frac{1}{z - d_k} E_k^*, \tag{108}$$

hence

$$\frac{1}{2\pi i} \int_{\Gamma_j} (zE - A)^\dagger dz = \sum_{k=1}^{r} \left(\frac{1}{2\pi i} \int_{\Gamma_j} \frac{dz}{z - d_k} \right) E_k^*$$

$$= \sum_{\{d_k = \hat{d}_j\}} E_k^*,$$

by the assumptions on Γ_j and Cauchy's integral theorem

$$= \hat{E}_j^*, \quad \text{by (104)}.$$

(b) Similarly we calculate

$$\frac{1}{2\pi i}\int_\Gamma f(z)(zE-A)^\dagger dz = \sum_{j=1}^q \sum_{k=1}^r \left(\frac{1}{2\pi i}\int_{\Gamma_j}\frac{f(z)}{z-d_k}dz\right)E_k^*$$

$$= \sum_{j=1}^q f(\hat{d}_j)\hat{E}_j^*$$

$$= \sum_{j=1}^r f(d_j)E_j^*, \qquad \text{proving (106)}.$$

Finally, (107) follows from (106) and Corollary 3. ∎

COROLLARY 4. Let A, E, Γ, and f be as in Lemma 1. Then

$$f(A)=E\left(\frac{1}{2\pi i}\int_\Gamma f(z)(zE-A)^\dagger dz\right)E. \qquad (109)$$

PROOF. Using (96) and (106) we calculate

$$E\left(\frac{1}{2\pi i}\int_\Gamma f(z)(zE-A)^\dagger dz\right)E=\left(\sum_{i=1}^r E_i\right)\left(\sum_{j=1}^r f(d_j)E_j^*\right)\left(\sum_{k=1}^r E_k\right)$$

$$= \sum_{j=1}^r f(d_j)E_j, \quad \text{by (94) and Theorem 3(c)},$$

$$=f(A). \quad ∎$$

The *generalized resolvent* of a matrix $A\in C^{m\times n}$ is the function $R(z,A):C\to C^{n\times m}$ given by

$$R(z,A)=(zE-A)^\dagger, \qquad (110)$$

where the partial isometry E is given as in Theorem 6. This definition is suggested by the classical definition of the resolvent of a square matrix as

$$R(z,A)=(zI-A)^{-1}, \qquad \text{for } z\notin\sigma(A).$$

In analogy to the classical case—see, e.g., Dunford and Schwartz [1], p. 568 —we state the following identity, known as the *(first) resolvent equation*.

LEMMA 2.   Let $A \in C_r^{m \times n}$ and let $d(A)$ and E be as in Theorem 6. Then

$$R(\lambda, A) - R(\mu, A) = (\mu - \lambda) R(\lambda, A) E R(\mu, A) \tag{111}$$

for any scalars $\lambda, \mu \notin d(A)$.

PROOF.

$$R(\lambda, A) - R(\mu, A) = (\lambda E - A)^\dagger - (\mu E - A)^\dagger, \qquad \text{by (110)}$$

$$= \sum_{k=1}^{r} \left\{ \frac{1}{\lambda - d_k} - \frac{1}{\mu - d_k} \right\} E_k^*, \qquad \text{by (108)}$$

$$= \sum_{k=1}^{r} \left\{ \frac{\mu - \lambda}{(\lambda - d_k)(\mu - d_k)} \right\} E_k^*$$

$$= (\mu - \lambda) \left(\sum_{k=1}^{r} \frac{1}{\lambda - d_k} E_k^* \right) E \left(\sum_{l=1}^{r} \frac{1}{\mu - d_l} E_l^* \right),$$

$$\text{by (94), (96) and Theorem 3(c)},$$

$$= (\mu - \lambda) R(\lambda, A) E R(\mu, A), \qquad \text{by (108).} \quad \blacksquare$$

The resolvent equation, (111), is used in the following lemma, based on Lancaster [1], p. 552.

LEMMA 3.   Let $A \in C^{m \times n}$, let $d(A)$ and E be given as in Theorem 6, and let the scalar functions $f, g: C \rightarrow C$ be analytic in a domain D containing $d(A)$. If Γ is a contour surrounding $d(A)$ and lying in the interior of D, then

$$\left(\frac{1}{2\pi i} \int_\Gamma f(\lambda) R(\lambda, A) \, d\lambda \right) E \left(\frac{1}{2\pi i} \int_\Gamma g(\lambda) R(\lambda, A) \, d\lambda \right)$$

$$= \frac{1}{2\pi i} \int_\Gamma f(\lambda) g(\lambda) R(\lambda, A) \, d\lambda. \tag{112}$$

PROOF.   Let Γ_1 be a contour surrounding Γ and still lying in the interior of D. Then

$$\frac{1}{2\pi i} \int_\Gamma g(\lambda) R(\lambda, A) \, d\lambda = \frac{1}{2\pi i} \int_{\Gamma_1} g(\mu) R(\mu, A) \, d\mu,$$

which when substituted in the left-hand side of (112) gives

$$\left(\frac{1}{2\pi i}\int_\Gamma f(\lambda)R(\lambda,A)\,d\lambda\right)E\left(\frac{1}{2\pi i}\int_{\Gamma_1}g(\mu)R(\mu,A)\,d\mu\right)$$

$$=-\frac{1}{4\pi^2}\int_{\Gamma_1}\!\int_\Gamma f(\lambda)g(\mu)R(\lambda,A)ER(\mu,A)\,d\lambda\,d\mu$$

$$=\frac{1}{4\pi^2}\int_{\Gamma_1}\!\int_\Gamma f(\lambda)g(\mu)\frac{R(\lambda,A)-R(\mu,A)}{\lambda-\mu}\,d\lambda\,d\mu,\qquad \text{by (111)}$$

$$=\frac{1}{4\pi^2}\int_\Gamma f(\lambda)R(\lambda,A)\left(\int_{\Gamma_1}\frac{g(\mu)}{\lambda-\mu}\,d\mu\right)d\lambda$$

$$-\frac{1}{4\pi^2}\int_{\Gamma_1}\left(\int_\Gamma\frac{f(\lambda)}{\lambda-\mu}\,d\lambda\right)g(\mu)R(\mu,A)\,d\mu$$

$$=\frac{1}{2\pi i}\int f(\lambda)g(\lambda)R(\lambda,A)\,d\lambda,\qquad \text{since}\qquad \int_{\Gamma_1}\frac{g(\mu)}{\lambda-\mu}\,d\mu=-2\pi i g(\lambda)$$

$$\text{and}\qquad \int_\Gamma\frac{f(\lambda)}{\lambda-\mu}\,d\lambda=0,\qquad \text{by our assumptions on}\,\Gamma,\Gamma_1.\quad\blacksquare$$

We illustrate now the application of the above concepts to the solution of the matrix equation

$$AXB=D \tag{113}$$

studied in Theorem 2.1. Here the matrices $A\in C^{m\times n}$, $B\in C^{k\times l}$, and $D\in C^{m\times l}$ are given, and, in addition, the matrices A and B have spectral representations, given by Theorem 6 as follows:

$$A=\sum_{i=1}^p d_i^A E_i^A,\qquad E^A=\sum_{i=1}^p E_i^A,\qquad p=\text{rank }A \tag{114}$$

and

$$B=\sum_{j=1}^q d_j^B E_j^B,\qquad E^B=\sum_{j=1}^q E_j^B,\qquad q=\text{rank }B. \tag{115}$$

THEOREM 8. Let A, B, D be as above, and let Γ_1 and Γ_2 be contours surrounding $d(A) = \{d_1^A, d_2^A, \ldots, d_p^A\}$ and $d(B) = \{d_1^B, d_2^B, \ldots, d_q^B\}$, respectively. If (113) is consistent, then it has the following solution:

$$X = -\frac{1}{4\pi^2} \int_{\Gamma_1} \int_{\Gamma_2} \frac{R(\lambda, A) DR(\mu, B)}{\lambda \mu} d\mu \, d\lambda. \tag{116}$$

PROOF.   From (109) it follows that

$$A = E^A \left(\frac{1}{2\pi i} \int_{\Gamma_1} \lambda R(\lambda, A) \, d\lambda \right) E^A$$

and

$$B = E^B \left(\frac{1}{2\pi i} \int_{\Gamma_2} \mu R(\mu, B) \, d\mu \right) E^B.$$

Therefore,

$$AXB = E^A \left[\frac{1}{2\pi i} \int_{\Gamma_1} \lambda R(\lambda, A) \, d\lambda \right]$$

$$\times E^A \left[\frac{1}{2\pi i} \int_{\Gamma_1} \frac{R(\lambda, A)}{\lambda} D \left(\frac{1}{2\pi i} \int_{\Gamma_2} \frac{R(\mu, B)}{\mu} d\mu \right) d\lambda \right]$$

$$\times E^B \left[\frac{1}{2\pi i} \int_{\Gamma_2} \mu R(\mu, B) \, d\mu \right] E^B$$

$$= E^A \left[\frac{1}{2\pi i} \int_{\Gamma_1} R(\lambda, A) \, d\lambda \right] D \left[\frac{1}{2\pi i} \int_{\Gamma_2} R(\mu, B) \, d\mu \right] E^B,$$

by a double application of Lemma 3

$$= E^A (E^A)^* D (E^B)^* E^B, \quad \text{by (106) with } f \equiv 1$$

$$= P_{R(A)} D P_{R(B^*)}, \quad \text{by (65)}$$

$$= AA^\dagger DB^\dagger B$$

$$= D \quad \text{if and only if (113) is consistent, by Theorem 2.1.}$$

Alternatively, it follows from (107) that X in (116) is $X = A^\dagger D B^\dagger$, a solution of (113) if it is consistent. ∎

For additional results along these lines see Lancaster [1] and Wimmer and Ziebur [1].

EXERCISES AND EXAMPLES

61. *Matrix functions: The classical definition.* For any $A \in C^{n \times n}$ with spectrum $\sigma(A)$, let $F(A)$ denote the class of all functions $f: C \to C$ which are analytic in some open set containing $\sigma(A)$. For any scalar function $f: C \to C$ which is analytic in some open set, the *corresponding matrix function* $f: C^{n \times n} \to C^{n \times n}$ is defined, at those $A \in C^{n \times n}$ such that $f \in F(A)$, by

$$f(A) = p(A), \tag{117}$$

where $p(A)$ is any polynomial such that, for each $\lambda \in \sigma(A)$,

$$p^{(i)}(\lambda) = f^{(i)}(\lambda), \qquad i = 0, 1, \ldots, \nu(\lambda) - 1 \tag{118}$$

where $\nu(\lambda)$ is the index (see Definition 4.1) of the matrix $A - \lambda I$, also called the *index of the eigenvalue* λ.

For other definitions of matrix functions, and their relations to the one given here, see Rinehart [1]. Additional results and references on matrix functions are Dunford and Schwartz ([1], pp. 556–565), Gantmacher [1], Frame, [1], and Lancaster [1].

62. If $A \in C^{n \times n}$ is normal with a spectral representation

$$A = \sum_{i=1}^{r} \lambda_i P_i, \tag{101}$$

then, for any $f \in F(A)$, definition (117) gives

$$f(A) = \sum_{i=1}^{r} f(\lambda_i) P_i + f(0) P_{N(A)}, \tag{119}$$

since the eigenvalues of a normal matrix have index one.

63. *Generalized powers.* The matrix function $f: C^{m \times n} \to C^{m \times n}$ corresponding to the scalar function

$$f(z) = z^k, \qquad k \text{ any integer},$$

is denoted by

$$f(A) = A^{\langle k \rangle}$$

and called the *generalized* kth *power* of $A \in C^{m \times n}$. Definition 1 shows that

$$A^{\langle k \rangle} = \sum_{i=1}^{r} d_i^k E_i, \quad \text{by (102)} \tag{120}$$

or equivalently

$$A^{\langle k \rangle} = U_{(r)} D_{(r)}^k V_{(r)}^*, \quad \text{by (103).} \tag{121}$$

The generalized powers of A satisfy

$$A^{\langle k \rangle} = \begin{cases} E, & k = 0 \\ A^{\langle k-1 \rangle} E^* A, & k \geqslant 1, \\ A^{\langle k+1 \rangle} E^* A^{\langle -1 \rangle}, & k \leqslant -1 \end{cases} \quad \text{in particular } A^{\langle 1 \rangle} = A \tag{122}$$

64. If in Theorem 6 the scalars $d(A)$ are chosen by (7), i.e., if $d(A) = \alpha(A)$, then for any integer k

$$A^{* \langle k \rangle} = A^{\langle k \rangle *}, \tag{123}$$

$$A^{\langle 2k+1 \rangle} = A(A^*A)^k = (AA^*)^k A; \tag{124}$$

in particular,

$$A^{\langle -1 \rangle} = A^{*\dagger} \tag{125}$$

65. If $A \in C_r^{n \times n}$ is normal, and if the scalars $d(A)$ are chosen by (57); i.e., if $d(A) = \sigma(A)$, then

$$A^{\langle k \rangle} = \begin{cases} A^k, & k \geqslant 1 \\ P_{R(A)}, & k = 0 \\ (A^\dagger)^k, & k \leqslant -1 \end{cases} . \tag{126}$$

66. *Ternary powers.* From (124) follows the definition of a polynomial in ternary powers of $A \in C^{m \times n}$, as a polynomial

$$\sum_k p_k A^{\langle 2k+1 \rangle} = \sum_k p_k (AA^*)^k A.$$

Such polynomials were studied by Hestenes [5] in the more general context of ternary algebras.

In (129) below, we express $A^\dagger$ as a polynomial in ternary powers of A^*. First we require the following.

67. Let $A \in C^{n \times n}$ be Hermitian, and let a vanishing polynomial of A, i.e., a polynomial $m(\lambda)$ satisfying $m(A) = O$, be given in the form

$$m(\lambda) = c\lambda^l (1 - \lambda q(\lambda)) \tag{4.26}$$

where $c \neq 0$, $l \geqslant 0$, and the leading coefficient of q is 1.
Then

$$A^\dagger = q(A) + q(O)[Aq(A) - I], \qquad (127)$$

and in particular

$$A^{-1} = q(A), \qquad \text{if } A \text{ is nonsingular (Albert [3], p. 75).}$$

PROOF.   From (4.26) it follows that

$$A^l = A^{l+1}q(A)$$

and since A is Hermitian

$$A^\dagger = (A^\dagger)^{l+1}A^l = AA^\dagger q(A)$$

$$= AA^\dagger[q(A) - q(O)] + AA^\dagger q(O)$$

$$= q(A) - q(O) + AA^\dagger q(O) \qquad (128)$$

since $q(A) - q(O)$ contains only positive powers of A. Postmultiplying (128) by A gives

$$A^\dagger A = [q(A) - q(O)]A + Aq(O)$$

$$= q(A)A = Aq(A),$$

which when substituted in (128), gives (127).   ∎

Alternatively, (127) can be shown to follow from the results of Section 4.6, since here $A^{(d)} = A^\dagger$.

68.   Let $A \in C^{m \times n}$ and let

$$m(\lambda) = c\lambda^l(1 - \lambda q(\lambda)) \qquad (4.26)$$

be a vanishing polynomial of A^*A, as in Ex. 67. Then

$$A^\dagger = q(A^*A)A^* \qquad (129)$$

(Penrose [1], Hestenes [5], Ben-Israel and Charnes [1]).

PROOF.   From (127) it follows that

$$(A^*A)^\dagger = q(A^*A) + q(O)[A^*Aq(A^*A) - I],$$

so, by Ex. 1.17(d),

$$A^\dagger = (A^*A)^\dagger A^* = q(A^*A)A^*.   ∎$$

A computational method based on (129) is given in Decell [1] and in Albert [3].

69. *Partial isometries.* Let $W \in C^{m \times n}$. Then W is a partial isometry if and only if

$$W = e^{iA}$$

for some $A \in C^{m \times n}$.

PROOF.   Follows from (103) and Exs. 37–38.

70.   Let $U \in C^{n \times n}$. Then U is a unitary matrix if and only if

$$U = e^{iH} + P_{N(H)} \tag{130}$$

for some Hermitian matrix $H \in C^{n \times n}$. Note that the exponential in (130) is defined according to Definition 1. For the classical definition given in Ex. 61, Eq. 130 should be replaced by

$$U = e^{iH}. \tag{130'}$$

71. *Polar decompositions.*   Let $A \in C_r^{m \times n}$ and let

$$A = GE = EH \tag{66}$$

be a polar decomposition of A, given as in Corollary 2. Then for any function f, Definition 1 gives

$$f(A) = f(G)E = Ef(H), \tag{131}$$

in particular

$$A^{\langle k \rangle} = G^k E = EH^k, \qquad \text{for any integer } k. \tag{132}$$

SUGGESTED FURTHER READING

Section 2.   Businger and Golub [2], Golub [3], Golub and Reinsch [1], Good [1], Hartwig [2], Hestenes [2], Lanczos [1], and Wedin [2].

Section 3.   Erdelyi ([1], [5], [6], [7]), Erdelyi and Miller [1], Halmos and Wallen [1], Hearon ([2], [3]), Hestenes ([2], [3], [4], [5]), and Poole and Boullion [1].

7

COMPUTATIONAL ASPECTS OF GENERALIZED INVERSES

1. INTRODUCTION

There are three principal situations in which it is required to obtain numerically a generalized inverse of a given matrix: (i) the case in which any $\{1\}$-inverse will suffice, (ii) the cases in which any $\{1,3\}$-inverse (or sometimes any $\{1,4\}$-inverse) will do, and (iii) the case in which a $\{2\}$-inverse having a specified range and null space is required.

The inverse desired in case (iii) is, in the vast majority of cases, the Moore–Penrose inverse, which may be described as the unique $\{2\}$-inverse (of the given matrix A) having the same range and null space as A^*. The Drazin inverse can also be fitted into this pattern, being the unique $\{2\}$-inverse of A having the same range and null space as A^l, where l is any integer not less than the index of A. When $l=1$, this is the group inverse.

Within each of these three situations, there are again two cases to be considered: (a) that in which the matrices involved are small and the calculations are made by hand, and can, if desired, be performed exactly using rational arithmetic, and (b) that in which a computer is employed. In case (a) the paramount consideration in comparing different methods is the number of arithmetic operations required; in case (b), "conditioning" of matrices and accumulation of rounding error must also be considered. The purpose of this chapter is merely to offer some useful suggestions to those who wish to do numerical work; the treatment is far from exhaustive and no error analysis is attempted.

Iterative methods for generalized inversion are discussed in Section 5. The remaining sections deal with direct methods.

283

2. COMPUTATION OF UNRESTRICTED {1}-INVERSES AND {1,2}-INVERSES

Let A be a given matrix for which a {1}-inverse is desired, when any {1}-inverse will suffice. If it should happen that A is of such structure, or has arisen in such a manner, that a nonsingular submatrix of maximal order is known, we can write

$$PAQ = \begin{bmatrix} A_{11} & A_{12} \\ A_{21} & A_{22} \end{bmatrix}, \tag{5.19}$$

where A_{11} is nonsingular and P and Q are permutation matrices used to bring the nonsingular submatrix into the upper left position. (If A is of full (column or row) rank, some of the submatrices in (5.19) will be absent.) Since rank A is the order of A_{11}, this implies that

$$A_{22} = A_{21} A_{11}^{-1} A_{12} \qquad (\text{Brand [1]}) \tag{5.34}$$

and a {1,2}-inverse of A is

$$A^{(1,2)} = Q \begin{bmatrix} A_{11}^{-1} & O \\ O & O \end{bmatrix} P \qquad (\text{C. R. Rao [2]}). \tag{5.25}$$

In the more usual case in which a nonsingular submatrix of maximal order is not known, and likewise, rank A is not known, perhaps the simplest method is that of Section 1.2, using Gaussian elimination to bring A to Hermite normal form. (It should be noted, however, that we are employing the "looser" definition of the Hermite normal form given in Section 1.2 of this book, and not the strict definition used in some texts, e.g., Marcus and Minc [1].) Thus if

$$EAP = \begin{bmatrix} I_r & K \\ O & O \end{bmatrix} \tag{1.6}$$

(with modifications in the case where A is of full rank), where E is nonsingular and P is a permutation matrix, then

$$A^{(1)} = P \begin{bmatrix} I_r & O \\ O & L \end{bmatrix} E \tag{1}$$

is a {1}-inverse of A for arbitrary L. Of course, the simplest choice is $L = O$, which gives the {1,2}-inverse

$$A^{(1,2)} = P \begin{bmatrix} I_r & O \\ O & O \end{bmatrix} E.$$

On the other hand, when A is square, a nonsingular {1}-inverse may sometimes be desired. This is obtained by taking L in (1) to be nonsingular. The simplest choice for L is a unit matrix, which gives

$$A^{(1)} = PE.$$

If the calculations are performed on a computer, then, as in the non-singular case, the accuracy may depend critically on the choice of pivots used in the Gaussian elimination. (For a discussion of pivoting see, e.g., Pennington [1]; for a simple illustration, see Ex. 2 below.)

EXERCISES

1. Show that (1) gives a {1,2}-inverse of A if and only if $L = O$.
2. Consider the two nonsingular matrices

$$A = \begin{bmatrix} \epsilon & 1 \\ 0 & 1 \end{bmatrix}, \quad B = \begin{bmatrix} \epsilon & 1 \\ 1 & 1 \end{bmatrix},$$

where ϵ is nonzero, but very close to zero. Compare the various ways (i.e., choices of pivots) of transforming A and B to their Hermite normal forms. The objective is a numerically stable process, which here means to avoid, or to postpone, division by ϵ.

3. COMPUTATION OF UNRESTRICTED {1,3}-INVERSES.

Let $A \in C_r^{m \times n}$ and let

$$A = FG \tag{2}$$

be a full-rank factorization. Then, by Ex. 1.26(b),

$$X = G^{(1)} F^\dagger, \tag{3}$$

where $G^{(1)}$ is an arbitrary element of $G\{1\}$, is a $\{1,2,3\}$-inverse of A. If the factorization (2) has been obtained from the Hermite normal form of A by the procedure described in Section 1.7, then

$$F = AP_1, \tag{4}$$

where P_1 denotes the first r columns of the permutation matrix P. Moreover, we may take $G^{(1)} = P_1$, and (3) gives

$$X = P_1 F^\dagger. \tag{5}$$

Since F is of full column rank,

$$F^\dagger = (F^*F)^{-1}F^* \tag{6}$$

by (1.25). Thus, (4), (6), and (5), in that order, give a $\{1,2,3\}$-inverse of A.

Note that (4) shows that F is a submatrix of A consisting of r linearly independent columns. In fact, the only purpose served by the computation of the Hermite normal form is in the selection of the r columns. Thus, the method is equally valid if r linearly independent columns have been determined in some other manner; see, e.g., Exs. 4–5 below.

Observe also that (5) shows that each of the r rows of $F^\dagger$ is a row of X (in general, not the corresponding row), while the remaining $n - r$ rows of X are rows of zeros. Thus, in the language of lineaer programming, X is a "basic" $\{1,2,3\}$-inverse of A.

EXERCISES

3. Use (4), (6), and (5) to obtain a $\{1\}$-inverse of

$$A = \begin{bmatrix} 1 & 0 & 0 & 1 \\ 1 & 1 & 0 & 0 \\ 0 & 1 & 1 & 0 \\ 0 & 0 & 1 & 1 \end{bmatrix}.$$

4. *Gram–Schmidt orthogonalization.* Given a nonzero $A \in C^{m \times n}$, a full column rank submatrix F can be found by the *Gram–Schmidt orthogonalization process* (abbreviated GSO) as follows.

Applying GSO (without normalization) to the columns $[a_1, a_2, \ldots, a_n]$ of A

gives an orthogonal basis $\{v_1, v_2, \ldots, v_r\}$ of $R(A)$, where

$$\left.\begin{aligned}
v_1 &= a_{c_1} \quad \text{if } a_{c_1} \neq 0 = a_j \quad \text{for } 1 \leqslant j < c_1. \\
x_j &= a_j - \sum_{l=1}^{k-1} \frac{(a_j, v_l)}{\|v_l\|^2} v_l, \quad j = c_{k-1}+1,\ c_{k-1}+2, \ldots, c_k,
\end{aligned}\right\} \tag{7}$$

and

$$v_k = x_{c_k} \quad \text{if } x_{c_k} \neq 0 = x_j \quad \text{for } c_{k-1}+1 \leqslant j < c_k, \quad k = 2, \ldots, r.$$

Then

$$F = [a_{c_1}, a_{c_2}, \ldots, a_{c_r}], \qquad P_1 = [e_{c_1}, e_{c_2}, \ldots, e_{c_r}] \tag{8}$$

are two matrices satisfying (4).

5. If F is given by (8) and (7), then

$$F^\dagger = \begin{bmatrix} a_{c_1}^* / \|a_{c_1}\|^2 \\ a_{c_2}^* / \|a_{c_2}\|^2 \\ \vdots \\ a_{c_r}^* / \|a_{c_r}\|^2 \end{bmatrix} \tag{9}$$

Thus the GSO gives everything needed in (5) to compute a $\{1,2,3\}$-inverse. See also Ex. 9 below.

6. Use (7), (8), (9), and (5) to calculate a $\{1,2,3\}$-inverse of the matrix given in Ex. 3.

4. COMPUTATION OF {2}-INVERSES WITH PRESCRIBED RANGE AND NULL SPACE

Let $A \in C_r^{m \times n}$, let $A\{2\}_{S,T}$ contain a nonzero matrix X, and let U and V be such that $R(U) = R(X)$, $N(V) = N(X)$, and the product VAU is defined. Then, by Theorems 2.11 and 2.12, rank U = rank V = rank VAU, and

$$X = U(VAU)^{(1)}V, \tag{10}$$

where $(VAU)^{(1)}$ is an arbitrary element of $(VAU)\{1\}$. This is the basic formula for the case considered in this section. Zlobec's formula

$$A^\dagger = A^*(A^*AA^*)^{(1)}A^* \tag{11}$$

(see Ex. 2.33) and Greville's formula

$$A^{(d)} = A^l (A^{2l+1})^{(1)} A^l, \qquad (4.30)$$

where l is a positive integer not less than the index of A, are particular cases. Formula (10) has the advantage that it does not require inversion of any nonsingular matrix. Aside from matrix multiplication, only the determination of a $\{1\}$-inverse of VAU is needed, and this can be obtained by the method of Section 1.2.

It should be noted, however, that when ill-conditioning of A is a problem, this is accentuated by forming products like A^*AA^* or A^{2l+1}, and in such cases, other methods are preferable.

In the case of the Moore–Penrose inverse, Noble's formula

$$A^\dagger = Q \begin{bmatrix} I_r \\ T^* \end{bmatrix} (I_r + TT^*)^{-1} A_{11}^{-1} (I_r + S^*S)^{-1} [I_r \quad S^*] P \qquad (5.28)$$

is available, if a maximal nonsingular (and well-conditioned) submatrix A_{11} is known, where the permutation matrices P and Q and the "multipliers" S and T are defined by

$$A = P^T \begin{bmatrix} A_{11} & A_{12} \\ A_{21} & A_{22} \end{bmatrix} Q^T$$

$$= P^T \begin{bmatrix} I_r \\ S \end{bmatrix} A_{11} [I_r \quad T] Q^T; \qquad \text{see Ex. 7.} \qquad (5.22)$$

Otherwise, it is probably best to use the method of Section 1.7 to obtain a full-rank factorization

$$A = FG. \qquad (1.20)$$

Then, the Moore–Penrose inverse is

$$A^\dagger = G^* (F^*AG^*)^{-1} F^*, \qquad (1.23)$$

while the group inverse is

$$A^\# = F(GF)^{-2} G, \qquad (4.12)$$

whenever GF is nonsingular.

In the computation of $A^{(d)}$ when the index of A exceeds 1, it is not easy to avoid raising A to a power. When ill-conditioning of A is serious, perhaps the best method is the sequential procedure of Cline [3], which involves full-rank factorization of matrices of successively smaller order, until a nonsingular matrix is reached. Thus, we take

$$A = B_1 G_1, \tag{12}$$

$$G_i B_i = B_{i+1} G_{i+1} \qquad (i = 1, 2, \ldots, k-1), \tag{13}$$

where k is the index of A. Then

$$A^{(d)} = B_1 B_2 \cdots B_k (G_k B_k)^{-1} G_k G_{k-1} \cdots G_1. \tag{14}$$

EXERCISES

7. *Noble's method.*   Let the nonzero matrix $A \in C_r^{m \times n}$ be transformed to a column-permuted Hermite normal form

$$PEAQ = \left[\begin{array}{c|c} I_r & T \\ \hline O & O \end{array} \right] = (PEP^T)(PAQ), \tag{15}$$

where P and Q are permutation matrices and E is a product of elementary row matrices of types (i) and (ii) (see Section 1.2),

$$E = E_k E_{k-1} \cdots E_2 E_1,$$

which does not involve permutation of rows.

Then, E can be chosen so that

$$PE\left[\begin{array}{c|c} A & I_m \end{array} \right] \left[\begin{array}{c|c} Q & O \\ \hline O & P^T \end{array} \right] = \left[\begin{array}{c|c|c|c} I_r & T & A_{11}^{-1} & O \\ \hline O & O & -S & I_{m-r} \end{array} \right] \tag{16}$$

giving all the matrices P, Q, T, S, and A_{11}^{-1} which appear in (5.28). Note that after the left-hand portion of the right member of (16) has been brought to the form (15), still further row operations may be needed to bring the right-hand portion to the required form (Noble [1]).

8. *Singular value decomposition.*   Let

$$A = U_{(r)} D_{(r)} V_{(r)}^* \tag{6.54}$$

be a singular value decomposition of $A \in C^{m \times n}$. Then

$$A^\dagger = V_{(r)} D_{(r)}^{-1} U^*_{(r)}$$

$$= V_{(r)} (U^*_{(r)} A V_{(r)})^{-1} U^*_{(r)} \qquad (6.55)$$

is shown to be a special case of (10) by taking

$$U = V_{(r)}, \qquad V = U^*_{(r)}.$$

A method for computing the Moore–Penrose inverse, based on (6.55), has been developed by Golub and Kahan [1]. See also Businger and Golub [1], [2] and Golub and Reinsch [1].

9. *Gram–Schmidt orthogonalization.*   The GSO of Exs. 4–5 can be modified to compute the Moore–Penrose inverse. This method is due to Rust, Burrus, and Schneeburger [1]; see also Albert [3], Chapter V.

10.   For the matrix A of Ex. 3, calculate $A^\dagger$ by:

(i)   Zlobec's formula (11).
(ii) Noble's formula (5.28).
(iii) MacDuffee's formula (1.23).
(iv) Greville's method, Section 5.5.

11.   For the matrix A of Ex. 3, calculate $A^\#$ by:

(i)   Formula (4.17).
(ii) Cline's formula (4.12).

12.   Show that, for a matrix A of index k that is not nilpotent, with B_i and G_i defined by (12) and (13), $G_k B_k$ is nonsingular. [*Hints*: Express A^k and A^{k+1} in terms of B_i and G_i ($i = 1, 2, \ldots, k$), and let r_k denote the number of columns of B_k, which is also the number of rows of G_k. Show that rank $A^k = r_k$, while rank $A^{k+1} = $ rank $G_k B_k$. Therefore, rank $A^{k+1} = $ rank A^k implies that $G_k B_k$ is nonsingular.]

13.   Use Theorem 4.7 to verify (14).

5.   ITERATIVE METHODS FOR COMPUTING $A^\dagger$

An iterative method for computing $A^\dagger$ is a set of instructions for generating a sequence $\{X_k : k = 0, 1, \ldots\}$ converging to $A^\dagger$. The instructions specify how to select the initial approximation X_0, how to proceed from X_k to X_{k+1} for each $k = 0, 1, \ldots$, and when to stop, having obtained a reasonable approximation of $A^\dagger$.

The rate of convergence of such an iterative method is determined in terms of the corresponding sequence of *residuals* $\{R_k : k = 0, 1, \ldots\}$:

$$R_k = P_{R(A)} - AX_k, \qquad k = 0, 1, \ldots, \tag{17}$$

which clearly converges to O as $X_k \to A^\dagger$. An iterative method is called a *pth-order iterative method* for some positive integer p, if there is a positive constant c such that

$$\|R_{k+1}\| \leqslant c \|R_k\|^p, \qquad k = 0, 1, \ldots, \tag{18}$$

for any multiplicative matrix norm; see, e.g., Ex. 1.33.

In analogy with the nonsingular case—see, e.g., Householder [1], pp. 94–95—we consider iterative methods of the type

$$X_{k+1} = X_k + C_k R_k, \qquad k = 0, 1, \ldots, \tag{19}$$

where $\{C_k : k = 0, 1, \ldots\}$ is a suitable sequence, and X_0 is the initial approximation (to be specified).

One objection to (19) as an iterative method for computing $A^\dagger$ is that (19) requires at each iteration the computation of the residual R_k, for which one needs the projection $P_{R(A)}$, whose computation is a task comparable to computing $A^\dagger$. This difficulty will be overcome here by choosing the sequence $\{C_k : k = 0, 1, \ldots\}$ in (19) to satisfy

$$C_k = C_k P_{R(A)}, \qquad k = 0, 1, \ldots. \tag{20}$$

For such a choice we have

$$C_k R_k = C_k (P_{R(A)} - AX_k), \qquad \text{by } (17)$$

$$= C_k (I - AX_k), \qquad \text{by } (20), \tag{21}$$

and (19) can therefore be rewritten as

$$X_{k+1} = X_k + C_k T_k, \qquad k = 0, 1, \ldots, \tag{22}$$

where

$$T_k = I - AX_k, \qquad k = 0, 1, \ldots. \tag{23}$$

The iterative method (19), or (22), is suitable for the case where A is an $m \times n$ matrix with $m \leqslant n$, for then R_k and T_k are $m \times m$ matrices. However, if $m > n$ the following dual version of (19) is preferable to it:

$$X'_{k+1} = X'_k + R'_k C'_k, \tag{19'}$$

where

$$R_k' = P_{R(A^*)} - X_k'A \qquad (17')$$

and $\{ C_k' : k = 0, 1, \ldots \}$ is a suitable sequence, satisfying

$$C_k' = P_{R(A^*)}C_k', \qquad k = 0, 1, \ldots, \qquad (20')$$

a condition which allows rewriting (19') as

$$X_{k+1}' = X_k' + T_k'C_k', \qquad k = 0, 1, \ldots, \qquad (22')$$

where

$$T_k' = I - X_k'A, \qquad k = 0, 1, \ldots . \qquad (23')$$

Indeed, if $m > n$ then (22') is preferable to (22), for the former method uses the $n \times n$ matrix T_k' while the latter uses T_k, which is an $m \times m$ matrix.

Since all the results and proofs pertaining to the iterative method (19) or (22) hold true, with obvious modifications, for the dual method (19') or (22'), we will, for the sake of convenience, restrict the discussion to the case

$$m \leqslant n, \qquad (24)$$

leaving to the reader the details of the complementary case.

A *first-order iterative method for computing* $A^\dagger$, of type (22), is presented in the following

THEOREM 1.   Let $O \neq A \in C^{m \times n}$ and let the initial approximation X_0 and its residual R_0 satisfy

$$X_0 \in R(A^*, A^*) \qquad (25)$$

(i.e., $X_0 = A^*BA^*$, for some $B \in C^{m \times n}$; see, e.g., Ex. 3.25), and

$$\rho(R_0) < 1, \qquad (26)$$

respectively. Then the sequence

$$\begin{aligned} X_{k+1} &= X_k + X_0 T_k \\ &= X_k + X_0(I - AX_k), \qquad k = 0, 1, \ldots, \end{aligned} \qquad (27)$$

converges to $A^\dagger$ as $k \to \infty$, and the corresponding sequence of residuals satisfies

$$\|R_{k+1}\| \leqslant \|R_0\| \, \|R_k\|, \qquad k = 0, 1, \ldots, \qquad (28)$$

for any multiplicative matrix norm.

PROOF. The sequence (27) is obtained from (22) by choosing

$$C_k = X_0, \qquad k = 0, 1, \ldots, \tag{29}$$

a choice which, by (25), satisfies (20), and allows rewriting (27) as

$$X_{k+1} = X_k + X_0 R_k$$
$$= X_k + X_0 (P_{R(A)} - AX_k), \qquad k = 0, 1, \ldots. \tag{30}$$

From (30) we compute the residual

$$R_{k+1} = P_{R(A)} - AX_{k+1}$$
$$= P_{R(A)} - AX_k - AX_0 R_k$$
$$= R_k - AX_0 R_k$$
$$= P_{R(A)} R_k - AX_0 R_k, \qquad \text{by (17)}$$
$$= R_0 R_k, \qquad k = 0, 1, \ldots,$$
$$= R_0^{k+2}, \qquad \text{by repeating the argument.} \tag{31}$$

For any multiplicative matrix norm, it follows from (31) that

$$\|R_{k+1}\| \leqslant \|R_0\| \, \|R_k\|. \tag{28}$$

From

$$R_{k+1} = R_0^{k+2}, \qquad k = 0, 1, \ldots, \tag{31}$$

it also follows, by using (26) and Ex. 1.45, that the sequence of residuals converges to the zero matrix:

$$P_{R(A)} - AX_k \to O \quad \text{as} \quad k \to \infty. \tag{32}$$

We will prove now that the sequence (27) converges. Rewriting the sequence (27) as

$$X_{k+1} = X_k + X_0 R_k, \tag{30}$$

it follows from (31) that

$$X_{k+1} = X_k + X_0 R_0^{k+1}$$
$$= X_{k-1} + X_0 R_0^k + X_0 R_0^{k+1}$$
$$= X_0 (I + R_0 + R_0^2 + \cdots + R_0^{k+1}), \qquad k = 0, 1, \ldots, \tag{33}$$

which, by (26) and Exs. 1.45–1.46, converges to a limit X_∞.

Finally we will show that $X_\infty = A^\dagger$. From (32) it follows that

$$AX_\infty = P_{R(A)},$$

and in particular, that X_∞ is a $\{1\}$-inverse of A. From (25) and (27) it is obvious that all X_k lie in $R(A^*,A^*)$, and therefore

$$X_\infty \in R(A^*,A^*),$$

proving that $X_\infty = A^\dagger$, since $A^\dagger$ is the unique $\{1\}$-inverse of A which lies in $R(A^*,A^*)$; see Ex. 3.28. ∎

For any integer $p \geqslant 2$, a pth-*order iterative method for computing* $A^\dagger$, of type (22), is described in the following.

THEOREM 2. Let $\bigcirc \neq A \in C^{m \times n}$ and let the initial approximation X_0 and its residual R_0 satisfy (25) and (26), respectively. Then for any integer $p \geqslant 2$, the sequence

$$X_{k+1} = X_k(I + T_k + T_k^2 + \cdots + T_k^{p-1})$$

$$= X_k(I + (I - AX_k) + (I - AX_k)^2 + \cdots + (I - AX_k)^{p-1}),$$

$$k = 0, 1, \ldots \tag{34}$$

converges to $A^\dagger$ as $k \to \infty$, and the corresponding sequence of residuals satisfies

$$\|R_{k+1}\| \leqslant \|R_k\|^p, \qquad k = 0, 1, \ldots. \tag{35}$$

PROOF. The sequence (34) is obtained from (22) by choosing

$$C_k = X_k(I + T_k + T_k^2 + \cdots + T_k^{p-2}). \tag{36}$$

From (25) and (34) it is obvious that all X_k lie in $R(A^*,A^*)$, and therefore the sequence $\{C_k\}$, given by (36), satisfies (20), proving that the sequence (34) can be rewritten in the form (19)

$$X_{k+1} = X_k(I + R_k + R_k^2 + \cdots + R_k^{p-1}), \qquad k = 0, 1, \ldots. \tag{37}$$

From (37) we compute

$$R_{k+1} = P_{R(A)} - AX_{k+1}$$

$$= P_{R(A)} - AX_k(I + R_k + \cdots + R_k^{p-1})$$

$$= R_k - AX_k(R_k + R_k^2 + \cdots + R_k^{p-1}). \tag{38}$$

Now for any $j = 1, \ldots, p - 1$

$$R_k^j - AX_k R_k^j = P_{R(A)} R_k^j - AX_k R_k^j$$

$$= R_k R_k^j = R_k^{j+1},$$

and therefore, the last line in (38) collapses to

$$R_{k+1} = R_k^p, \tag{39}$$

which implies (35). The remainder of the proof, namely, that the sequence (37) converges to $A^\dagger$, can be given analogously to the proof of Theorem 1.
∎

The iterative methods (27) and (34) are related by the following:

THEOREM 3. Let $O \neq A \in C^{m \times n}$ and let the sequence $\{X_k : k = 0, 1, \ldots\}$ be constructed as in Theorem 1. Let p be any integer $\geqslant 2$, and let a sequence $\{\tilde{X}_j : j = 0, 1, \ldots\}$ be constructed as in Theorem 2 with the same initial approximation X_0 as the first sequence:

$$\tilde{X}_0 = X_0,$$

$$\tilde{X}_{j+1} = \tilde{X}_j \big(I + \tilde{T}_j + \tilde{T}_j^2 + \cdots + \tilde{T}_j^{p-1} \big), \qquad j = 0, 1, \ldots, \tag{34}$$

where

$$\tilde{T}_j = I - A\tilde{X}_j, \qquad j = 0, 1, \ldots. \tag{23}$$

Then

$$\tilde{X}_j = X_{p^j - 1}, \qquad j = 0, 1, \ldots. \tag{40}$$

PROOF. We use induction on j to prove (40), which obviously holds for $j = 0$. Assuming

$$\tilde{X}_j = X_{p^j - 1}, \tag{40}$$

we will show that

$$\tilde{X}_{j+1} = X_{p^{j+1} - 1}.$$

From

$$X_k = X_0 \big(I + R_0 + R_0^2 + \cdots + R_0^k \big) \tag{33}$$

and (40), it follows that

$$\tilde{X}_j = X_0 \big(I + R_0 + R_0^2 + \cdots + R_0^{p^j - 1} \big). \tag{41}$$

Rewriting (34) as

$$\tilde{X}_{j+1} = \tilde{X}_j\left(I + \tilde{R}_j + \tilde{R}_j^2 + \cdots + \tilde{R}_j^{p-1}\right), \tag{37}$$

it follows from

$$\tilde{R}_j = P_{R(A)} - A\tilde{X}_j$$

$$= P_{R(A)} - AX_{p^j-1}, \quad \text{by (40)},$$

$$= R_{p^j-1}$$

$$= R_0^{p^j}, \quad \text{by (31)},$$

that

$$\tilde{X}_{j+1} = \tilde{X}_j\left(I + R_0^{p^j} + R_0^{2p^j} + \cdots + R_0^{(p-1)p^j}\right)$$

$$= X_0\left(I + R_0 + R_0^2 + \cdots + R_0^{p^j-1}\right)\left(I + R_0^{p^j} + R_0^{2p^j} + \cdots + R_0^{(p-1)p^j}\right),$$

$$\text{by (41)},$$

$$= X_0\left(I + R_0 + R_0^2 + \cdots + R_0^{p^{j+1}-1}\right)$$

$$= X_{p^{j+1}-1}, \quad \text{by (33)}. \quad \blacksquare$$

Theorem 3 shows that an approximation $\tilde{X}_j$ obtained by the pth-order method (34) in j iterations, will require $p^j - 1$ iterations of the first-order method (27), both methods using the same initial approximation. For any two iterative methods of different orders, the higher-order method will, in general, require fewer iterations but more computations per iteration. A discussion of the optimal order p for methods of type (34) is given in Ex. 20.

EXERCISES AND EXAMPLES

14. Show that the condition

$$X_0 \in R(A^*,A^*) \tag{25}$$

is necessary for the convergence of the iterative methods (27) and (34).

EXAMPLE. Let

$$A = \frac{1}{2}\begin{bmatrix} 1 & 1 \\ 1 & 1 \end{bmatrix}, \qquad B = \epsilon\begin{bmatrix} 1 & -1 \\ -1 & 1 \end{bmatrix}, \qquad \epsilon \neq 0,$$

and let

$$X_0 = A + B.$$

Then

$$R_0 = P_{R(A)} - AX_0 = O$$

and in particular (26) holds, but

$$X_0 \notin R(A^*, A^*)$$

and both sequences (27) and (34) reduce to

$$X_k = X_0, \qquad k = 0, 1, \ldots,$$

without converging to $A^\dagger$.

15. Let $O \neq A \in C^{m \times n}$, and let X_0 and $R_0 = P_{R(A)} - AX_0$ satisfy

$$X_0 \in R(A^*, A^*), \tag{25}$$

$$\rho(R_0) < 1. \tag{26}$$

Then

$$A^\dagger = X_0 (I - R_0)^{-1}. \tag{42}$$

PROOF. The proof of Theorem 1 shows $A^\dagger$ to be the limit of

$$X_k = X_0 (I + R_0 + R_0^2 + \cdots + R_0^k) \tag{33}$$

as $k \to \infty$. But the sequence (33) converges, by Ex. 1.46, to the right member of (42). ■

The Special Case $X_0 = \beta A^*$

A frequent choice of the initial approximation X_0, in the iterative methods (27) and (34), is

$$X_0 = \beta A^* \tag{43}$$

for a suitable real scalar β. This special case is treated in the following three exercises.

16. Let $O \neq A \in C_r^{m \times n}$, let β be a real scalar, and let

$$R_0 = P_{R(A)} - \beta AA^*,$$

$$T_0 = I - \beta AA^*.$$

Then the following are equivalent.

(a) The scalar β satisfies

$$0 < \beta < \frac{2}{\lambda_1(AA^*)}, \tag{44}$$

where

$$\lambda_1(AA^*) \geqslant \lambda_2(AA^*) \geqslant \cdots \geqslant \lambda_r(AA^*) > 0$$

are the nonzero eigenvalues of AA^*.

(b) $\rho(R_0) < 1$.

(c) $\rho(T_0) \leqslant 1$ and $\lambda = -1$ is not an eigenvalue of T_0.

PROOF.   The nonzero eigenvalues of R_0 and T_0 are among

$$\{1 - \beta\lambda_i(AA^*) : i = 1, \dots, r\}$$

and

$$\{1 - \beta\lambda_i(AA^*) : i = 1, \dots, m\},$$

respectively. The equivalence of (a), (b), and (c) then follows from the observation that (44) is equivalent to

$$|1 - \beta\lambda_i(AA^*)| < 1, \qquad i = 1, \dots, r. \quad \blacksquare$$

17.   Let $O \neq A \in C_r^{m \times n}$, and let the real scalar β satisfy

$$0 < \beta < \frac{2}{\lambda_1(AA^*)}. \tag{44}$$

Then:

(a) The sequence

$$X_0 = \beta A^*, \qquad X_{k+1} = X_k(I - \beta AA^*) + \beta A^*, \qquad k = 0, 1, \dots, \tag{45}$$

or equivalently

$$X_k = \beta \sum_{j=0}^{k} A^*(I - \beta AA^*)^j, \qquad k = 0, 1, \dots, \tag{46}$$

is a first-order iterative method for computing $A^\dagger$.

(b) The corresponding residuals $R_k = P_{R(A)} - AX_k$ are given by

$$R_k = (P_{R(A)} - \beta AA^*)^{k+1}, \qquad k = 0, 1, \ldots. \tag{47}$$

(c) For any k, the spectral norm of R_k, $\|R_k\|_2$, is minimized by choosing

$$\beta = \frac{2}{\lambda_1(AA^*) + \lambda_r(AA^*)}, \tag{48}$$

in which case the minimal $\|R_k\|_2$ is

$$\|R_k\|_2 = \left(\frac{\lambda_1(AA^*) - \lambda_r(AA^*)}{\lambda_1(AA^*) + \lambda_r(AA^*)} \right)^{k+1}, \qquad k = 0, 1, \ldots. \tag{49}$$

PROOF.   (a) Substituting (43) in (27) results in (45) or equivalently in (46).

    (b) Follows from (46).

    (c) R_k is Hermitian and therefore

$$\|R_k\|_2 = \rho(R_k), \qquad \text{by Ex. 1.45}$$

$$= \rho(R_0^{k+1}), \qquad \text{by (31)}$$

$$= \rho^{k+1}(R_0), \qquad \text{by Ex. 1.44.}$$

Thus, $\|R_k\|_2$ is minimized by the same β that minimizes $\rho(R_0)$. Since the nonzero eigenvalues of $R_0 = P_{R(A)} - \beta AA^*$ are

$$1 - \beta\lambda_i(AA^*), \qquad i = 1, \ldots, r$$

it is clear that β minimizes

$$\rho(R_0) = \max\{|1 - \beta\lambda_i(AA^*)| : i = 1, \ldots, r\}$$

if, and only if,

$$-(1 - \beta\lambda_1(AA^*)) = 1 - \beta\lambda_r(AA^*), \tag{50}$$

which is (48). Finally (49) is obtained by substituting (48) in

$$\rho(R_k) = \max\{|1 - \beta\lambda_i(AA^*)|^{k+1} : i = 1, \ldots, r\}, \qquad \text{by (47)}$$

$$= |1 - \beta\lambda_r(AA^*)|^{k+1}, \qquad \text{for } \beta \text{ satisfying (50).} \quad \blacksquare$$

18. Let A, β be as in Ex. 17. Then for any integer $p \geqslant 2$, the sequence

$$X_{k+1} = X_k (I + T_k + T_k^2 + \cdots + T_k^{p-1}) \tag{34}$$

with

$$X_0 = \beta A^* \tag{43}$$

is a pth-order iterative method for computing $A^\dagger$. The corresponding residuals are

$$R_k = (P_{R(A)} - \beta A A^*)^{p^{k+1}}$$

and their spectral norms are minimized by β of (48). The iterative methods of Exs. 17 and 18 were studied by Ben-Israel and Cohen [1], Petryshyn [1], and Zlobec [1].

19. *A second-order iterative method.* An important special case of Theorem 2 is the case $p = 2$, resulting in the following second-order iterative method for computing $A^\dagger$. Let $O \neq A \in C^{m \times n}$ and let the initial approximation X_0 and its residual R_0 satisfy (25) and (26), respectively. Then the sequence

$$X_{k+1} = X_k (2I - A X_k), \qquad k = 0, 1, \ldots \tag{51}$$

converges to $A^\dagger$ as $k \to \infty$, and the corresponding sequence of residuals satisfies

$$\| R_{k+1} \| \leqslant \| R_k \|^2, \qquad k = 0, 1, \ldots. \quad \blacksquare \tag{52}$$

The iterative method (51) studied by Ben-Israel [4], Ben-Israel and Cohen [1], Petryshyn [1], and Zlobec [1] is a generalization of the well-known method of Schultz [1] for the iterative inversion of a nonsingular matrix; see, e.g., Householder [1], p. 95. A detailed error analysis of (51) is given in Söderström and Stewart [1].

20. *Discussion of the optimum order p.* As in Theorem 3 we denote by $\{ X_k \}$ and $\{ \tilde{X}_k \}$ the sequences generated by the first-order method (27) and by the pth-order method (34), respectively, using the same initial approximation $X_0 = \tilde{X}_0$. Taking the sequence $\{ X_k \}$ as the standard for comparing different orders p in (34), we use (40) to conclude that for each $k = 0, 1, \ldots$, the smallest $\tilde{k}$ such that the iterant $\tilde{X}_{\tilde{k}}$ is beyond X_k, is the smallest integer $\tilde{k}$ satisfying

$$p^{\tilde{k}} - 1 \geqslant k$$

and therefore

$$\tilde{k} = \langle \ln(k+1)/\ln p \rangle, \tag{53}$$

where the brackets $\langle \ \ \rangle$ indicate the smallest integer equaling or exceeding their contents.

In assessing the amount of computational effort per iteration, we assume that the amount of computational effort required to add or subtract an identity matrix is negligible compared to the effort to perform a matrix multiplication. Assuming (24) and hence the usage of the methods (27) and (34), rather than their duals based on (22'), we define a *unit of computational effort* as the effort required to multiply two $m \times m$ matrices. Accordingly, premultiplying an $n \times m$ matrix by an $m \times n$ matrix requires n/m units, as does the premultiplication of an $m \times m$ matrix by an $n \times m$ matrix. The iteration

$$X_{k+1} = X_k(I + T_k + T_k^2 + \cdots + T_k^{p-1})$$

$$= X_k(I + T_k(I + \cdots + T_k(I + T_k)\cdots)) \tag{34}$$

thus requires:

n/m units of effort to compute T_k,
$p-2$ units of effort to compute $T_k(I + \cdots + T_k(I + T_k)\cdots)$
n/m units of effort to multiply $X_k(I + \cdots + T_k^{p-1})$ adding to

$$p - 2 + 2\frac{n}{m} \tag{54}$$

units of effort.

The figure (54) can be improved for certain p. For example, the iteration (34) can be written for $p = 2^q$, $q = 1, 2, \ldots$, as

$$X_{k+1} = X_k \prod_{j=1}^{2^{q-1}} (I + T_k^j)$$

$$= X_k(I + T_k)(I + T_k^2)(I + T_k^4)\cdots(I + T_k^{2^{q-1}}), \tag{55}$$

requiring only

$$2(q-1) + 2\frac{n}{m} \tag{56}$$

units of effort, improving on (54) for all $q \geqslant 3$; see also Lonseth [1].

In comparing the first-order iterative method (27) and the second-order method (51) (obtained from (34) for $p = 2$) one sees that both methods require $2(n/m)$ units of effort per iteration. Therefore, by Theorem 3, the pth-order method (34) with $p=2$, is superior to the first-order method (27).

For a given integer $k=1,2,\ldots$ we define the *optimum order p* as the order of the iterative method (34) which, starting with an initial approximation X_0, minimizes the computational effort required to obtain, or go beyond, the approximation X_k, obtained by the first-order method (27) in k iterations.

Combining (53), (54), and (55) it follows that for a given k, the optimum p is the integer p minimizing

$$\left(p-2+2\frac{n}{m}\right)<\frac{\ln(k+1)}{\ln p}>, \qquad p=2,3,\ldots, \quad p\neq 2^q, \quad q=1,2,\ldots \quad (57)$$

or

$$\left(2q-2+2\frac{n}{m}\right)<\frac{\ln(k+1)}{q\ln 2}>, \qquad p=2^q, \quad q=1,2,\ldots. \quad (58)$$

Lower bounds for (57) and (58) are

$$\ln(k+1)\frac{p-2+2(n/m)}{\ln p}, \qquad p=2,3,\ldots, \quad p\neq 2^q, \quad q=1,2,\ldots \quad (57')$$

and

$$\ln(k+1)\frac{2q-2+2(n/m)}{q\ln 2}, \qquad p=2^q, \quad q=1,2,\ldots, \quad (58')$$

respectively, suggesting the following definition which is independent of k. The *approximate optimum order p* is the integer $p \geqslant 2$ minimizing

$$f(p)=\begin{cases} \dfrac{p-2+2(n/m)}{\ln p}, & p\neq 2^q, \quad p=1,2,\ldots \\[2mm] \dfrac{2(q-1+(n/m))}{q\ln 2}, & p=2^q, \quad q=1,2,\ldots. \end{cases} \quad (59)$$

The approximate optimum order p depends on the ratio n/m.

21. Iterative methods for computing projections. Since $AA^\dagger = P_{R(A)}$, it follows that for any sequence $\{X_k\}$, the sequence $\{Y_k = AX_k\}$ satisfies

$$Y_k \to P_{R(A)} \qquad \text{if} \quad X_k \to A^\dagger.$$

Thus, for any iterative method for computing $A^\dagger$ defined by a sequence of succcessive approximations $\{X_k\}$, there is an associated iterative method for computing $P_{R(A)}$ defined by the sequence $\{Y_k = AX_k\}$. Similarly, an iterative method for computing $P_{R(A^*)}$, is given by the sequence $\{Y_k' = X_k A\}$, since $A^\dagger A = P_{R(A^*)}$.

The residuals R_k, $k = 0, 1, \ldots$, of the sequence $\{Y_k\}$ will still be defined by (1), or equivalently

$$R_k = P_{R(A)} - Y_k, \qquad k = 0, 1, \ldots. \tag{60}$$

Therefore, the iterative method $\{Y_k = AX_k\}$ for computing $P_{R(A)}$ is of the same order as the iterative method $\{X_k\}$ for computing $A^\dagger$.

In particular, a pth-order iterative method for computing $P_{R(A)}$, based on Theorem 2, is given as follows.

Let $O \neq A \in C^{m \times n}$ and let the initial approximation Y_0 and its residual R_0 satisfy

$$Y_0 \in R(A, A^*) \tag{61}$$

and

$$\rho(R_0) < 1, \tag{26}$$

respectively. Then for any integer $p \geqslant 2$, the sequence

$$Y_{k+1} = Y_k(I + T_k + T_k^2 + \cdots + T_k^{p-1}) \tag{62}$$

with

$$T_k = I - Y_k, \qquad k = 0, 1, \ldots$$

converges to $P_{R(A)}$, as $k \to \infty$, and the corresponding sequence of residuals (60) satisfies

$$\|R_{k+1}\| \leqslant \|R_k\|^p, \qquad k = 0, 1, \ldots, \tag{35}$$

for any multiplicative matrix norm.

22. *A monotone property of* (62). Let $O \neq A \in C^{m \times n}$, let p be an even positive integer, and let the sequence $\{Y_k\}$ be given by (62), (61), (26), and the additional condition that Y_0 be Hermitian. Then the sequence $\{\text{trace } Y_k : k = 1, 2, \ldots\}$ is monotone increasing and converges to rank A.

PROOF. From (60), (31), and Theorem 3 it follows that

$$Y_k = P_{R(A)} - R_0^{p^k}. \tag{63}$$

From the fact that the trace of a matrix equals the sum of its eigenvalues, it follows that

$$\text{trace } P_{R(A)} = \dim R(A) = \text{rank } A$$

and

$$\text{trace } R_0^{p^k} = \sum_{i=1}^{m} \lambda_i \left(R_0^{p^k} \right)$$

$$= \sum_{i=1}^{m} \lambda_i^{p^k}(R_0), \qquad k = 0, 1, \ldots$$

which is a monotone decreasing sequence converging to 0, since p is even, R_0 is Hermitian (by (60) and the assumption that Y_0 is Hermitian), and therefore its eigenvalues $\lambda_i(R_0)$, which by (26) have moduli less than 1, are real. The proof is completed by noting that, by (63),

$$\text{trace } Y_k = \text{trace } P_{R(A)} - \text{trace } R_0^{p^k}$$

$$= \text{rank } A - \sum_{i=1}^{m} \lambda_i^{p^k}(R_0). \qquad \blacksquare$$

23. *A lower bound on rank A.* Let $O \neq A \in C^{m \times n}$ and let the sequence $\{ Y_k : k = 0, 1, \ldots \}$ be as in Ex. 22. Then

$$\text{rank } A \geqslant \langle \text{trace } Y_k \rangle, \qquad k = 1, 2, \ldots \tag{64}$$

where the brackets $\langle \ \rangle$ indicate the smallest integer equaling or exceeding their contents.

24. *Iterative methods for computing matrix products involving generalized inverses.* In many applications one has to compute a matrix product $A^\dagger B$ or $BA^\dagger$, where $A \in C^{m \times n}$ is given and B is a given matrix or vector. The iterative methods for computing $A^\dagger$ given above can be adapted for computing such products.

Consider, for example, the iterative method

$$X_{k+1} = X_k \left(I + T_k + T_k^2 + \cdots + T_k^{p-1} \right), \qquad k = 0, 1, \ldots \tag{34}$$

where p is an integer $\geqslant 2$,

$$T_k = I - AX_k, \qquad k = 0, 1, \ldots \tag{23}$$

and the initial approximation X_0 satisfies (25) and (26). A corresponding iterative method for computing $BA^\dagger$, for a given $B \in C^{q \times n}$, is given as follows.

Let $X_0 \in C^{n \times m}$ satisfy (25) and (26) and let the sequence $\{ Z_k : k$

$=0, 1, \ldots\}$ be given by

$$Z_0 = BX_0, \tag{65}$$

$$Z_{k+1} = Z_k M_k, \qquad k = 0, 1, \ldots \tag{66}$$

where

$$M_k = I + T_k + T_k^2 + \cdots + T_k^{p-1}, \qquad k = 0, 1, \ldots \tag{67}$$

$$T_{k+1} = I + M_k(T_k - I), \qquad k = 0, 1, \ldots \tag{68}$$

and

$$T_0 = I - AX_0.$$

Then the sequence $\{Z_k\}$ converges to $BA^\dagger$ as $k \to \infty$. (Garnett, Ben-Israel, and Yau [1]).

SUGGESTED FURTHER READING

Section 4. Albert [3], Businger and Golub [2], Decell [1], Germain-Bonne [1], Golub and Reinsch [1], Graybill, Meyer, and Painter [1], Ijiri [1], Kublanovskaya [1], Noble ([1], [3]), Pereyra and Rosen [1], Peters and Wilkinson [1], Pyle [1], Shinozaki, Sibuya, and Tanabe [1], Stallings and Boullion [1], Tewarson ([1], [2], [3]), Urquhart [1], and Willner [1].

Section 5. Kammerer and Nashed ([1], [2], [3], [4]), Nashed ([1], [2]), Showalter [1], Showalter and Ben-Israel [1], Whitney and Meany [1], and Zlobec ([3], [4]).

8

GENERALIZED INVERSES OF LINEAR OPERATORS BETWEEN HILBERT SPACES*

1. INTRODUCTION

The observation that generalized inverses are like prose ("Good Heavens! For more than forty years I have been speaking prose without knowing it" —Molière, *Le Bourgeois Gentilhomme*) is nowhere truer than in the literature of linear operators. In fact, generalized inverses of integral and differential operators were studied by Fredholm, Hilbert, Schmidt, Bounitzky, Hurwitz, and others, before E. H. Moore introduced generalized inverses in an algebraic setting; see, e.g., the historical survey in Reid [7].

This chapter is a brief and biased introduction to generalized inverses of linear operators between Hilbert spaces, with special emphasis on the similarities to the finite-dimensional case. Thus the spectral theory of such operators is omitted.

Following the preliminaries in Section 2, generalized inverses are introduced and studied in Section 3. Applications to integral and differential operators are sampled in Exs. 18–36. The minimization properties of generalized inverses are studied in Section 4. Integral and series representations of generalized inverses, and iterative methods for their computation, are given in Section 5.

*This chapter requires familiarity with the basic concepts of linear functional analysis, in particular, the theory of linear operators in Hilbert space.

2. HILBERT SPACES AND OPERATORS: PRELIMINARIES AND NOTATION

In this section we have collected, for convenience, some preliminary results needed later. These are well-known results, which can be found, in the form stated here or in a more general form, in the standard texts on functional analysis; see, e.g., Taylor [1] and Yosida [1].

(A) Our *Hilbert spaces* will be denoted by $\mathcal{H}, \mathcal{H}_1, \mathcal{H}_2,$ etc. In each space, the *inner product* of two vectors x and y is denoted by (x,y) and the *norm* is denoted by $\| \quad \|$. The *closure* of a subset L of $\mathcal{H}$ will be denoted by $\overline{L}$ and its orthogonal complement by $L^\perp$. $L^\perp$ is a closed subspace of $\mathcal{H}$, and

$$L^\perp = \overline{L}^\perp.$$

The *sum*, $M + N$, of two subsets $M, N \subset \mathcal{H}$ is

$$M + N = \{ x + y : x \in M, y \in N \}.$$

If M, N are subspaces of $\mathcal{H}$ and $M \cap N = \{0\}$, then $M + N$ is called the *direct sum* of M and N, and denoted by $M \oplus N$. If in addition $M \subset N^\perp$ we denote their sum by $M \overset{\perp}{\oplus} N$ and call it the *orthogonal direct sum* of M and N. Even if the subspaces M, N are closed, their sum $M + N$ need not be closed; see, e.g., Ex. 1. An orthogonal direct sum of two closed subspaces is closed. Conversely, if L, M are closed subspaces of $\mathcal{H}$ and $M \subset L$, then

$$L = M \overset{\perp}{\oplus} (L \cap M^\perp). \tag{1}$$

If (1) holds for two subspaces $M \subset L$, we say that L is *decomposable with respect to* M. See Exs. 5–6.

(B) The *(Cartesian) product* of $\mathcal{H}_1, \mathcal{H}_2$ will be denoted by

$$\mathcal{H}_{1,2} = \mathcal{H}_1 \times \mathcal{H}_2 = \{ \{ x,y \} : x \in \mathcal{H}_1, y \in \mathcal{H}_2 \}$$

where $\{x,y\}$ is an ordered pair. $\mathcal{H}_{1,2}$ is a Hilbert space with inner product

$$(\{ x_1,y_1 \}, \{ x_2,y_2 \}) = (x_1,x_2) + (y_1,y_2).$$

Let $J_i : \mathcal{H}_i \to \mathcal{H}_{1,2}$, $i = 1, 2$, be defined by

$$J_1 x = \{ x,0 \} \qquad \text{for all } x \in \mathcal{H}_1$$

and

$$J_2 y = \{ 0,y \} \qquad \text{for all } y \in \mathcal{H}_2.$$

The transformations J_1 and J_2 are isometric isomorphisms, mapping $\mathcal{H}_1$ and $\mathcal{H}_2$ onto

$$\mathcal{H}_{1,0} = J_1 \mathcal{H}_1 = \mathcal{H}_1 \times \{ 0 \}$$

and

$$\mathcal{K}_{0,2} = J_2 \mathcal{K}_2 = \{0\} \times \mathcal{K}_2,$$

respectively. Here $\{0\}$ is an appropriate zero space.

(C) Let $\mathcal{L}(\mathcal{K}_1, \mathcal{K}_2)$ denote the *class of linear operators from* $\mathcal{K}_1$ *to* $\mathcal{K}_2$. In what follows we will use *operator* to mean a linear operator. For any $T \in \mathcal{L}(\mathcal{K}_1, \mathcal{K}_2)$ we denote the *domain* of T by $D(T)$, the *range* of T by $R(T)$, the *null space* of T by $N(T)$, and the *carrier* of T by $C(T)$ where

$$C(T) = D(T) \cap N(T)^{\perp}.$$

The *graph*, $G(T)$, of a $T \in L(\mathcal{K}_1, \mathcal{K}_2)$ is

$$G(T) = \{\{x, Tx\} : x \in D(T)\}.$$

Clearly, $G(T)$ is a subspace of $\mathcal{K}_{1,2}$, and $G(T) \cap \mathcal{K}_{0,2} = \{0,0\}$. Conversely, if G is a subspace of $\mathcal{K}_{1,2}$ and $G \cap \mathcal{K}_{0,2} = \{0,0\}$, then G is the graph of a unique $T \in \mathcal{L}(\mathcal{K}_1, \mathcal{K}_2)$, defined for any point x in its domain

$$D(T) = J_1^{-1} P_{\mathcal{K}_{1,0}} G(T)$$

by

$$Tx = y,$$

where y is the unique vector in $\mathcal{K}_2$ such that $\{x, y\} \in G$.

Similarly, for any $T \in \mathcal{L}(\mathcal{K}_2, \mathcal{K}_1)$ the *inverse graph* of $T, G^{-1}(T)$, is defined by

$$G^{-1}(T) = \{\{Ty, y\} : y \in D(T)\}.$$

A subspace G in $\mathcal{K}_{1,2}$ is an inverse graph of some $T \in \mathcal{L}(\mathcal{K}_2, \mathcal{K}_1)$ if and only if $G \cap \mathcal{K}_{1,0} = \{0,0\}$, in which case T is uniquely determined by G (von Neumann [4]).

(D) An operator $T \in \mathcal{L}(\mathcal{K}_1, \mathcal{K}_2)$ is called *closed* if $G(T)$ is a closed subspace of $\mathcal{K}_{1,2}$. Equivalently T is closed if

$$x_n \in D(T), x_n \to x_0, Tx_n \to y_0 \Rightarrow x_0 \in D(T) \text{ and } Tx_0 = y_0$$

where $\to$ denotes strong convergence. A closed operator has a closed null space. The subclass of *closed operators* in $\mathcal{L}(\mathcal{K}_1, \mathcal{K}_2)$ will be denoted by $\mathcal{C}(\mathcal{K}_1, \mathcal{K}_2)$.

(E) An operator $T \in \mathcal{L}(\mathcal{K}_1, \mathcal{K}_2)$ is called *bounded* if its *norm* $\|T\|$ is finite, where

$$\|T\| = \sup_{0 \neq x \in \mathcal{K}_1} \frac{\|Tx\|}{\|x\|}.$$

The subclass of *bounded operators* in $\mathcal{L}(\mathcal{K}_1, \mathcal{K}_2)$ is denoted by $\mathcal{B}(\mathcal{K}_1, \mathcal{K}_2)$. If $T \in \mathcal{B}(\mathcal{K}_1, \mathcal{K}_2)$, then it may be assumed, without loss of generality, that $D(T)$ is closed or even that $D(T) = \mathcal{K}_1$. A bounded $T \in \mathcal{B}(\mathcal{K}_1, \mathcal{K}_2)$ is closed if and only if $D(T)$ is closed. Thus we may write $\mathcal{B}(\mathcal{K}_1, \mathcal{K}_2) \subset \mathcal{C}(\mathcal{K}_1, \mathcal{K}_2)$. Conversely, a closed $T \in \mathcal{C}(\mathcal{K}_1, \mathcal{K}_2)$ is bounded if $D(T) = \mathcal{K}_1$. This statement is the *closed graph theorem*.

(F) Let $T_1, T_2 \in \mathcal{L}(\mathcal{K}_1, \mathcal{K}_2)$ with $D(T_1) \subset D(T_2)$. If $T_2 x = T_1 x$ for all $x \in D(T_1)$, then T_2 is called an *extension* of T_1, and T_1 is called a *restriction* of T_2. These relations are denoted by

$$T_1 \subset T_2$$

or by

$$T_1 = T_{2[D(T_1)]}.$$

Let $T \in \mathcal{L}(\mathcal{K}_1, \mathcal{K}_2)$ and let the restriction of T to $C(T)$ be denoted by T_0

$$T_0 = T_{[C(T)]}.$$

Then

$$G(T_0) = \{ \{x, Tx\} : x \in C(T) \}$$

satisfies

$$G(T_0) \cap \mathcal{K}_{1,0} = \{0,0\}$$

and hence is the inverse graph of an operator $S \in \mathcal{L}(\mathcal{K}_2, \mathcal{K}_1)$ with

$$D(S) = R(T_0).$$

Clearly,

$$STx = x \qquad \text{for all } x \in C(T),$$

and

$$TSy = y \qquad \text{for all } y \in R(T_0).$$

Thus, if T_0 is considered as an operator in $\mathcal{L}(\overline{C(T)}, \overline{R(T_0)})$, then T_0 is invertible in its domain. The inverse T_0^{-1} is closed if and only if T_0 is closed. For $T \in \mathcal{L}(\mathcal{K}_1, \mathcal{K}_2)$, both $C(T)$ and T_0 may be trivial; see, e.g.,

Exs. 2 and 4.

(G) An operator $T \in \mathcal{L}(\mathcal{K}_1, \mathcal{K}_2)$ is called *dense* (or *densely defined*) if $\overline{D(T)} = \mathcal{K}_1$. Since any $T \in \mathcal{L}(\mathcal{K}_1, \mathcal{K}_2)$ can be considered to be an element of $\mathcal{L}(\overline{D(T)}, \mathcal{K}_2)$, any operator can be assumed to be dense without loss of generality.

For any $T \in \mathcal{L}(\mathcal{K}_1, \mathcal{K}_2)$, the condition $\overline{D(T)} = \mathcal{K}_1$ is equivalent to

$$G(T)^{\perp} \cap \mathcal{K}_{1,0} = \{0,0\},$$

where

$$G(T)^{\perp} = \{\{y,z\} : (y,x) + (z,Tx) = 0 \text{ for all } x \in D(T)\} \subset \mathcal{K}_{1,2}.$$

Thus for any dense $T \in \mathcal{L}(\mathcal{K}_1, \mathcal{K}_2)$, $G(T)^{\perp}$ is the inverse graph of a unique operator in $\mathcal{C}(\mathcal{K}_2, \mathcal{K}_1)$. This operator is $-T^*$, where T^*, the *adjoint* of T, satisfies

$$(T^*y, x) = (y, Tx) \qquad \text{for all } x \in D(T).$$

(H) For any dense $T \in \mathcal{L}(\mathcal{K}_1, \mathcal{K}_2)$,

$$\overline{N(T)} = R(T^*)^{\perp}, \qquad N(T^*) = R(T)^{\perp}. \tag{2}$$

In particular, $T[T^*]$ has a dense range if and only if $T^*[T]$ is one-to-one.

(I) Let $T \in \mathcal{L}(\mathcal{K}_1, \mathcal{K}_2)$ be dense.
If both T and T^* have inverses, then $(T^{-1})^* = (T^*)^{-1}$.
T has a bounded inverse if and only if $R(T^*) = \mathcal{K}_1$.
T^* has a bounded inverse if $R(T) = \mathcal{K}_2$. The converse holds if T is closed.
T^* has a bounded inverse and $R(T^*) = \mathcal{K}_1$ if and only if T has a bounded inverse and $\overline{R(T)} = \mathcal{K}_1$ (Taylor [1], Goldberg [1]).

(J) An operator $T \in \mathcal{L}(\mathcal{K}_1, \mathcal{K}_2)$ is called *closable* (or *preclosed*) if T has a closed extension. Equivalently, T is closable if

$$\overline{G(T)} \cap \mathcal{K}_{0,2} = \{0,0\},$$

in which case $\overline{G(T)}$ is the graph of an operator $\overline{T}$, called the *closure* of T. $\overline{T}$ is the minimal closed extension of T.

Since $G(T)^{\perp\perp} = \overline{G(T)}$ it follows that for a dense T, T^{**} is defined only if T is closable, in which case

$$T \subset T^{**} = \overline{T}$$

and

$$T = T^{**}$$

if and only if T is closed.

(K) A dense operator $T \in \mathcal{L}(\mathcal{H}, \mathcal{H})$ is called *symmetric* if

$$T \subset T^*$$

and *self-adjoint* if

$$T = T^*,$$

in which case it is called *non-negative*, and denoted by $T \geq 0$, if

$$(Tx, x) \geq 0 \quad \text{for all } x \in D(T).$$

If $T \in \mathcal{C}(\mathcal{H}_1, \mathcal{H}_2)$ is dense, then T^*T and TT^* are non-negative, and $I + TT^*$ and $I + T^*T$ have bounded inverses (von Neumann [1]).

(L) An operator $P \in \mathcal{B}(\mathcal{H}, \mathcal{H})$ is an *orthogonal projector* if

$$P = P^* = P^2,$$

in which case $R(P)$ is closed and

$$\mathcal{H} = R(P) \overset{\perp}{\oplus} N(P).$$

Conversely, if L is a closed subspace of $\mathcal{H}$, then there is a unique orthogonal projector P_L such that

$$L = R(P_L) \quad \text{and} \quad L^\perp = N(P_L).$$

(M) An operator $T \in \mathcal{C}(\mathcal{H}_1, \mathcal{H}_2)$ is called *normally solvable* if $R(T)$ is closed, which, by (2), is equivalent to the following condition: The equation

$$Tx = y$$

is consistent if and only if y is orthogonal to any solution u of

$$T^*u = 0.$$

This condition accounts for the name "normally solvable."

For any $T \in \mathcal{C}(\mathcal{H}_1, \mathcal{H}_2)$, the following statements are equivalent:

(a) T is normally solvable.

(b) The restriction $T_0 = T_{[C(T)]}$ has a bounded inverse.

(c) The non-negative number

$$\gamma(T) = \inf \left\{ \frac{\|Tx\|}{\|x\|} : 0 \neq x \in C(T) \right\} \tag{3}$$

is positive (Hestenes [4], Theorem 3.3).

EXERCISES AND EXAMPLES

1. *A nonclosed sum of closed subspaces.* Let $T \in \mathcal{B}(\mathcal{H}_1, \mathcal{H}_2)$, and let

$$D = J_1 D(T) = \{ \{x, 0\} : x \in D(T) \}.$$

Without loss of generality we assume that $D(T)$ is closed. Then D is closed. Also $G(T)$ is closed since T is bounded. But

$$G(T) + D$$

is nonclosed if $R(T)$ is nonclosed, since

$$\{x, y\} \in G(T) + D \Leftrightarrow y \in R(T) \qquad \text{(Halmos [2], p. 26)}.$$

2. *Unbounded linear functionals.* Let T be an unbounded linear functional on $\mathcal{H}$. Then $N(T)$ is dense in $\mathcal{H}$, and consequently $N(T)^{\perp} = \{0\}$, $C(T) = \{0\}$.

An example of such a functional on $L^2[0, \infty]$ is

$$Tx = \int_0^{\infty} tx(t)\, dt.$$

To show that $N(T)$ is dense, let $x_0 \in L^2[0, \infty]$ with $Tx_0 = \alpha$. Then a sequence $\{x_n\} \subset N(T)$ converging to x_0 is

$$x_n(t) = \begin{cases} x_0(t) & \text{if } t < 1 \text{ or } t > n+1 \\ x_0(t) - \dfrac{\alpha}{nt} & \text{if } 1 \leqslant t \leqslant n+1 \end{cases}.$$

Indeed,

$$\|x_n - x_0\|^2 = \int_1^{n+1} \frac{\alpha^2}{(nt)^2}\, dt = \frac{\alpha^2}{n(n+1)} \to 0.$$

3. Let D be a dense subspace of $\mathcal{H}$, and let F be a closed subspace such that $F^{\perp}$ is finite dimensional. Then

$$\overline{D \cap F} = F$$

(Erdelyi and Ben-Israel [1], Lemma 5.1).

4. *An operator with a trivial carrier.* Let D be any proper dense subspace of $\mathcal{H}$ and choose $x \notin D$. Let $F = [x]^{\perp}$, where $[x]$ is the line generated by x. Then $\overline{F \cap D} = F$, by Ex. 3. However $D \not\subset F$, so we can choose a subspace $A \neq \{0\}$ in D such that

$$D = A \oplus (D \cap F).$$

Define $T \in \mathcal{L}(\mathcal{H}, \mathcal{H})$ by

$$D(T) = D$$

and

$$T(y + z) = y \qquad \text{if } y \in A, z \in D \cap F.$$

Then

$$N(T) = D \cap F,$$

$$\overline{N(T)} = \overline{D \cap F} = F,$$

$$N(T)^{\perp} = F^{\perp} = [x],$$

$$C(T) = D(T) \cap N(T)^{\perp} = \{0\}.$$

5. Let L, M be subspaces of $\mathcal{H}$, and let $M \subset L$. Then

$$L = M \overset{\perp}{\oplus} (L \cap M^{\perp}) \qquad (1)$$

if and only if

$$P_{\overline{M}} x \in M \qquad \text{for all } x \in L.$$

In particular, a space is decomposable with respect to any closed subspace (Arghiriade [3]).

6. Let L, M, N be subspaces of $\mathcal{H}$ such that

$$L = M \overset{\perp}{\oplus} N.$$

Then

$$M = L \cap N^{\perp}, \qquad N = L \cap M^{\perp}.$$

Thus an orthogonal direct sum is decomposable with respect to each summand.

7. *A bounded operator with nonclosed range.* Let l^2 denote the Hilbert space of square summable sequences and let $T \in \mathcal{B}(l^2, l^2)$ be defined, for some $0 < k < 1$, by

$$T(\alpha_0, \alpha_1, \alpha_2, \ldots, \alpha_n, \ldots) = (\alpha_0, k\alpha_1, k^2\alpha_2, \ldots, k^n\alpha_n, \ldots).$$

The vector $y = \left(1, \dfrac{1}{2}, \dfrac{1}{3}, \ldots, \dfrac{1}{n}, \ldots\right)$ is in $\overline{R(T)}$, since $y = \lim_{n \to \infty} Tx_n$,

$x_n = \left(1, \dfrac{1}{2k}, \dfrac{1}{3k^2}, \ldots, \dfrac{1}{nk^{n-1}}, 0, 0, \ldots\right)$. However, $y \notin R(T)$.

8. *Linear integral operators.* Let $L^2 = L^2[a,b]$, the Lebesgue square integrable functions on the finite interval $[a,b]$. Let $K(s,t)$ be an L^2-*kernel* on $a \leqslant s, t \leqslant b$, meaning that the Lebesgue integral

$$\int_a^b \int_a^b |K(s,t)|^2 \, ds \, dt$$

exists and is finite; see, e.g., Smithies ([2], Section 1.6).

Consider the two operators $T_1, T_2 \in \mathfrak{B}(L^2, L^2)$ defined by

$$(T_1 x)(s) = \int_a^b K(s,t) x(t) \, dt, \qquad a \leqslant s \leqslant b,$$

$$(T_2 x)(s) = x(s) - \int_a^b K(s,t) x(t) \, dt, \qquad a \leqslant s \leqslant b,$$

called *Fredholm integral operators of the first kind* and *the second kind,* respectively.

Then

(a) $R(T_2)$ is closed.

(b) $R(T_1)$ is nonclosed unless it is finite dimensional.

More generally, if $T \in \mathcal{L}(\mathcal{K}_1, \mathcal{K}_2)$ is completely continuous then $R(T)$ is nonclosed unless it is finite dimensional (Kammerer and Nashed [3], Proposition 2.5).

9. Let $T \in \mathcal{C}(\mathcal{K}_1, \mathcal{K}_2)$ be dense. Then T is normally solvable if and only if T^* is. Also, T is normally solvable if and only if TT^* or T^*T is.

3. GENERALIZED INVERSES OF LINEAR OPERATORS BETWEEN HILBERT SPACES

A natural definition of generalized inverses in $\mathcal{L}(\mathcal{K}_1, \mathcal{K}_2)$ is the following one due to Tseng [3].

DEFINITION 1. Let $T \in \mathcal{L}(\mathcal{K}_1, \mathcal{K}_2)$. Then an operator $T^g \in \mathcal{L}(\mathcal{K}_2, \mathcal{K}_1)$ is a *Tseng generalized inverse* (abbreviated g.i.) of T if

$$R(T) \subset D(T^g) \tag{4}$$

$$R(T^g) \subset D(T) \tag{5}$$

$$T^g T x = P_{\overline{R(T^g)}} x \qquad \text{for all } x \in D(T) \tag{6}$$

$$TT^g y = P_{\overline{R(T)}} y \qquad \text{for all } y \in D(T^g). \tag{7}$$

This definition is symmetric in T and T^g, thus T is a g.i. of T^g.

An operator $T \in \mathcal{L}(\mathcal{K}_1, \mathcal{K}_2)$ may have a unique g.i., or infinitely many g.i.'s or it may have none. We will show in Theorem 1 that T has a g.i. if and only if its domain is decomposable with respect to its null space,

$$D(T) = N(T) \overset{\perp}{\oplus} \left(D(T) \cap N(T)^{\perp} \right)$$

$$= N(T) \overset{\perp}{\oplus} C(T). \tag{8}$$

By Ex. 5, this condition is satisfied if $N(T)$ is closed. Thus it holds for all closed operators, and in particular for bounded operators. If T has g.i.'s, then it has a *maximal g.i.*, some of whose poperties are collected in Theorem 2. For bounded operators with closed range, the maximal g.i. coincides with the Moore–Penrose inverse, and will likewise be denoted by $T^\dagger$. See Theorem 3.

For operators T without g.i.'s, the maximal g.i. $T^\dagger$ can be "approximated" in several ways, with the objective of retaining as many of its useful properties as possible. One such approach, due to Erdelyi [9] is described in Definition 3 and Theorem 4.

Some properties of g.i.'s, when they exist, are given in the following three lemmas, due to Arghiriade [3], which are needed later.

LEMMA 1. If $T^g \in \mathcal{L}(\mathcal{K}_2, \mathcal{K}_1)$ is a g.i. of $T \in \mathcal{L}(\mathcal{K}_1, \mathcal{K}_2)$, then $D(T)$ is decomposable with respect to $R(T^g)$.

PROOF. Follows from Ex. 5 since, for any $x \in D(T)$

$$P_{\overline{R(T^g)}} x = T^g T x, \qquad \text{by (6).} \quad \blacksquare$$

LEMMA 2. If $T^g \in \mathcal{L}(\mathcal{K}_2, \mathcal{K}_1)$ is a g.i. of $T \in \mathcal{L}(\mathcal{K}_1, \mathcal{K}_2)$, then T is a one-to-one mapping of $R(T^g)$ onto $R(T)$.

PROOF. Let $y \in R(T)$. Then

$$y = P_{\overline{R(T)}} y = T T^g y, \qquad \text{by (7),}$$

proving that $T(R(T^g)) = R(T)$.

Now we prove that T is one-to-one on $R(T^g)$. Let $x_1, x_2 \in R(T^g)$ satisfy

$$T x_1 = T x_2.$$

Then

$$x_1 = P_{\overline{R(T^g)}} x_1 = T^g T x_1 = T^g T x_2 = P_{\overline{R(T^g)}} x_2 = x_2. \quad \blacksquare$$

LEMMA 3. If $T^g \in \mathcal{L}(\mathcal{K}_2, \mathcal{K}_1)$ is a g.i. of $T \in \mathcal{L}(\mathcal{K}_1, \mathcal{K}_2)$, then:

$$N(T) = D(T) \cap R(T^g)^\perp \qquad (9)$$

and

$$C(T) = R(T^g). \qquad (10)$$

PROOF. Let $x \in D(T)$. Then, by Lemma 1,

$$x = x_1 + x_2, \quad x_1 \in R(T^g), \quad x_2 \in D(T) \cap R(T^g)^\perp, \quad x_1 \perp x_2. \quad (11)$$

Now

$$x_1 = P_{\overline{R(T^g)}}\, x = T^g T(x_1 + x_2) = T^g T x_1$$

and therefore,

$$T^g T x_2 = 0,$$

which, by Lemma 2 with T and T^g interchanged, implies that

$$Tx_2 = 0, \qquad (12)$$

hence

$$D(T) \cap R(T^g)^\perp \subset N(T).$$

Conversely, let $x \in N(T)$ be decomposed as in (11). Then

$$0 = Tx = T(x_1 + x_2)$$

$$= Tx_1, \qquad \text{by (12)},$$

which, by Lemma 2, implies that $x_1 = 0$ and therefore

$$N(T) \subset D(T) \cap R(T^g)^\perp,$$

completing the proof of (9).

Now

$$D(T) = R(T^g) \overset{\perp}{\oplus} \left(D(T) \cap R(T^g)^\perp \right), \qquad \text{by Lemma 1,}$$

$$= R(T^g) \overset{\perp}{\oplus} N(T),$$

which, by Ex. 6, implies that

$$R(T^g) = D(T) \cap N(T)^\perp,$$

proving (10). ∎

The existence of g.i.'s is settled in the following theorem announced, without proof, by Tseng [3]. Our proof follows that of Arghiriade [3].

THEOREM 1. Let $T \in \mathcal{L}(\mathcal{H}_1, \mathcal{H}_2)$. Then T has a g.i. if and only if

$$D(T) = N(T) \stackrel{\perp}{\oplus} C(T), \tag{8}$$

in which case, for any subspace $L \subset R(T)^\perp$, there is a g.i. T_L^g of T, with

$$D(T_L^g) = R(T) \stackrel{\perp}{\oplus} L \tag{13}$$

and

$$N(T_L^g) = L. \tag{14}$$

PROOF.   If T has a g.i., then (8) follows from Lemmas 1 and 3.

Conversely, suppose that (8) holds. Then

$$R(T) = T(D(T)) = T(C(T)) = R(T_0), \tag{15}$$

where $T_0 = T_{[C(T)]}$ is the restriction of T to $C(T)$. The inverse T_0^{-1} exists, by Section 2(F), and satisfies

$$R(T_0^{-1}) = C(T)$$

and, by (15),

$$D(T_0^{-1}) = R(T).$$

For any subspace $L \subset R(T)^\perp$, consider the extension T_L^g of T_0^{-1} with domain

$$D(T_L^g) = R(T) \stackrel{\perp}{\oplus} L \tag{13}$$

and null space

$$N(T_L^g) = L. \tag{14}$$

From its definition, it follows that T_L^g satisfies

$$D(T_L^g) \supset R(T)$$

and

$$R(T_L^g) = R(T_0^{-1}) = C(T) \subset D(T). \tag{16}$$

For any $x \in D(T)$

$$T_L^g T x = T_L^g T P_{\overline{C(T)}} x, \qquad \text{by (8)}$$

$$= T_0^{-1} T_0 P_{\overline{C(T)}} x, \qquad \text{by Ex. 5}$$

$$= P_{\overline{R(T_L^g)}} x, \qquad \text{by (16).}$$

Finally, any $y \in D(T_L^g)$ can be written, by (13), as

$$y = y_1 + y_2, \qquad y_1 \in R(T), \qquad y_2 \in L, \qquad y_1 \perp y_2,$$

and therefore

$$T T_L^g y = T T_L^g y_1, \qquad \text{by (14)}$$

$$= T_0 T_0^{-1} y_1$$

$$= y_1$$

$$= P_{\overline{R(T)}} y.$$

Thus T_L^g is a g.i. of T. ∎

The g.i. T_L^g is uniquely determined by its domain (13) and null space (14); see Ex. 10.

The maximal choice of the subspace L in (13) and (14) is $L = R(T)^\perp$. For this choice we have the following

DEFINITION 2. Let $T \in L(\mathcal{H}_1, \mathcal{H}_2)$ satisfy (8). Then the *maximal g.i.* of T, denoted by $T^\dagger$, is the g.i. of T with domain

$$D(T^\dagger) = R(T) \overset{\perp}{\oplus} R(T)^\perp \tag{17}$$

and null space

$$N(T^\dagger) = R(T)^\perp. \tag{18}$$

By Ex. 10, the g.i. $T^\dagger$ so defined is unique. It is maximal in the sense that any other g.i. of T is a restriction of $T^\dagger$.

Moreover, $T^\dagger$ is dense, by (17), and has a closed null space, by (18). Choosing L as a dense subspace of $R(T)^\perp$ shows that an operator T may have infinitely many dense g.i.'s T_L^g. Also, T may have infinitely many g.i.'s

T_L^g with closed null space, each obtained by choosing L as a closed subspace of $R(T)^\perp$. However, $T^\dagger$ is the unique dense g.i. with closed null space; see Ex. 11.

For closed operators, the maximal g.i. can be alternatively defined, by means of the following construction due to Hestenes [4], see also Landesman [1].

Let $T \in \mathcal{C}(\mathcal{K}_1, \mathcal{K}_2)$ be dense. Since $N(T)$ is closed, it follows, from Ex. 5, that

$$D(T) = N(T) \overset{\perp}{\oplus} C(T) \tag{8}$$

and therefore

$$G(T) = N \overset{\perp}{\oplus} C, \tag{19}$$

where, using the notation of Section 2(B), (C), and (F),

$$N = J_1 N(T) = G(T) \cap \mathcal{K}_{1,0}, \tag{20}$$

$$C = \{ \{x, Tx\} : x \in C(T) \}. \tag{21}$$

Similarly, since T^* is closed, it follows from Section 2(G), that

$$G(T)^\perp = N^* \overset{\perp}{\oplus} C^* \tag{22}$$

with

$$N^* = J_2 N(T^*) = G(T)^\perp \cap \mathcal{K}_{0,2}, \tag{23}$$

$$C^* = \{ \{-T^* y, y\} : y \in C(T^*) \}. \tag{24}$$

Now

$$\mathcal{K}_{1,2} = G(T) \overset{\perp}{\oplus} G(T)^\perp, \qquad \text{since } T \text{ is closed}$$

$$= \left(N \overset{\perp}{\oplus} C \right) \overset{\perp}{\oplus} \left(N^* \overset{\perp}{\oplus} C^* \right), \qquad \text{by (19) and (22)}$$

$$= \left(C \overset{\perp}{\oplus} N^* \right) \overset{\perp}{\oplus} \left(C^* \overset{\perp}{\oplus} N \right)$$

$$= G^\dagger \overset{\perp}{\oplus} G^{\dagger *}, \tag{25}$$

where

$$G^\dagger = C \overset{\perp}{\oplus} N^*, \tag{26}$$

$$G^{\dagger *} = C^* \overset{\perp}{\oplus} N. \tag{27}$$

Since

$$G^\dagger \cap \mathcal{K}_{1,0} = \{0,0\}, \qquad \text{by Section } 2(F),$$

it follows that $G^\dagger$ is the inverse graph of an operator $T^\dagger \in \mathcal{C}(\mathcal{K}_2, \mathcal{K}_1)$, with domain

$$J_2^{-1} P_{\mathcal{K}_{0,2}} G^\dagger = T(C(T)) \overset{\perp}{\oplus} N(T^*)$$

$$= R(T) \overset{\perp}{\oplus} R(T)^\perp, \qquad \text{by (15) and (2),}$$

and null space

$$J_2^{-1} N^* = N(T^*) = R(T)^\perp$$

and such that

$$T^\dagger T x = P_{\overline{C(T)}} x, \qquad \text{for any } x \in N(T) \overset{\perp}{\oplus} C(T),$$

and

$$TT^\dagger y = P_{\overline{R(T)}} y, \qquad \text{for any } y \in R(T) \overset{\perp}{\oplus} R(T)^\perp.$$

Thus $T^\dagger$ is the maximal g.i. of Definition 2.

Similarly, $G^{\dagger *}$ is the graph of the operator $-T^{*\dagger} \in \mathcal{C}(\mathcal{K}_1, \mathcal{K}_2)$, which is the maximal g.i. of $-T^*$.

This elegant construction makes obvious the properties of the maximal g.i., collected in the following:

THEOREM 2 (Hestenes [4]). Let $T \in \mathcal{C}(\mathcal{K}_1, \mathcal{K}_2)$ be dense. Then

(a) $T^\dagger \in \mathcal{C}(\mathcal{K}_2, \mathcal{K}_1)$

(b) $D(T^\dagger) = R(T) \overset{\perp}{\oplus} N(T^*), \qquad N(T^\dagger) = N(T^*),$

(c) $R(T^\dagger) = C(T),$

(d) $T^\dagger T x = P_{\overline{R(T^\dagger)}} x \qquad \text{for any } x \in D(T)$

(e) $TT^\dagger y = P_{\overline{R(T)}} y \qquad \text{for any } y \in D(T^\dagger)$

(f) $T^{\dagger\dagger} = T.$

(g) $T^{*\dagger} = T^{\dagger *},$

(h) $N(T^{*\dagger}) = N(T),$

(i) T^*T and $T^\dagger T^{*\dagger}$ are non-negative and

$$(T^*T)^\dagger = T^\dagger T^{*\dagger}, \qquad N(T^*T) = N(T)$$

(j) TT^* and $T^{*\dagger}T^\dagger$ are non-negative and $(TT^*)^\dagger = T^{*\dagger}T^\dagger$,
$N(TT^*) = N(T^*)$. ■

For bounded operators with closed range, the various characterizations of the maximal g.i. are collected in the following

THEOREM 3 (Petryshyn [1]). If $T \in \mathcal{B}(\mathcal{K}_1, \mathcal{K}_2)$ and $R(T)$ is closed, then $T^\dagger$ is characterized as the unique solution X of each of the following equivalent systems:

(a) $\quad TXT = T, XTX = X, \qquad (TX)^* = TX, (XT)^* = XT,$

(b) $\quad TX = P_{R(T)}, \qquad N(X^*) = N(T),$

(c) $\quad TX = P_{R(T)}, \qquad XT = P_{R(T^*)}, \qquad XTX = X,$

(d) $\quad XTT^* = T^*, \qquad XX^*T^* = X,$

(e) $\quad XTx = x \qquad$ for all $x \in R(T^*),$

$\quad\quad\quad Xy = 0 \qquad$ for all $y \in N(T^*),$

(f) $\quad XT = P_{R(T^*)}, \qquad N(X) = N(T^*)$

(g) $\quad TX = P_{R(T)}, \qquad XT = P_{R(X)}.$ ■

The notation $T^\dagger$ is justified by Theorem 3 (a), which lists the four Penrose equations (1.1)–(1.4).

If $T \in \mathcal{L}(\mathcal{K}_1, \mathcal{K}_2)$ does not satisfy (8), then it has no g.i., by Theorem 1. In this case one can still approximate $T^\dagger$ by an operator which has some properties of $T^\dagger$, and reduces to it if $T^\dagger$ exists. Such an approach, due to Erdelyi [9], is described in the following

DEFINITION 3. Let $T \in \mathcal{L}(\mathcal{K}_1, \mathcal{K}_2)$ and let T_r be the restriction of T defined by

$$D(T_r) = N(T) \overset{\perp}{\oplus} C(T), N(T_r) = N(T). \tag{28}$$

The *(Erdelyi) g.i.* of T is defined as $T_r^\dagger$, which exists since T_r satisfies (8). The inverse graph of $T_r^\dagger$ is

$$G^{-1}(T_r^\dagger) = \left\{ \{x, Tx + z\} : x \in C(T), z \in (T(C(T)))^\perp \right\}, \tag{29}$$

from which the following properties of $T_r^\dagger$ can be easily deduced.

THEOREM 4 (Erdelyi [9]). Let $T \in \mathcal{L}(\mathcal{H}_1, \mathcal{H}_2)$, and let its restriction T_r be defined by (28). Then

(a) $T_r^\dagger = T^\dagger$ if $T^\dagger$ exists,

(b) $D(T_r^\dagger) = T(C(T)) \overset{\perp}{\oplus} (T(C(T)))^\perp$,

   and in general, $R(T) \not\subset D(T_r^\dagger)$,

(c) $R(T_r^\dagger) = C(T)$, $\overline{R(T_r^\dagger)} = N(T)^\perp$,

(d) $T_r^\dagger T x = P_{\overline{R(T_r^\dagger)}} x$, for all $x \in D(T_r)$,

(e) $TT_r^\dagger y = P_{\overline{R(T)}} y$, for all $y \in D(T_r^\dagger)$,

(f) $D\left((T_r^\dagger)_r^\dagger\right) = \overline{N(T)} \overset{\perp}{\oplus} C(T)$,

(g) $R\left((T_r^\dagger)_r^\dagger\right) = T(C(T))$,

(h) $N\left((T_r^\dagger)_r^\dagger\right) = \overline{N(T)}$,

(i) $T \subset (T_r^\dagger)_r^\dagger$ if (8) holds,

(j) $T = (T_r^\dagger)_r^\dagger$ if and only if $N(T)$ is closed.

(k) If T is dense and closable, then

$$T_r^{\dagger*} \subset (T^*)_r^\dagger. \quad \blacksquare$$

See also Ex. 15.

EXERCISES AND EXAMPLES

10. Let $T \in \mathcal{L}(\mathcal{H}_1, \mathcal{H}_2)$ have g.i.'s and let L be a subspace of $R(T)^\perp$. Then the conditions

$$D(T^g) = R(T) \overset{\perp}{\oplus} L, \qquad (13)$$

$$N(T^g) = L \qquad (14)$$

determine a unique g.i., which is thus equal to T_L^g as constructed in the proof of Theorem 1.

PROOF.   Let T^g be a g.i. of T satisfying (13) and (14), and let $y \in D(T^g)$ be written as

$$y = y_1 + y_2, \quad y_1 \in R(T), \quad y_2 \in L.$$

Then

$$T^g y = T^g y_1, \quad \text{by (14)}$$

$$= T^g T x_1, \quad \text{for some } x_1 \in D(T)$$

$$= P_{\overline{R(T^g)}} x_1, \quad \text{by (6)}$$

$$= P_{\overline{C(T)}} x_1, \quad \text{by (10)}.$$

We claim that this determines T^g uniquely. For, suppose there is an $x_2 \in D(T)$ with $y_1 = Tx_2$. Then, as above,

$$T^g y = P_{\overline{C(T)}} x_2$$

and therefore

$$P_{\overline{C(T)}} x_1 - P_{\overline{C(T)}} x_2 = P_{\overline{C(T)}} (x_1 - x_2)$$

$$= 0 \quad \text{since } x_1 - x_2 \in N(T). \quad \blacksquare$$

11.   Let $T \in \mathcal{L}(\mathcal{K}_1, \mathcal{K}_2)$ have g.i.'s. Then $T^\dagger$ is the unique dense g.i. with closed null space.

PROOF.   Let T^g be any dense g.i. with closed null space. Then

$$D(T^g) = N(T^g) \overset{\perp}{\oplus} C(T^g), \quad \text{by Theorem 1}$$

$$= N(T^g) \overset{\perp}{\oplus} R(T), \quad \text{by (10)},$$

which, together with the assumptions $\overline{D(T^g)} = \mathcal{K}_2$ and $N(T^g) = \overline{N(T^g)}$, implies that

$$N(T^g) = R(T)^\perp.$$

Thus, T^g has the same domain and null space as $T^\dagger$, and therefore $T^g = T^\dagger$, by Ex. 10.

12.   Let $T \in \mathcal{B}(\mathcal{K}_1, \mathcal{K}_2)$ have a closed range $R(T)$ and let $T_1 \in \mathcal{B}(\mathcal{K}_1, R(T))$ be defined by

$$T_1 x = Tx \quad \text{for all } x \in \mathcal{K}_1.$$

Then

(a) T_1^* is the restriction of T^* to $R(T)$.

(b) The operator $T_1 T_1^* \in \mathcal{B}(R(T), R(T))$ is invertible.

(c) $$T^\dagger = P_{R(T^*)} T_1^* (T_1 T_1^*)^{-1} P_{R(T)} \qquad (\text{Kurepa } [1]).$$

13. Let $T \in \mathcal{C}(\mathcal{H}_1, \mathcal{H}_2)$. Then $R(T)$ is closed if and only if $T^\dagger$ is bounded (Landesman [1]).

PROOF. Follows from Section 2(M).

14. Let $T \in \mathcal{B}(\mathcal{H}_1, \mathcal{H}_2)$ have closed range. Then

$$T^\dagger = (T^*T)^\dagger T^* = T^*(TT^*)^\dagger \qquad (\text{Desoer and Whalen } [1]).$$

15. For arbitrary $T \in \mathcal{L}(\mathcal{H}_1, \mathcal{H}_2)$ consider its extension $\tilde{T}$ with

$$D(\tilde{T}) = D(T) + \overline{N(T)}, \qquad N(\tilde{T}) = \overline{N(T)}, \qquad \tilde{T} = T \text{ on } D(T), \quad (30)$$

which coincides with T if $N(T)$ is closed. Since $D(\tilde{T})$ is decomposable with respect to $N(\tilde{T})$, it might seem that $\tilde{T}$ can be used to obtain $\tilde{T}^\dagger$, a substitute for (the possibly nonexisting) $T^\dagger$.

Show that $\tilde{T}$ is not well defined by (30) if

$$D(T) \cap \overline{N(T)} \neq N(T) \qquad \text{and} \qquad N(\tilde{T}) \neq D(\tilde{T}), \quad (31)$$

which is the only case of interest since otherwise $D(T)$ is decomposable with respect to $N(T)$ or $\tilde{T}$ is identically $\bigcirc$ in its domain.

PROOF. By (31) there exist x_0 and y such that

$$x_0 \in D(T) \cap \overline{N(T)}, \qquad x_0 \notin N(T)$$

and

$$y \in D(T), \qquad y \notin \overline{N(T)}.$$

Then

$$\tilde{T}(x_0 + y) = \tilde{T}y, \qquad \text{since } x_0 \in N(\tilde{T})$$

and on the other hand

$$\tilde{T}(x_0 + y) = T(x_0 + y), \qquad \text{since } x_0, y \in D(T)$$

$$\neq Ty, \qquad \text{since } x_0 \notin N(T). \quad \blacksquare$$

16. Let $T \in \mathcal{B}(\mathcal{K}_1, \mathcal{K}_2)$ have closed range. Then

$$\| T^{\dagger} \| = \frac{1}{\gamma(T)},$$

where $\gamma(T)$ is defined in (3) (Petryshyn [1], Lemma 2).

17. Let $F \in \mathcal{B}(\mathcal{K}_3, \mathcal{K}_2)$ and $G \in \mathcal{B}(\mathcal{K}_1, \mathcal{K}_3)$ with $R(G) = \mathcal{K}_3 = R(F^*)$, and define $A \in \mathcal{B}(\mathcal{K}_1, \mathcal{K}_2)$ by $A = FG$. Then

$$A^{\dagger} = G^*(GG^*)^{-1}(F^*F)^{-1}F^*$$

$$= G^{\dagger}F^{\dagger} \qquad (\text{Holmes [1], p. 223}).$$

Compare with Theorem 1.5 and Ex. 1.16.

18. *Generalized inverses of linear integral operators.* In this exercise and in Exs. 19–24 below we consider the *Fredholm integral equation of the second kind*

$$x(s) - \lambda \int_a^b K(s,t)x(t)\,dt = y(s), \qquad a \leqslant s \leqslant b, \qquad (32)$$

written for short as

$$(I - \lambda K)x = y,$$

where all functions are complex, $[a,b]$ is a bounded interval, λ is a complex scalar and $K(s,t)$ is an L^2-*kernel* on $[a,b] \times [a,b]$; i.e., the (Lebesgue) integral $\int_a^b \int_a^b |K(s,t)|^2\,ds\,dt$ exists and is finite. Writing L^2 for $L^2[a,b]$, we need the following facts from the Fredholm theory of integral equations; see, e.g. Smithies [2]. For any λ, K as above

(a) $(I - \lambda K) \in \mathcal{B}(L^2, L^2)$,

(b) $(I - \lambda K)^* = I - \bar{\lambda}K^*$,   where $K^*(s,t) = \overline{K(t,s)}$.

(c) The null spaces $N(I - \lambda K)$ and $N(I - \bar{\lambda}K^*)$ have equal finite dimensions.

$$\dim N(I - \lambda K) = \dim N(I - \bar{\lambda}K^*) = n(\lambda), \quad \text{say}. \qquad (33)$$

(d) A scalar λ is called a *regular value* of K if $n(\lambda) = 0$, in which case the operator $I - \lambda K$ has an inverse $(I - \lambda K)^{-1} \in \mathcal{B}(L^2, L^2)$ written as

$$(I - \lambda K)^{-1} = I + \lambda R, \qquad (34)$$

where $R = R(s,t,\lambda)$ is an L^2-kernel called the *resolvent* of K.

(e) A scalar λ is called an *eigenvalue* of K if $n(\lambda) > 0$, in which case any nonzero $x \in N(I - \lambda K)$ is called an *eigenfunction* of K corresponding to λ. For any λ and, in particular, for any eigenvalue λ, both range spaces $R(I - \lambda K)$ and $R(I - \bar{\lambda} K^*)$ are closed and, by (2),

$$R(I - \lambda K) = N(I - \bar{\lambda} K^*)^{\perp}, \qquad R(I - \bar{\lambda} K^*) = N(I - \lambda K)^{\perp}. \qquad (35)$$

Thus, if λ is a regular value of K then (32) has, for every $y \in L^2$, a unique solution given by

$$x = (I + \lambda R)y,$$

that is

$$x(s) = y(s) + \lambda \int_a^b R(s,t,\lambda) y(t) \, dt, \qquad a \leqslant s \leqslant b. \qquad (36)$$

If λ is an eigenvalue of K then (32) is consistent if and only if y is orthogonal to any $u \in N(I - \bar{\lambda} K^*)$, in which case the general solution of (32) is

$$x = x_0 + \sum_{i=1}^{n(\lambda)} c_i x_i, \qquad c_i \text{ arbitrary scalars}, \qquad (37)$$

where x_0 is a particular solution of (32), and $\{x_1, \ldots, x_{n(\lambda)}\}$ is a basis of $N(I - \lambda K)$.

19. Pseudo resolvents. Let λ be an eigenvalue of K. Following Hurwitz [1], an L^2-kernel $R = R(s,t;\lambda)$ is called a *pseudo resolvent* of K if for any $y \in R(I - \lambda K)$, the function

$$x(s) = y(s) + \lambda \int_a^b R(s,t;\lambda) y(t) \, dt \qquad (36)$$

is a solution of (32).

A pseudo resolvent was constructed by Hurwitz as follows.

Let λ_0 be an eigenvalue of K, and let $\{x_1, \ldots, x_n\}$ and $\{u_1, \ldots, u_n\}$ be orthonormal bases of $N(I - \lambda_0 K)$ and $N(I - \bar{\lambda}_0 K^*)$ respectively. Then λ_0 is a regular value of the kernel

$$K_0(s,t) = K(s,t) - \frac{1}{\lambda_0} \sum_{i=1}^{n} u_i(s) \overline{x_i(t)}, \qquad (38)$$

written for short as

$$K_0 = K - \frac{1}{\lambda_0} \sum_{i=1}^{n} u_i x_i^*,$$

and the resolvent R_0 of K_0 is a pseudo resolvent of K, satisfying

$$(I+\lambda_0 R_0)(I-\lambda_0 K)x = x, \quad \text{for all } x \in R(I-\bar{\lambda}K^*)$$

$$(I-\lambda_0 K)(I+\lambda_0 R_0)y = y, \quad \text{for all } y \in R(I-\lambda K) \tag{39}$$

$$(I+\lambda_0 R_0)u_i = x_i, \quad i=1,\ldots,n.$$

PROOF. Follows from the matrix case, Ex. 2.42. ∎

20. A comparison with Theorem 2.2 shows that $I+\lambda R$ is a $\{1\}$-inverse of $I-\lambda K$, if R is a pseudo resolvent of K. As with $\{1\}$-inverses, the pseudo resolvent is nonunique. Indeed, for R_0, u_i, x_i as above, the kernel

$$R_0 + \sum_{i,j=1}^{n} c_{ij} x_i u_j^* \tag{40}$$

is a pseudo resolvent of K for any choice of scalars c_{ij}.

The pseudo resolvent constructed by Fredholm [1], who called the resulting operator $I+\lambda R$ a *pseudo inverse* of $I-\lambda K$, is the first explicit application, known to us, of a generalized inverse.

The class of all pseudo resolvents of a given kernel K is characterized as follows.

Let K be an L^2-kernel, let λ_0 be an eigenvalue of K and let $\{x_1,\ldots,x_n\}$ and $\{u_1,\ldots,u_n\}$ be orthonormal bases of $N(I-\lambda_0 K)$ and $N(I-\bar{\lambda}_0 K^*)$, respectively. An L^2-kernel R is a pseudo resolvent of K if and only if

$$R = K + \lambda KR - \frac{1}{\lambda_0} \sum_{i=1}^{n} \beta_i u_i^*,$$

$$R = K + \lambda RK - \frac{1}{\lambda_0} \sum_{i=1}^{n} x_i \alpha_i^* \tag{41}$$

where $\alpha_i, \beta_i \in L^2$ satisfy

$$(\alpha_i, x_j) = \delta_{ij}, (\beta_i, u_j) = \delta_{ij}, \quad i,j=1,\ldots,n. \tag{42}$$

Here KR stands for the kernel $KR(s,t) = \int_a^b K(s,u)R(u,t)du$, etc.

If λ is a regular value of K then (41) reduces to

$$R = K + \lambda KR, \quad R = K + \lambda RK, \tag{43}$$

which uniquely determines the resolvent $R(s,t;\lambda)$ (Hurwitz [1]).

21. Let K, λ_0, x_i, u_i, and R_0 be as above. Then the maximal g.i. of $I - \lambda_0 K$ is

$$(I - \lambda_0 K)^\dagger = I + \lambda_0 R_0 - \sum_{i=1}^{n} x_i u_i^*, \tag{44}$$

corresponding to the pseudo resolvent

$$R = R_0 - \frac{1}{\lambda_0} \sum_{i=1}^{n} x_i u_i^*. \tag{45}$$

22. Let $K(s,t) = u(s)\overline{v(t)}$, where

$$\int_a^b u(s)\overline{v(s)}\, ds = 0.$$

Then every scalar λ is a regular value of K.

23. Consider the equation

$$x(s) - \lambda \int_{-1}^{1} (1 + 3st)x(t)\, dt = y(s) \tag{46}$$

with $K(s,t) = 1 + 3st$. The resolvent is

$$R(s,t;\lambda) = \frac{1 + 3st}{1 - 2\lambda}.$$

K has a single eigenvalue $\lambda = \frac{1}{2}$ and an orthonormal basis of $N(I - \frac{1}{2}K)$ is

$$\left\{ x_1(s) = \frac{1}{\sqrt{2}},\, x_2(s) = \frac{\sqrt{3}}{\sqrt{2}}s \right\}$$

which, by symmetry, is also an orthonormal basis of $N(I - \frac{1}{2}K^*)$. From (38) we get

$$K_0(s,t) = K(s,t) - \frac{1}{\lambda_0}\sum u_i(s)\overline{x_i(t)}$$

$$= (1 + 3st) - 2\left(\frac{1}{\sqrt{2}}\frac{1}{\sqrt{2}} + \frac{\sqrt{3}}{\sqrt{2}}s\frac{\sqrt{3}}{\sqrt{2}}t \right)$$

$$= 0,$$

and the resolvent of $K_0(s,t)$ is therefore

$$R_0(s,t;\lambda) = 0.$$

If $\lambda \neq \frac{1}{2}$, then for each $y \in L^2[-1,1]$ Eq. (46) has a unique solution

$$x(s) = y(s) + \lambda \int_{-1}^{1} \frac{1+3st}{1-2\lambda} y(t)\, dt.$$

If $\lambda = \frac{1}{2}$, then (46) is consistent if and only if

$$\int_{-1}^{1} y(t)\, dt = 0, \qquad \int_{-1}^{1} ty(t)\, dt = 0,$$

in which case the general solution is

$$x(s) = y(s) + c_1 + c_2 s; \qquad c_1, c_2 \text{ arbitrary.}$$

23A.   Let

$$K(s,t) = 1 + s + 3st, \qquad -1 \leqslant s, t \leqslant 1.$$

Then $\lambda = \frac{1}{2}$ is the only eigenvalue and

$$\dim N(I - \tfrac{1}{2}K) = 1.$$

An orthonormal basis of $N(I - \frac{1}{2}K)$ is the single vector

$$x_1(s) = \frac{\sqrt{3}}{\sqrt{2}} s, \qquad -1 \leqslant s \leqslant 1.$$

An orthonormal basis of $N(I - \frac{1}{2}K^*)$ is

$$u_1(s) = \frac{1}{\sqrt{2}}, \qquad -1 \leqslant s \leqslant 1.$$

The Hurwitz kernel (38) is

$$K_0(s,t) = (1 + s + 3st) - 2\left(\frac{1}{\sqrt{2}} \frac{\sqrt{3}}{\sqrt{2}} t \right)$$

$$= 1 + s - \sqrt{3}\, t + 3st, \qquad -1 \leqslant s, t \leqslant 1.$$

Compute the resolvent R_0 of K_0, which is a pseudo resolvent of K. (*Hint:* Use the following exercise.)

24. *Degenerate kernels.* A kernel $K(s,t)$ is called *degenerate* if it is a finite sum of products of L^2 functions, as follows:

$$K(s,t) = \sum_{i=1}^{m} f_i(s)\,\overline{g_i(t)}. \tag{47}$$

Degenerate kernels are convenient because they reduce the integral equation (32) to a finite system of linear equations. Also, any L^2-kernel can be approximated, arbitrarily close, by a degenerate kernel; see, e.g., Smithies [2], p. 40, and Halmos [2], Problem 137.

Let $K(s,t)$ be given by (47). Then

(a) The scalar λ is an eigenvalue of (47) if and only if $1/\lambda$ is an eigenvalue of the $m \times m$ matrix

$$B = [b_{ij}],$$

where

$$b_{ij} = \int_a^b f_j(s)\,\overline{g_i(s)}\ ds.$$

(b) Any eigenfunction of $K[K^*]$ corresponding to an eigenvalue λ $[\bar{\lambda}]$ is a linear combination of the m functions $f_1,\ldots,f_m$ $[g_1,\ldots,g_m]$.

(c) If λ is a regular value of (47), then the resolvent at λ is

$$R(s,t;\lambda) = \frac{\det\begin{bmatrix} 0 & \vdots & f_1(s) & \cdots & f_m(s) \\ \hline -\overline{g_1(t)} & \vdots & & & \\ \vdots & \vdots & & I-\lambda B & \\ -\overline{g_m(t)} & \vdots & & & \end{bmatrix}}{\det(I-\lambda B)}.$$

See also Kantorovich and Krylov [1], Chapter II.

Generalized Inverses of Linear Differential Operators

The following 12 exercises deal with generalized inverses of closed dense operators $L \in \mathcal{C}(\mathcal{S}_1, \mathcal{S}_2)$ with $\overline{D(L)} = \mathcal{S}_1$, where:

(i) S_1, S_2 are spaces of (scalar or vector) functions which are either the Hilbert space $L^2[a,b]$ or the *space of continuous functions* $C[a,b]$, where $[a,b]$ is a given finite real interval. Since $C[a,b]$ is a dense subspace of $L^2[a,b]$, a closed dense linear operator mapping $C[a,b]$ into S_2 may be considered as a dense operator in $C(L^2[a,b], S_2)$.

(ii) L is defined for all x in its domain $D(L)$ by

$$Lx = lx, \tag{47}$$

where l is a *differential expression*, for example, in the vector case,

$$lx = A_1(t)\frac{d}{dt}x + A_0(t)x, \tag{48}$$

where $A_0(t), A_1(t)$ are $n \times n$ matrix coefficients, with suitable regularity conditions; see, e.g., Ex. 30 below.

(iii) The *domain* of L consists of those functions x in S_1 for which l makes sense and $lx \in S_2$, and which satisfy certain given conditions, such as *initial* or *boundary conditions*.

If a differential operator L is invertible and there is a kernel (function, or matrix in the vector case)

$$G(s,t), \qquad a \leqslant s, t \leqslant b,$$

such that for all $y \in R(L)$

$$(L^{-1}y)(s) = \int_a^b G(s,t)y(t)\,dt, \qquad a \leqslant s \leqslant b,$$

then $G(s,t)$ is called the *Green's function* (or *matrix*) of L. In this case, for any $y \in R(L)$, the unique solution of

$$Lx = y \tag{49}$$

is given by

$$x(s) = \int_a^b G(s,t)y(t)\,dt, \qquad a \leqslant s \leqslant b. \tag{50}$$

If L is not invertible, but there is a kernel $G(s,t)$ such that, for any $y \in R(L)$, a particular solution of (49) is given by (50), then $G(s,t)$ is called a *generalized Green's function* (or *matrix*) of L. A generalized Green's function of L is therefore a kernel of an integral operator which is a generalized inverse of L.

Generalized Green's functions were introduced by Hilbert [1] in 1904, and consequently studied by Myller, Westfall, and Bountitzky [1], Elliott [1], [2], and Reid [1]; see, e.g., the historical survey in Reid [4].

25. *Derivatives.* Let

$\mathbb{S}$ = the real space $L^2[0,\pi]$ of real valued functions,

$\mathbb{S}^1$ = the absolutely continuous functions $x(t), 0 \leqslant t \leqslant \pi$, whose derivatives x' are in $\mathbb{S}$,

$$\mathbb{S}^2 = \{ x \in \mathbb{S}^1 : x' \in \mathbb{S}^1 \},$$

and let L be the operator d/dt with

$$D(L) = \{ x \in \mathbb{S}^1 : x(0) = x(\pi) = 0 \}.$$

Then

(a) $L \in \mathcal{C}(\mathbb{S}, \mathbb{S})$, $\overline{D(L)} = \mathbb{S}$, $C(L) = D(L)$,

$$R(L) = \left\{ y \in \mathbb{S} : \int_0^\pi y(t) \, dt = 0 \right\} = \overline{R(L)}.$$

(b) The adjoint L^* is the operator $-d/dt$ with

$$D(L^*) = \mathbb{S}^1, \quad C(L^*) = \mathbb{S}^1 \cap R(L), \quad R(L^*) = \mathbb{S}.$$

(c) $L^*L = -\dfrac{d^2}{dt^2}$ with $D(L^*L) = \{ x \in \mathbb{S}^2 : x(0) = x(\pi) = 0 \}$

and $R(L^*L) = \mathbb{S}$.

(d) $LL^* = -\dfrac{d^2}{dt^2}$ with $D(LL^*) = \{ x \in \mathbb{S}^2 : x'(0) = x'(\pi) = 0 \}$

and $R(LL^*) = R(L)$.

(e) $L^\dagger$ is defined on $D(L^\dagger) = \mathbb{S}$ by

$$(L^\dagger y)(t) = \int_0^t y(s) \, ds - \frac{t}{\pi} \int_0^\pi y(s) \, ds, \quad 0 \leqslant t \leqslant \pi$$

(Hestenes [4], Example 1).

26. For L of Ex. 25, determine which of the following equations hold and interpret your results:

(a) $L^{\dagger *} = L^{* \dagger}$,

(b) $L^\dagger = (L^*L)^\dagger L^* = L^* (LL^*)^\dagger$,

(c) $L^{\dagger \dagger} = L$.

27. Gradients. Let

$\mathcal{S}$ = the real space $L^2[[0,\pi]\times[0,\pi]]$ of real valued functions $x(t_1,t_2)$,

$$0 \leqslant t_1, t_2 \leqslant \pi,$$

$\mathcal{S}^1$ = the subclass of $\mathcal{S}$ with the properties
   (i) $x(t_1,t_2)$ is absolutely continuous in $t_1[t_2]$ for almost all t_2 $[t_1]$, $0 \leqslant t_1, t_2 \leqslant \pi$;
   (ii) the partial derivatives $\partial x/\partial t_1$, $\partial x/\partial t_2$ which exist almost everywhere are in $\mathcal{S}$,

and let L be the gradient operator

$$lx = \begin{bmatrix} \dfrac{\partial x}{\partial t_1} \\[2mm] \dfrac{\partial x}{\partial t_2} \end{bmatrix}$$

with domain

$$D(L) = \{ x \in \mathcal{S}^1 : x(0,t_2) = x(\pi,t_2) = 0 \text{ for almost all } t_2,$$

$$x(t_1,0) = x(t_1,\pi) = 0 \text{ for almost all } t_1, 0 \leqslant t_1, t_2 \leqslant \pi \}$$

Then:

(a) $L \in \mathcal{C}(\mathcal{S}, \mathcal{S} \times \mathcal{S})$, $\overline{D(L)} = \mathcal{S}$.

(b) The adjoint L^* is the negative of the divergence operator

$$l^*y = l^* \begin{bmatrix} y_1 \\ y_2 \end{bmatrix} = -\frac{\partial y_1}{\partial t_1} - \frac{\partial y_2}{\partial t_2}$$

with

$$D(L^*) = \{ y \in \mathcal{S} \times \mathcal{S} : y \in C^1 \}.$$

(c) $L^*L = -\text{Laplacian} = -\left[\dfrac{\partial^2}{\partial t_1^2} + \dfrac{\partial^2}{\partial t_2^2} \right]$

(d) The Green's function of L^*L is

$$G(s_1,s_2,t_1,t_2) = \frac{4}{\pi^2} \sum_{m,n=1}^{\infty} \frac{1}{m^2+n^2} \sin(ms_1)\sin(ns_2)\sin(mt_1)\sin(nt_2),$$

$$0 \leqslant s_i, t_j \leqslant \pi.$$

(e) If

$$y = \begin{bmatrix} y_1 \\ y_2 \end{bmatrix} \in \mathbb{S} \times \mathbb{S},$$

then

$$(L^\dagger y)(t_1, t_2) = \sum_{j=1}^{2} \int_0^\pi \int_0^\pi \frac{\partial}{\partial s_j} G(s_1, s_2, t_1, t_2) y_j(s_1, s_2) \, ds_1 \, ds_2$$

(Landesman [1], Section 5).

28. *Ordinary linear differential equations with homogeneous boundary conditions.* Let

$\mathbb{S}$ = the real space $L^2[a,b]$ of real valued functions,

$C^k[a,b]$ = the real valued functions on $[a,b]$ with k derivatives and

$$x^{(k)} = \frac{d^k x}{dt^k} \in C[a,b],$$

$$\mathbb{S}^k = \left\{ x \in C^{k-1}[a,b] : x^{(k-1)} \text{ absolutely continuous, } x^{(k)} \in \mathbb{S} \right\}$$

and let L be the operator

$$l = \sum_{i=0}^{n} a_i(t) \left(\frac{d}{dt} \right)^i, \qquad a_i \in C^i[a,b], i = 0, 1, \ldots, n,$$

$$a_n(t) \neq 0, \quad a \leqslant t \leqslant b, \tag{51}$$

with domain $D(L)$ consisting of all $x \in \mathbb{S}^n$ which satisfy

$$M\hat{x} = 0, \tag{52}$$

where $M \in R_m^{m \times 2n}$ is a matrix with a specified null space $N(M)$, and $\hat{x} \in R^{2n}$ is the boundary vector

$$\hat{x}^T = \left[x(a), x'(a), \cdots, x^{(n-1)}(a); x(b), x'(b), \cdots, x^{(n-1)}(b) \right].$$

Finally let $\tilde{L}$ be the operator l of (51) with $D(\tilde{L}) = \mathbb{S}^n$. Then

(a) $L \in \mathcal{C}(\mathbb{S}, \mathbb{S}), \qquad \overline{D(L)} = \mathbb{S}.$

(b) $\dim N(\tilde{L}) = n = \dim N(\tilde{L}^*).$

(c) $N(L) \subset N(\tilde{L})$, $N(L^*) \subset N(\tilde{L}^*)$, hence $\dim N(L) \leqslant n$ and $\dim N(L^*) \leqslant n$.

(d) $R(L)$ is closed.

(e) The restriction $L_0 = L_{[C(L)]}$ of L to its carrier is a one-to-one mapping of $C(L)$ onto $R(L)$; $L_0 \in \mathcal{C}(C(L), R(L))$.

(f) $L_0^{-1} \in \mathcal{B}(R(L), C(L))$.

(g) $L^\dagger$, the extension of L_0^{-1} to all of $\mathcal{S}$ with $N(L^\dagger) = R(L)^\perp$ is bounded and satisfies

$$LL^\dagger y = P_{R(L)} y, \quad \text{for all } y \in \mathcal{S}$$

$$L^\dagger L x = P_{N(L)^\perp} x, \quad \text{for all } x \in D(L).$$

For proofs of (a) and (d) see Halperin [1] and Schwartz [1]. The proof of (e) is contained in Section 2(F), and (f) follows from the closed graph theorem (Locker [2]).

29. For L as in Ex. 28, find the generalized Green's function which corresponds to $L^\dagger$, i.e., find the kernel $L^\dagger(s, t)$ such that

$$(L^\dagger y)(s) = \int_a^b L^\dagger(s, t) y(t) dt \quad \text{for all } y \in D(L^\dagger) = \mathcal{S}.$$

Hint: A generalized Green's function of $\tilde{L}$ is

$$\tilde{G}(s, t) = \begin{cases} \displaystyle\sum_{j=1}^n \frac{x_j(s) \det(X_j(t))}{a_n(t) \det(X(t))}, & a \leqslant t \leqslant s \leqslant b \\ 0, & a \leqslant s \leqslant t \leqslant b \end{cases} \tag{53}$$

where

$\{x_1, \ldots, x_k\}$ is an orthonormal basis of $N(L)$,

$\{x_1, \ldots, x_k, x_{k+1}, \ldots, x_n\}$ is an orthomormal basis of $N(\tilde{L})$,

$$X(t) = \left[x_j^{(i-1)}(t) \right], \quad i, j = 1, \ldots, n,$$

$X_j(t)$ is the matrix obtained from $X(t)$ by replacing the jth column by $[0, 0, \ldots, 0, 1]^T$.

This is proved in Coddington and Levinson ([1], Theorem 6.4).

Since $R(L) \subset R(\tilde{L})$ it follows, for any $y \in R(L)$, that the general solution of

$$Lx = y$$

is

$$x(s) = \int_z^b \tilde{G}(s,t)y(t)\,dt + \sum_{i=1}^n c_i x_i(s),$$

c_i arbitrary.   (54)

Writing the particular solution $L^{\dagger}y$ in the form (54)

$$L^{\dagger}y = x_0 + \sum_{i=1}^n c_i x_i,$$

$$x_0(s) = \int_a^b G(s,t)y(t)\,dt,$$ (55)

we determine its coefficients $\{c_1,\ldots,c_n\}$ as follows:

(a) The coefficients $\{c_1,\ldots,c_k\}$ are determined by $L^{\dagger}y \in N(L)^{\perp}$, since, by (55),

$$\left(L^{\dagger}y, x_j\right) = 0 \Rightarrow c_j = -\left(x_0, x_j\right), \qquad j = 1,\ldots,k.$$

(b) The remaining coefficients $\{c_{k+1},\ldots,c_n\}$ are determined by the boundary condition (52). Indeed, writing (55) as

$$L^{\dagger}y = x_0 + Xc, \qquad c^T = [c_1,\ldots,c_n],$$

it follows from (52) that

$$M\hat{x}_0 + M\hat{X}c = 0, \qquad \text{where } \hat{X} = \begin{bmatrix} X(a) \\ X(b) \end{bmatrix}.$$ (56)

A solution of (56) is

$$c = -\left(M\hat{X}\right)^{(1)} M\hat{x}_0,$$ (57)

where $(M\hat{X})^{(1)} \in R^{n \times m}$ is any $\{1\}$ inverse of $M\hat{X} \in R^{m \times n}$. Now $\{x_1,\ldots,x_k\} \subset D(L)$, and therefore

$$M\hat{X} = [\,\mathbf{O} \quad B\,], \qquad B \in R_{n-k}^{m \times (n-k)}.$$

Thus, we may use in (57),

$$\left(M\hat{X}\right)^{(1)} = \begin{bmatrix} \mathbf{O} \\ B^{(1)} \end{bmatrix}, \qquad \text{for any } B^{(1)} \in B\{1\},$$

obtaining

$$c = -\begin{bmatrix} O \\ B^{(1)} \end{bmatrix} M\hat{x}_0,$$

which uniquely determines $\{c_{k+1},\ldots,c_n\}$.

Substituting these coefficients $\{c_1,\ldots,c_n\}$ in (54) finally gives $L^{\dagger}(s,t)$ (Locker [2]).

30. *The vector case.* Let $\mathcal{S}_n$ and $\mathcal{S}_n^k$ denote the spaces of n-dimensional vector functions whose components belong to $\mathcal{S}$ and $\mathcal{S}^k$, respectively, of Ex. 28. Let L be the differential operator

$$lx = A_1(t)\frac{dx}{dt} + A_0(t)x, \qquad a \leqslant t \leqslant b \tag{48}$$

where A_0, A_1 are $n \times n$ matrix functions satisfying*

(i) $A_0(t)$ is continuous on $[a,b]$.

(ii) $A_1(t)$ is continuously differentiable and nonsingular on $[a,b]$, with domain $D(L)$ consisting of those vector function $x \in \mathcal{S}_n^1$ which satisfy

$$M\hat{x} = 0, \tag{52}$$

where $M \in R_m^{m \times 2n}$ is a matrix with a specified null space $N(M)$ and $\hat{x} \in R^{2n}$ is the boundary vector

$$\hat{x} = \begin{bmatrix} x(a) \\ x(b) \end{bmatrix}. \tag{58}$$

Let $\tilde{L}$ be the differential operator (48) with domain $D(\tilde{L}) = \mathcal{S}_n^1$. Then

(a) $L \in \mathcal{C}(\mathcal{S}_n, \mathcal{S}_n), \qquad \overline{D(L)} = \mathcal{S}_n.$

(b) The adjoint of L is the operator L^*, defined by

$$l^*y = -\frac{d}{dt}(A_1^*(t)y) + A_0^*(t)y \tag{59}$$

*Much weaker regularity conditions will do; see, e.g., Reid [3] and [5], Chapter III.

on its domain

$$D(L^*) = \{ y \in S_n^1 : y^*(b)x(b) - y^*(a)x(a) = 0 \text{ for all } x \in D(L) \}$$

$$= \left\{ y \in S_n^1 : P^* \begin{bmatrix} I & O \\ O & -I \end{bmatrix} \hat{y} = 0 \text{ for any} \right.$$

$$\left. P \in R_{2n-m}^{(2n-m) \times 2n} \text{ with } MP = O \right\}. \tag{60}$$

(c) $\dim N(\tilde{L}) = n$.

(d) Let

$$k = \dim N(L) \quad \text{and} \quad k^* = \dim N(L^*).$$

Then

$$\max\{0, n-m\} \leqslant k \leqslant \min\{n, 2n-m\}$$

and

$$k + m = k^* + n.$$

(e) $R(L) = N(L^*)^\perp, \qquad R(L^*) = N(L)^\perp,$

hence both $R(L)$ and $R(L^*)$ are closed.

(f) Let

$$X(t) = [x_1(t), \ldots, x_n(t)]$$

be a *fundamental matrix* of $\tilde{L}$, i.e., let the vectors $\{x_1, \ldots, x_n\}$ form a basis of $N(\tilde{L})$. Then

$$\tilde{G}(s,t) = \tfrac{1}{2} \text{sign}(s-t) X(s) X(t)^{-1} \tag{61}$$

is a generalized Green's matrix of $\tilde{L}$.

(g) Let $(M\hat{X})^{(1)}$ be any {1}-inverse of $M\hat{X}$ where $\hat{X} = \begin{bmatrix} X(a) \\ X(b) \end{bmatrix}$. Then

$$G(s,t) = \tfrac{1}{2} X(s) \left[\text{sign}(s-t)I - (M\hat{X})^{(1)} M \begin{bmatrix} I & O \\ O & -I \end{bmatrix} \hat{X} \right] X(t)^{-1}$$

$$\tag{62}$$

is a generalized Green's matrix of L (Reid [4] and [5], Chapter III).

PROOF OF (g). For any $y \in R(L)$, the general solution of

$$Lx = y \tag{49}$$

is

$$x(s) = \int_a^b \tilde{G}(s,t)y(t)\,dt + \sum_{i=1}^n c_i x_i(s) \tag{54}$$

or

$$x = x_0 + Xc, \qquad c^T = [c_1, \ldots, c_n]$$

and from (52) it follows that

$$c = -(M\hat{X})^{(1)} M\hat{x}_0 \tag{57}$$

and (62) follows by substituting (57) in (54).

31. The differential expression

$$lx = \sum_{i=1}^n a_i(t) \frac{d^i x}{dt^i}, \qquad x \text{ scalar function} \tag{51}$$

is a special case of

$$lx = A_1(t) \frac{dx}{dt} + A_0(t)x, \qquad x \text{ vector function.} \tag{48}$$

32. *The class of all generalized Green's matrices.* Let L be as in Ex. 30 and let $X_0(t)$ and $Y_0(t)$ be $n \times k$ and $n \times k^*$ matrix functions whose columns are bases of $N(L)$ and $N(L^*)$, respectively. Then a kernel $H(s,t)$ is a generalized Green's matrix of L if and only if

$$H(s,t) = G(s,t) + X_0(s)A^*(t) + B(s)Y_0^*(t), \tag{63}$$

where $G(s,t)$ is any generalized Green's matrix of L (in particular (62)), and $A(t)$ and $B(t)$ are $n \times k$ and $n \times k^*$ matrix functions which are Lebesgue measurable and essentially bounded (Reid [3]).

33. Let $X_0(t)$ and $Y_0(t)$ be as in Ex. 32. If $\Theta(t)$ and $\Psi(t)$ are Lebesgue measurable and essentially bounded matrix functions such that the matrices

$$\int_a^b \Theta^*(t) X_0(t)\,dt, \qquad \int_a^b Y_0^*(t)\Psi(t)\,dt$$

are nonsingular, then L has a unique generalized Green's function $G_{\Theta,\Psi}$ such that

$$\int_a^b \Theta^*(s)\,G(s,t)\,ds = \mathrm{O},$$

$$\int_a^b G(s,t)\Psi(t)\,dt = \mathrm{O}, \qquad a \leqslant s, t \leqslant b \qquad (\text{Reid [3]}). \qquad (64)$$

Thus the generalized inverse determined by $G_{\Theta,\Psi}$ has null space spanned by the columns of Ψ and range which is the orthogonal complement of the columns of Θ. Compare with Section 2.5.

34. *Existence and properties of $L^\dagger$.* If in Ex. 33 we take

$$\Theta = X_0, \qquad \Psi = Y_0,$$

then we get a generalized inverse of L which has the same range and null space as L^*. This generalized inverse is the analog of the Moore–Penrose inverse of L and will likewise be denoted by $L^\dagger$.

Show that $L^\dagger$ satisfies the four Penrose equations (1.1)–(1.4) as far as can be expected.

(a) $LL^\dagger L = L,$

(b) $L^\dagger LL^\dagger = L^\dagger,$

(c) $LL^\dagger = P_{R(L)},$

$\qquad (LL^\dagger)^* = P_{R(L)} \qquad \text{on } D(L^*),$

(d) $L^\dagger L = P_{R(L^*)} \qquad \text{on } D(L),$

$\qquad (L^\dagger L)^* = P_{R(L^*)} \qquad (\text{Loud [1], [2]}).$

35. *Loud's construction of $L^\dagger$.* Just as in the matrix case (see Theorem 2.10(c) and Ex. 2.32) it follows here that

$$L^\dagger = P_{R(L^*)}GP_{R(L)}, \qquad (65)$$

where G is any generalized Green's matrix.

In computing $P_{R(L^*)}$ and $P_{R(L)}$ we use Ex. 30(e) to obtain

$$P_{R(L^*)} = I - P_{N(L)}, \qquad P_{R(L)} = I - P_{N(L^*)}. \qquad (66)$$

Here $P_{N(L)}$ and $P_{N(L^*)}$ are integral operators of the first kind with kernels

$$K_{N(L)}(s,t) = X_0(s)\left(\int_a^b X_0^*(u)X_0(u)\,du\right)^{-1} X_0^*(t) \qquad (67)$$

and

$$K_{N(L^*)}(s,t) = Y_0(s)\left(\int_a^b Y_0^*(u)Y_0(u)\,du\right)^{-1} Y_0^*(t), \qquad (68)$$

respectively, where X_0 and Y_0 are as in Ex. 32.

Thus, for any generalized Green's matrix $G(s,t)$, $L^\dagger$ has the kernel

$$L^\dagger(s,t) = G(s,t) - \int_a^b K_{N(L)}(s,u)G(u,t)\,du - \int_a^b G(s,u)K_{N(L^*)}(u,t)\,du$$

$$+ \int_a^b \int_a^b K_{N(L)}(s,u)G(u,v)K_{N(L^*)}(v,t)\,du\,dv \qquad (69)$$

(Loud [2]).

36. Let L be the differential operator given by

$$lx = x' - B(t)x, \qquad 0 \leqslant t \leqslant 1$$

with boundary conditions

$$x(0) = x(1) = 0.$$

Then the adjoint L^* is given by

$$l^*y = -y' - B(t)^*y$$

with no boundary conditions.

Let $X(t)$ be a fundamental matrix for

$$lx = 0.$$

Then $X(t)^{*-1}$ is a fundamental matrix for

$$l^*y = 0.$$

Now $N(L) = \{0\}$ and therefore $K_{N(L)} = O$. Also, $N(L^*)$ is spanned by the columns of $X(t)^{*-1}$, so by (68)

$$K_{N(L^*)}(s,t) = X(s)^{*-1}\left(\int_0^1 X(u)X(u)^{*-1}\,du\right)X(t)^{-1}. \qquad (70)$$

A generalized Green's matrix for L is

$$G(s,t) = \begin{cases} X(s)X(t)^{-1}, & 0 \leqslant s < t \leqslant 1 \\ O, & 0 \leqslant t < s \leqslant 1 \end{cases}. \tag{71}$$

Finally, by (69),

$$L^{\dagger}(s,t) = G(s,t) - \int_0^1 G(s,u) K_{N(L^*)}(u,t)\,du,$$

with G and $K_{N(L^*)}$ given by (71) and (70), respectively (Loud [2], pp. 201–202).

4. MINIMAL PROPERTIES OF GENERALIZED INVERSES

In this section, which is based on Erdelyi and Ben-Israel [1], we develop certain distinguishing minimal properties of generalized inverses of operators between Hilbert spaces. The matrix case appears in Chapter 3.

DEFINITION 4.   Let $T \in \mathcal{L}(\mathcal{K}_1, \mathcal{K}_2)$ and consider the linear equation

$$Tx = y. \tag{72}$$

If the infinum

$$\|Tx' - y\| = \inf_{x \in D(T)} \|Tx - y\| \tag{73}$$

is attained by a vector $x' \in D(T)$, then x' is called an *extremal solution* of (72). Among the extremal solutions there may exist a unique vector x_0 of least norm

$$\|x_0\| < \|x'\|,$$

for all extremal solutions $x' \neq x_0$. Then x_0 is called the *least extremal solution*.

Other names for extremal solutions are *virtual solutions* (Tseng [5]), and *approximate solutions*.

Example 37 shows that extremal solutions need not exist. Their existence is characterized in the following theorem.

THEOREM 5.   Let $T \in \mathcal{L}(\mathcal{K}_1, \mathcal{K}_2)$. Then

$$Tx = y \cdot \tag{72}$$

has an extremal solution if and only if

$$P_{\overline{R(T)}}y \in R(T). \tag{74}$$

PROOF.   For every $x \in D(T)$

$$\| Tx - y \|^2 = \| P_{\overline{R(T)}}(Tx - y) \|^2 + \| P_{R(T)^\perp}(Tx - y) \|^2$$

$$= \| P_{\overline{R(T)}}(Tx - y) \|^2 + \| P_{R(T)^\perp}y \|^2.$$

Thus

$$\| Tx - y \| \geqslant \| P_{R(T)^\perp}y \|, \qquad \text{for all } x \in D(T)$$

with equality if and only if

$$Tx = P_{\overline{R(T)}}y. \tag{75}$$

Clearly,

$$\inf_{x \in D(T)} \| Tx - y \| = \| P_{R(T)^\perp}y \|, \tag{76}$$

which is attained if and only if (75) is satisfied for some $x \in D(T)$.   ∎

See also Ex. 44.

The existence of extremal solutions does not guarantee the existence of a least extremal solution; see, e.g., Ex. 39. Before settling this issue we require

LEMMA 4.   Let x' and x'' be extremal solutions of (72). Then

(a) $\qquad P_{N(T)^\perp}x' = P_{N(T)^\perp}x''$

(b) $\qquad P_{\overline{N(T)}}x' \in N(T) \qquad$ if and only if $\quad P_{\overline{N(T)}}x'' \in N(T)$.

PROOF.   (a) From (75),

$$Tx' = Tx'' = P_{\overline{R(T)}}y$$

and hence

$$T(x' - x'') = 0, \tag{77}$$

proving (a).

(b) From (77),

$$x' - x'' = P_{\overline{N(T)}}(x' - x'')$$

and then

$$P_{\overline{N(T)}}x' = P_{\overline{N(T)}}x'' + (x' - x''),$$

proving (b).   ∎

The existence of the least extremal solution is characterized in the following:

THEOREM 6 (Erdelyi and Ben-Israel [1]). Let x be an extremal solution of (72). There exists a least extremal solution if and only if

$$P_{\overline{N(T)}}x \in N(T) \tag{78}$$

in which case, the least extremal solution is

$$x_0 = P_{N(T)^\perp}x. \tag{79}$$

PROOF. Let x' be an extremal solution of (72). Then

$$\|x'\|^2 = \|P_{\overline{N(T)}}x'\|^2 + \|P_{N(T)^\perp}x'\|^2$$

$$= \|P_{\overline{N(T)}}x'\|^2 + \|P_{N(T)^\perp}x\|^2, \qquad \text{by Lemma 4,}$$

proving that

$$\|x'\| \geqslant \|P_{N(T)^\perp}x\|$$

with equality if and only if

$$P_{\overline{N(T)}}x' = 0. \tag{80}$$

If: Let condition (78) be satisfied and define

$$x_0 = x - P_{\overline{N(T)}}x.$$

Then x_0 is an extremal solution since

$$Tx_0 = Tx.$$

Also

$$P_{\overline{N(T)}}x_0 = 0,$$

which, by (80), proves that x_0 is the least extremal solution.

Only if. Let x_0 be the least extremal solution of (72). Then, by (80),

$$x_0 = P_{\overline{N(T)}}x_0 + P_{N(T)^\perp}x_0 = P_{N(T)^\perp}x,$$

and hence

$$x_0 = x - P_{\overline{N(T)}}x.$$

But

$$Tx_0 = Tx,$$

since both x_0 and x are extremal solutions, and therefore

$$TP_{\overline{N(T)}}x = 0,$$

proving (78). ∎

As in the matrix case (see Corollary 3.3), here too a unique generalized inverse is characterized by the property that it gives the least extremal solution whenever it exists. We define this inverse as follows:

DEFINITION 5.   Let $T \in \mathcal{L}(\mathcal{H}_1, \mathcal{H}_2)$, let

$$C(T) = D(T) \cap N(T)^{\perp},$$
$$B(T) = D(T) \cap \overline{N(T)}, \tag{81}$$

and let $A(T)$ be a subspace satisfying

$$D(T) = A(T) \oplus \left(B(T) \overset{\perp}{\oplus} C(T) \right). \tag{82}$$

(Examples 42 and 43 below show that, in the general case, this complicated decomposition cannot be avoided.) Let

$$G_0 = \{ \{x, Tx\} : x \in C(T) \}, \qquad G_1 = G(T)^{\perp} \cap \mathcal{H}_{0,2} = J_2 R(T)^{\perp}.$$

The *extremal g.i.* of T, denoted by $T_e^{\dagger}$, is defined by its inverse graph

$$G_0 + G_1 = \{ \{x, Tx + z\} : x \in C(T), z \in R(T)^{\perp} \}.$$

The following properties of $T_e^{\dagger}$ are easy consequences of the above construction.

THEOREM 7   (Erdelyi and Ben-Israel [1]).   Let $T \in \mathcal{L}(\mathcal{H}_1, \mathcal{H}_2)$. Then

(a) $D(T_e^{\dagger}) = T(C(T)) \overset{\perp}{\oplus} R(T)^{\perp}$,   and in general, $R(T) \not\subset D(T_e^{\dagger})$.

(b) $R(T_e^{\dagger}) = C(T)$.

(c) $N(T_e^{\dagger}) = R(T)^{\perp}$.

(d) $T T_e^{\dagger} y = P_{\overline{R(T)}} y$,   for all $y \in D(T_e^{\dagger})$.

(e) $T_e^{\dagger} T x = P_{\overline{R(T_e^{\dagger})}} x$, for all $x \in N(T) \overset{\perp}{\oplus} C(T)$.   ∎

See also Exs. 40–41 below.

The extremal g.i. $T_e^{\dagger}$ is characterized, in terms of the least extremal solution, as follows:

THEOREM 8 (Erdelyi and Ben-Israel [1]). The least extremal solution x_0 of (72) exists if and only if

$$y \in D(T_e^\dagger), \tag{83}$$

in which case

$$x_0 = T_e^\dagger y. \tag{84}$$

PROOF. Assume (83). By Theorem 7(a)

$$P_{\overline{R(T)}} y = y_0 \in T(C(T)) \subset R(T),$$

and, by Theorem 5, extremal solutions do exist. Let x_0 be the unique vector in $C(T)$ such that

$$P_{\overline{R(T)}} y = y_0 = Tx_0.$$

Then, by Theorem 3(a), (c), and (e),

$$T_e^\dagger y = T_e^\dagger y_0 = T_e^\dagger Tx_0 = x_0,$$

and, by Theorem 3(d),

$$\| Tx_0 - y \| = \| TT_e^\dagger y - y \| = \| P_{\overline{R(T)}} y - y \| = \| P_{R(T)^\perp} y \|,$$

which, by (76) shows that x_0 is an extremal solution. Since

$$x_0 \in R(T_e^\dagger) \subset N(T)^\perp,$$

it follows, from Lemma 4, that

$$x_0 = P_{N(T)^\perp} x$$

for any extremal solution x of (72). By Theorem 6, x_0 is the least extremal solution.

Conversely, let x_0 be the least extremal solution whose existence we assume. By Theorem 2, $x_0 \in C(T)$, and by Theorem 3(e),

$$T_e^\dagger Tx_0 = x_0.$$

Since x_0 is an extremal solution, it follows from (75) that

$$Tx_0 = P_{\overline{R(T)}} y \in T(C(T))$$

and therefore

$$x_0 = T_e^\dagger Tx = T_e^\dagger P_{\overline{R(T)}} y$$

$$= T_e^\dagger y. \quad \blacksquare$$

If $N(T)$ is closed then $T_e^\dagger$ coincides with the maximal g.i. $T^\dagger$. Thus for closed operators, and in particular for bounded operators, $T_e^\dagger$ should be replaced by $T^\dagger$ in the statement of Theorem 8.

EXERCISES AND EXAMPLES

37. *An equation without extremal solution.* Let T and y be as in Ex. 7. Then

$$Tx = y$$

has no extremal solutions.

38. It was noted in Ex. 8, that, in general, the Fredholm integral operator of the first kind has a nonclosed range. Consider the kernel

$$G(s,t) = \begin{cases} s(1-t), & 0 \leqslant s \leqslant t \leqslant 1 \\ t(1-s), & 0 \leqslant t \leqslant s \leqslant 1 \end{cases},$$

which is a generalized Green's function of the operator

$$-\frac{d^2}{dt^2}, \quad 0 \leqslant t \leqslant 1.$$

Let $T \in \mathcal{B}\,(L^2[0,1], L^2[0,1])$ be defined by

$$(Tx)(s) = \int_0^1 G(s,t)x(t)\,dt.$$

Show that there exists a $y \in L^2[a,b]$ for which

$$Tx = y$$

has no extremal solution.

39. *An equation without a least extremal solution.* Consider the unbounded functional on $L^2[0,\infty]$

$$Tx = \int_0^\infty tx(t)\,dt$$

discussed in Ex. 2. Then the equation

$$Tx = 1$$

is consistent, and each of the functions

$$x_n(t) = \begin{cases} \dfrac{1}{nt}, & 1 \leqslant t \leqslant n+1 \\ 0, & \text{otherwise} \end{cases}$$

is a solution, $n = 1, 2, \ldots$.

Since

$$\|x_n\|^2 = \int_1^{n+1} \frac{1}{(nt)^2} \, dt = \frac{1}{n(n+1)} \to 0,$$

there is no extremal solution of least norm.

40. *Properties of* $(T_e^\dagger)^\dagger$. By Theorem 7(a) and (c), it follows that $D(T_e^\dagger)$ is decomposable with respect to $N(T_e^\dagger)$. Thus $T_e^\dagger$ has a maximal (Tseng) g.i., denoted by $T_e^{\dagger\dagger}$. Some of its properties are listed below.

(a) $\quad G(T_e^{\dagger\dagger}) = \{ \{x + z, Tx\} : x \in C(T), z \in C(T)^\perp \}$.

(b) $\quad D(T_e^{\dagger\dagger}) = C(T) \overset{\perp}{\oplus} C(T)^\perp$.

(c) $\quad R(T_e^{\dagger\dagger}) = T(C(T))$.

(d) $\quad N(T_e^{\dagger\dagger}) = C(T)^\perp$.

41. Let $T \in \mathcal{L}(\mathcal{H}_1, \mathcal{H}_2)$ and let

$$D_0(T) = N(T) \overset{\perp}{\oplus} C(T).$$

Then

(a) $\quad D(T_e^{\dagger\dagger}) = C(T) \overset{\perp}{\oplus} \overline{N(T)} \overset{\perp}{\oplus} D_0(T)^\perp$,

a refinement of Ex. 40(b).

(b) $\quad D_0(T) \subset D(T) \cap D(T_e^{\dagger\dagger})$

and

$$T_{[D_0(T)]} = T_{e[D_0(T)]}^{\dagger\dagger}.$$

(c) $T_e^{\dagger\dagger}$ is an extension of T if and only if $D(T)$ is decomposable with respect to $N(T)$, in which case $T_e^{\dagger\dagger}$ is an extension by zero to $\overline{N(T)} \overset{\perp}{\oplus} D(T)^\perp$.

42. *An example of* $A(T) \neq \{0\}$, $A(T) \subset D(T_e^{\dagger\dagger})$. Let T be the operator defined in Ex. 4. Then, by Ex. 4,

$$B(T) = D(T) \cap \overline{N(T)}$$
$$= D \cap (\overline{D \cap F})$$
$$= D \cap F$$
$$= N(T),$$

and

$$C(T) = \{0\},$$

showing that

$$A(T) \neq \{0\}, \qquad \text{by (82).}$$

Thus

$$A(T) = A \text{ of Ex. 4,}$$

and

$$D(T_e^\dagger) = A^\perp = N(T_e^\dagger).$$

Finally, from $C(T)^\perp = \mathcal{K}$,

$$D(T_e^{\dagger\dagger}) = \mathcal{K} \supset A$$

with

$$N(T_e^{\dagger\dagger}) = \mathcal{K}.$$

43. *An example of $A(T) \neq \{0\}$, $A(T) \cap D(T_e^{\dagger\dagger}) = \{0\}$.* Let $\mathcal{K}$ be a Hilbert space and let M, N be subspaces of $\mathcal{K}$ such that

$$M \neq \overline{M}, \qquad N \neq \overline{N} \subset M^\perp.$$

Choose

$$y \in \overline{M}/M \quad \text{and} \quad z \in M^\perp / \left(N \overset{\perp}{\oplus} (N^\perp \cap M^\perp) \right);$$

let

$$x = y + z$$

and

$$D = M \oplus N \oplus [x]$$

where $[x]$ is the line spanned by x. Define $T \in \mathcal{L}(\mathcal{K}, \mathcal{K})$ on $D(T) = D$ by

$$T(u + v + \alpha x) = v + \alpha x, \qquad u \in M, \quad v \in N, \quad \alpha x \in [x].$$

Then

$$C(T) = N, \qquad N(T) = M, \qquad A(T) = [x]$$

and

$$x \notin D(T_e^{\dagger\dagger}).$$

44. Let $T \in \mathcal{B}(\mathcal{H}_1, \mathcal{H}_2)$. Then

$$Tx = y \tag{72}$$

has an extremal solution if and only if there is a positive scalar β such that

$$|(y,z)|^2 \leqslant \beta(z, AA^*z), \qquad \text{for every } z \in N(AA^*)^{\perp}$$

(Tseng [5]; see also Holmes [1] Section 35).

45. Let $T \in \mathcal{B}(\mathcal{H}_1, \mathcal{H}_2)$, $S \in \mathcal{B}(\mathcal{H}_1, \mathcal{H}_3)$ be normally solvable, and let

$$T_S = T_{[N(S)]}$$

denote the restriction of T to $N(S)$. If T_S is also normally solvable, then $T_S^{\dagger}$ is called the *N(S)-restricted pseudoinverse of* T. It is the unique solution X of the following five equations:

$$SX = O,$$

$$XTX = X,$$

$$(TX)^* = TX,$$

$$TXT = T \qquad \text{on } N(S),$$

$$P_{N(S)}(XT)^* = XT \qquad \text{on } N(S) \qquad \text{(Minamide and Nakamura [2])}.$$

46. Let T, S, and $T_S^{\dagger}$ be as in Ex. 45. Then for any $y_0 \in \mathcal{H}_2$ and $z_0 \in R(S)$, the least extremal solution of

$$Tx = y_0$$

subject to

$$Sx = z_0$$

is given by

$$x_0 = T_S^{\dagger}(y_0 - TS^{\dagger}z_0) + S^{\dagger}z_0 \qquad \text{(Minamide and Nakamura [2])}.$$

47. Let $\mathcal{H}_1, \mathcal{H}_2, \mathcal{H}_3$ be Hilbert spaces, let $T \in \mathcal{B}(\mathcal{H}_1, \mathcal{H}_2)$ with $R(T) = \mathcal{H}_2$ and let $S \in \mathcal{B}(\mathcal{H}_1, \mathcal{H}_3)$. For any $y \in \mathcal{H}_2$, there is a unique $x_0 \in \mathcal{H}_1$

satisfying

$$Tx = y \qquad (72)$$

and which minimizes the functional

$$\|Sx\|^2 + \|x\|^2$$

over all solutions of (72). This x_0 is given by

$$x_0 = (I + S^*S)^{-1} T^\dagger y_0,$$

where y_0 is the unique vector in $\mathcal{H}_2$ satisfying

$$y = T(I + S^*S)^{-1} T^\dagger y_0$$

(Porter and Williams [2], Theorem 7).

48. Let $\mathcal{H}_1, \mathcal{H}_2, \mathcal{H}_3, T$, and S be as above. Then for any $y \in \mathcal{H}_2$, $x_1 \in \mathcal{H}_1$, and $y_1 \in \mathcal{H}_2$ there is a unique $x_0 \in \mathcal{H}_1$ which is a solution of

$$Tx = y \qquad (72)$$

and which minimizes

$$\|Sx - y_1\|^2 + \|x - x_1\|^2$$

from among all solutions of (72). This x_0 is given by

$$x_0 = (I + S^*S)^{-1}(T^\dagger y_0 + x_0 + S^* y_1),$$

where y_0 is the unique vector in $\mathcal{H}_2$ satisfying

$$y = T(I + S^*S)^{-1}(T^\dagger y_0 + x_1 + S^* y_1)$$

(Porter and Williams [2], Theorem 8).

5. SERIES AND INTEGRAL REPRESENTATIONS AND ITERATIVE COMPUTATION OF GENERALIZED INVERSES

Direct computational methods, in which the exact solution requires a ꞓnite number of steps (such as the elimination methods of Section 7.2–7.4) cannot be used, in general, for the computation of generalized inverses of operators. The exceptions are operators with nice algebraic properties, such as the integral and differential operators of Exs. 18–36 with their finite-dimensional null spaces. In the general case, the only computable representations of generalized inverses involve infinite series, or integrals, approximated by suitable iterative methods. Such representations and methods are sampled in this section, based on Showalter and Ben-Israel [1], where the proofs, omitted here, can be found.

To motivate the idea behind our development consider the problem of minimizing

$$f(x) = (Ax - y, Ax - y),\qquad(85)$$

where $A \in \mathfrak{B}\,(\mathfrak{K}_1, \mathfrak{K}_2)$ and $\mathfrak{K}_1, \mathfrak{K}_2$ are Hilbert spaces.

Treating x as a function $x(t)$, $t \geqslant 0$, with $x(0) = 0$, we differentiate (85):

$$\frac{d}{dt}f(x) = 2\operatorname{Re}(Ax - y, A\dot{x}), \qquad \dot{x} = \frac{d}{dt}x$$

$$= 2\operatorname{Re}(A^*(Ax - y), \dot{x})\qquad(86)$$

and setting

$$\dot{x} = -A^*(Ax - y),\qquad(87)$$

it follows from (86) that

$$\frac{d}{dt}f(x) = -2\|A^*(Ax - y)\|^2 < 0.\qquad(88)$$

This version of the steepest descent method, given in Rosenbloom [1], results in $f(x(t))$ being a monotone decreasing function of t, asymptotically approaching its infimum as $t \to \infty$. We expect $x(t)$ to approach asymptotically $A^\dagger y$, so by solving (87)

$$x(t) = \int_0^t \exp[-A^*A(t - s)]A^*y\,ds\qquad(89)$$

and observing that y is arbitrary we get

$$A^\dagger = \lim_{t \to \infty} \int_0^t \exp[-A^*A(t - s)]A^*\,ds,\qquad(90)$$

which is the essence of Theorem 9.

Here as elsewhere in this section, the convergence is in the strong operator topology. Thus the limiting expression

$$A^\dagger = \lim_{t \to \infty} B(t) \qquad \text{or} \qquad B(t) \to A^\dagger \quad \text{as} \quad t \to \infty\qquad(91)$$

means that for all $y \in D(A^\dagger)$

$$A^\dagger y = \lim_{t \to \infty} B(t)y$$

in the sense that

$$\lim_{t \to \infty} \|(A^\dagger - B(t))y\| = 0. \tag{92}$$

A numerical integration of (87) with suitably chosen step size similarly results in

$$A^\dagger = \sum_{k=0}^{\infty} (I - \alpha A^*A)^k \alpha A^*, \tag{93}$$

where

$$0 < \alpha < \frac{2}{\|A\|^2}, \tag{94}$$

which is the essence of Theorem 10.

In statements like (92) it is necessary to distinguish between points $y \in \mathcal{K}_2$ relative to the given $A \in \mathcal{B}(\mathcal{K}_1, \mathcal{K}_2)$. Indeed, the three cases

$$P_{\overline{R(A)}}y \in R(AA^*), \qquad P_{\overline{R(A)}}y \in (R(A)/R(AA^*)),$$

$$P_{\overline{R(A)}}y \in \left(\overline{R(A)}/R(A)\right)$$

have different rates of convergence in (92). Here $x \in (X/Y)$ means $x \in X$, $x \notin Y$. We abbreviate these as follows:

$$(y \in \mathrm{I}) \qquad \text{means} \qquad P_{\overline{R(A)}}y \in R(AA^*),$$

$$(y \in \mathrm{II}) \qquad \text{means} \qquad P_{\overline{R(A)}}y \in (R(A)/R(AA^*)), \tag{95}$$

$$(y \in \mathrm{III}) \qquad \text{means} \qquad P_{\overline{R(A)}}y \in \left(\overline{R(A)}/R(A)\right).$$

We note that $A^\dagger y$ is not defined for $(y \in \mathrm{III})$, a case which does not exist if $R(A)$ is closed.

THEOREM 9   (Showalter and Ben-Israel [1]).   Let $A \in \mathcal{B}(\mathcal{K}_1, \mathcal{K}_2)$ and define, for $t \geqslant 0$

$$L_1(t) = \int_0^t \exp[-A^*A(t-s)]ds,$$

$$L_2(t) = \int_0^t \exp[-AA^*(t-s)]ds, \tag{96}$$

$$B(t) = L_1(t)A^* = A^*L_2(t).$$

Then:

(a) $\|(A^\dagger - B(t))y\|^2 \leqslant \dfrac{\|A^\dagger y\|^2 \|(AA^*)^\dagger y\|^2}{\|(AA^*)^\dagger y\|^2 + 2\|A^\dagger y\|^2 t}$ if $(y \in I)$ and $t \geqslant 0$.

(b) $\|(A^\dagger - B(t))y\|^2$

is a decreasing function of $t \geqslant 0$, with limit zero as $t \to \infty$, if $(y \in II)$.

(c) $\|(P_{\overline{R(A)}} - AB(t))y\|^2 \leqslant \dfrac{\|y\|^2 \|A^\dagger y\|^2}{\|A^\dagger y\|^2 + 2\|y\|^2 t}$ if $(y \in I)$

$\qquad\qquad\qquad\qquad\qquad\qquad$ or $(y \in II)$, and $t \geqslant 0$.

(d) $\|(P_{\overline{R(A)}} - AB(t))y\|^2$

is a decreasing function of $t \geqslant 0$, with limit zero as $t \to \infty$, if $(y \in III)$.   ∎
Note that even though $A^\dagger y$ is not defined for $(y \in III)$, still

$$AB(t) \to P_{\overline{R(A)}} \qquad \text{as } t \to \infty.$$

The discrete version of Theorem 9 is the following theorem.

THEOREM 10   (Showalter and Ben-Israel [1]).   Let $A \in \mathscr{B}(\mathscr{K}_1, \mathscr{K}_2)$, let
c be a real number, $0 < c < 2$, and let

$$\alpha = \frac{c}{\|A\|^2}.$$

For any $y \in \mathscr{K}_2$ define

$$x = T^\dagger y \qquad \text{if } (y \in I) \text{ or } (y \in II)$$

and define the sequences $\{y_N\}, \{x_N\}$ by

$$y_0 = 0, \qquad x_0 = 0,$$

$$(y - y_{N+1}) = (I - \alpha AA^*)(y - y_N) \qquad \text{if } (y \in I) \text{ or } (y \in II) \text{ or } (y \in III)$$

$$(x - x_{N+1}) = (I - \alpha A^*A)(x - x_N) \qquad \text{if } (y \in I) \text{ or } (y \in II),$$

$$N = 1, 2, \ldots.$$

Then the sequence

$$B_N = \sum_{k=0}^{N} (I - \alpha A^*A)^k \alpha A^*, \qquad N = 0, 1, \ldots \tag{97}$$

converges to $A^\dagger$ as follows:

(a) $\|(A^\dagger - B_N)y\|^2 \leqslant \dfrac{\|A^\dagger y\|^2 \|(AA^*)^\dagger y\|^2}{\|(AA^*)^\dagger y\|^2 + N[(2-c)c/\|A\|^2]\|A^\dagger y\|^2}$ if $(y \in \mathrm{I})$

and $N = 1, 2, \ldots$.

(b) $\|(A^\dagger - B_N)y\|^2 = \|x - x_N\|^2$

converges monotonically to zero if $(y \in \mathrm{II})$.

(c) $\|(P_{\overline{R(A)}} - AB_N)y\|^2 \leqslant \dfrac{\|y\|^2 \|A^\dagger y\|^2}{\|A^\dagger y\|^2 + N[(2-c)c/\|A\|^2]\|y\|^2}$ if $(y \in \mathrm{I})$

or $(y \in \mathrm{II})$ and $N = 1, 2, \ldots$.

(d) $\|(P_{\overline{R(A)}} - AB_N)y\|^2 = \|y - y_N\|^2$

converges monotonically to zero if $(y \in \mathrm{III})$. ∎

The convergence $B_N \rightarrow A^\dagger$, in the uniform operator topology, was established by Petryshyn [1], restricting A to have closed range.

As in the matrix case, studied in Section 7.5, higher-order iterative methods are more efficient means of summing the series (93) than the first-order method (97). Two such methods, of order $p \geqslant 2$, are given in the following:

THEOREM 11   (Showalter and Ben-Israel [1]). Let A, α and $\{B_N : N = 0, 1, \ldots\}$ be as in Theorem 10. Let p be an integer

$$p \geqslant 2$$

and define the sequences $\{C_{N,p} : N = 0, 1, \ldots\}$ and $\{D_{N,p} : N = 0, 1, \ldots\}$ as follows:

$$C_{0,p} = \alpha A^*, \qquad C_{N+1,p} = C_{N,p} \sum_{k=0}^{p-1} (I - AC_{N,p})^k, \tag{98}$$

$$D_{0,p} = \alpha A^*, \qquad D_{N+1,p} = D_{N,p} \sum_{k=1}^{p} \binom{p}{k} (-AD_{N,p})^{k-1}. \tag{99}$$

Then, for all $N = 0, 1, \ldots$

$$B_{(p^{N+1}-1)} = C_{N+1,p} = D_{N+1,p}. \; \blacksquare \tag{100}$$

Consequently $\{C_{N,p}\}$ and $\{D_{N,p}\}$ are pth-order iterative methods for computing $A^\dagger$, with the convergence rates established in Theoem 10; e.g.,

$$\|(A^\dagger - C_{N,p})y\|^2 \leqslant \frac{\|A^\dagger y\|^2 \|(AA^*)^\dagger y\|^2}{\|(AA^*)^\dagger y\|^2 + (p^N - 1)[(2-c)c/\|A\|^2]\|A^\dagger y\|^2}$$

if $(y \in I)$ and $N = 1, 2, \ldots.$

The series (98) is somewhat simpler to use if the term $(I - AC_{N,p})^k$ can be evaluated by only $k-1$ operator multiplications, e.g. for matrices. The form (99) is preferable otherwise, e.g. for integral operators.

For other iterative methods and comprehensive bibliographies on the subject, see Kammerer and Nashed [1],[3] and [4] and Zlobec [4].

EXERCISES AND EXAMPLES

49. Let $A \in \mathfrak{B}(\mathcal{K}_1, \mathcal{K}_2)$ have closed range, let $b \in \mathcal{K}_2$ and let $B \in R(A^*, A^*).$* Then the sequence

$$x_{k+1} = x_k - B(Ax_k - b), \qquad k = 0, 1, \ldots \qquad (101)$$

converges to $A^\dagger b$ for all $x_0 \in R(A^*)$ if

$$\rho(P_{R(A^*)} - BA) < 1$$

where $\rho(T)$ denotes the spectral radius of T; see, e.g., Taylor [1], p. 262 (Zlobec [4]).

The choice $B = \alpha A^*$ in (101) reduces it to the iterative method (97). Other choices of B are given in the following exercise.

50. *Splitting methods.* Let A be as in Ex. 49, and write

$$A = M + N, \qquad (102)$$

where $M \in \mathfrak{B}(\mathcal{K}_1, \mathcal{K}_2)$ has closed range and $N(A) = N(M)$. Choosing

$$B = wM^\dagger, \qquad w \neq 0$$

in (101) gives

$$x_{k+1} = [(1-w)I - wM^\dagger N]x_k + wM^\dagger b, x_0 \in R(A^*), \qquad (103)$$

---

*For $S, T \in \mathfrak{B}(\mathcal{K}_1, \mathcal{K}_2)$ with closed ranges, $R(S, T) = \{Z : Z = SWT$ for some $W \in \mathfrak{B}(\mathcal{K}_2, \mathcal{K}_1)\}$.

in particular, for $w = 1$,

$$x_{k+1} = - M^\dagger N x_k + M^\dagger b, \qquad x_0 \in R(A^*). \qquad (104)$$

(Zlobec [4], Berman and Plemmons [1]).

SUGGESTED FURTHER READING

Section 1. For alternative or more general treatments of generalized inverses of operators see F. V. Atkinson ([1], [2]), Beutler ([1], [2]), Davis and Robinson [1], Hamburger [1], Hansen and Robinson [1], Hestenes [5], Holmes [1], Leach [1], Nashed ([2], [3]), Nashed and Votruba [1], Pietsch [1], Porter and Williams ([1], [2]), Przeworska-Rolewicz and Rolewicz [1], Sheffield [1], Votruba [1], Wyler [1], and Zarantonello [1].

Section 3. For integral equations see K. E. Atkinson [1], Courant and Hilbert [1], Kammerer and Nashed ([1], [2], [3]), Korganoff and Pavel-Parvu [1], Lonseth [1], and Rall [1].

For application to Wiener-Hopf operators see Lent [1].

For applications to differential operators see also Bradley ([1], [2]), Courant and Hilbert [1], Greub and Rheinboldt [1], Kallina [1], Locker [1], Tucker [1], and Wyler [2].

For application in bifurcation theory see Stakgold [1].

BIBLIOGRAPHY

Afriat, S. N.
[1] Orthogonal and oblique projectors and the characteristics of pairs of vector spaces, *Proc. Cambridge Phil. Soc.*, **53** (1957), 800–816.

Albert, A.
[1] *An Introduction and Beginner's Guide to Matrix Pseudo Inverses*, ARCON–Advanced Research Consultants, Lexington, Mass., 1964.
[2] Conditions for positive and nonnegative definiteness in terms of pseudo-inverses, *SIAM J. Appl. Math.*, **17**, (1969), 434–440.
[*3*] *Regression and the Moore–Penrose Pseudoinverse*, Academic Press, New York, 1972, xiii + 180 pp.
[4] The Gauss–Markov theorem for regression models with possibly singular covariances, *SIAM J. Appl. Math.*, **24** (1973), 182–187.

Albert, A.; Sittler, R. W.
[1] A method for computing least squares estimators that keep up with the data, *SIAM J. Control*, **3** (1966), 384–417.

Altman, M.
[1] A generalization of Newton's method, *Bull. Acad. Polon. Sci. Ser. Sci. Math. Astronom. Phys.*, **3** (1955), 189–193.
[2] On a generalization of Newton's method, *Bull. Acad. Polon. Sci. Ser. Sci. Math. Astronom. Phys.*, **5** (1957), 789–795.

Amir-Moéz, A. R.
[1] Geometry of generalized inverses, *Math. Mag.*, **43** (1970), 33–36.

Anderson, W. N. Jr.
[1] Shorted operators, *SIAM J. Appl. Math.*, **20** (1971), 520–525.

Anderson, W. N. Jr.; Duffin, R. J.
[1] Series and parallel addition of matrices, *J. Math. Anal. Appl.*, **26** (1969), 576–594.

Arghiriade, E.
[1] Sur les matrices qui sont permutables avec leur inverse généralisée, *Atti Accad. Naz. Lincei Rend. Cl. Sci. Fis. Mat. Natur. Ser. VIII*, **35** (1963), 244–251.
[2] Remarques sur l'inverse généralisée d'un produit de matrices, *Atti Accad. Naz. Lincei Rend. Cl. Sci. Fis. Mat. Natur. Ser. VIII*, **42** (1967), 621–625.
[3] Sur l'inverse généralisée d'un operateur lineaire dans les espaces de Hilbert, *Atti Accad. Naz. Lincei Rend. Cl. Sci. Fis. Mat. Natur. Ser. VIII*, **45** (1968), 471–477.

Arghiriade, E.; Dragomir, A.

[1] Remarques sur quelques théorèmes relatives à l'inverse généralisée d'un operatéur linéaire dans les espaces de Hilbert, *Atti Accad. Naz. Lincei Rend. Cl. Sci. Fis. Mat. Natur. Ser. VIII*, **46** (1969), 333–338.

Atkinson, F. V.

[1] The normal solubility of linear equations in normed spaces (Russian), *Mat. Sbornik N.S.*, **28(70)** (1951), 3–14 (*Math. Rev.*, **13** (1952), 46).

[2] On relatively regular operators, *Acta Sci. Math. Szeged*, **15** (1953), 38–56 (Reviewed in *Math. Rev.*, **15** (1954), 134–135).

Atkinson, K. E.

[1] The solution of non-unique linear integral equations, *Numer. Math.*, **10** (1967), 117–124.

Autonne, L.

[1] *Bull. Soc. Math. France*, **30** (1902), 121–133.

[2] Sur les matrices hypohermitiennes et sur les matrices unitaires, *Ann Univ. Lyon*, **38** (1917), 1–77.

Balakrishnan, A. V.

[1] An operator theoretic formulation of a class of control problems and a steepest descent method of solution, *J. Soc. Indust. Appl. Math. Ser. A: Control.*, **1** (1963), 109–127.

Banerjee, K. S.

[1] Singularity in Hotelling's weighing designs and generalized inverses, *Ann. Math. Statist.*, **37** (1966), 1021–1032. Erratum, *ibid.* **40** (1969), 719.

Banerjee, K. S.; Federer, W. T.

[1] On the structure and analysis of singular fractional replicates, *Ann. Math. Statist.*, **39** (1968), 657–663.

Barnett, S.

[1] *Matrices in Control Theory*, Van Nostrand Reinhold, London, 1971, xiv + 221 pp.

Baskett, Th.S.; Katz, I. J.

[1] Theorems on product of EP_r matrices, *Linear Algebra and Appl.*, **2** (1969), 87–103.

Bauer, F. L.

[1] Elimination with weighted row combinations for solving linear equations and least squares problems, *Numer. Math.*, **7** (1965), 338–352. Republished, pp. 119–133 in Wilkinson and Reinsch [1].

[2] Theory of norms, Tech. Report No. CS 75, Computer Science Dept., Stanford University, Stanford, California, 1967.

Beckenbach, E. F.; Bellman, R.

[1] *Inequalities* (3rd revised printing). Springer, New York, 1971, xi + 198 pp.

Bellman, R.

[1] *Introduction to Matrix Analysis*, 2nd edition, McGraw-Hill Book Co., New York, 1970, xxiii + 403 pp.

Beltrami, E. J.

[1] A constructive proof of the Kuhn-Tucker multiplier rule, *J. Math. Anal. Appl.*, **26** (1969), 297–306.

Ben-Israel, A.

[1] On direct sum decompositions of Hestenes algebras, *Israel J. Math.*, **2** (1964), 50–54.

[2] A modified Newton-Raphson method for the solution of systems of equations, *Israel J. Math.*, **3** (1965), 94–98.

[3] A Newton-Raphson method for the solution of systems of equations, *J. Math. Anal. Appl.*, **15** (1966), 243–252.

[4] A note on an iterative method for generalized inversion of matrices, *Math. Comp.*, **20** (1966), 439–440.
[5] On error bounds for the generalized inverse, *SIAM J. Numer. Anal.*, **3** (1966), 582–592.
[6] A note on the Cayley transform, *Notices Amer. Math. Soc.*, **13** (1966), 599.
[7] On the geometry of subspaces in Euclidean *n*-spaces, *SIAM J. Appl. Math.*, **15** (1967), 1184–1198.
[8] On iterative methods for solving nonlinear least squares problems over convex sets, *Israel J. Math.*, **5** (1967), 211–224.
[9] On applications of generalized inverses in nonlinear analysis, pp. 183–202 in Boullion and Odell [1].
[10] On matrices of index zero or one, *SIAM J. Appl. Math.*, **17** (1968), 1118–1121.
[11] On optimal solutions of 2-person 0-sum games, *Atti Accad. Naz. Lincei Rend. Cl. Sci. Fis. Mat. Natur. Ser. VIII*, **44** (1968), 274–278.
[12] On decompositions of matrix spaces with applications to matrix equations, *Atti Accad. Naz. Lincei Rend. Cl. Sci. Fis. Mat. Natur. Ser. VIII*, **45** (1968), 54–60.
[13] A note on partitioned matrices and equations, *SIAM Review*, **11** (1969), 247–250.
[14] On Newton's method in nonlinear programming, pp. 339–352 in *Proc. Princeton Sympos. Math. Prog.* (H. W. Kuhn, Editor). Princeton University Press, Princeton, N. J., 1970, vi+620 pp.

Ben-Israel, A.; Charnes, A.
[1] Contributions to the theory of generalized inverses, *J. Soc. Indust. Appl. Math.*, **11** (1963), 667–699.
[2] Generalized inverses and the Bott-Duffin network analysis, *J. Math. Anal. Appl.*, **7** (1963), 428–435. Erratum, *ibid.* **18** (1967), 393.
[3] An explicit solution of a special class of linear programming problems, *Operations Research*, **16** (1968), 1167–1175.

Ben-Israel, A.; Charnes, A.; Robers, P. D.
[1] On generalized inverses and interval linear programming, pp. 53–70 in Boullion and Odell [1].

Ben-Israel, A.; Cohen, D.
[1] On iterative computation of generalized inverses and associated projections, *SIAM J. Numer. Anal.*, **3** (1966), 410–419.

Ben-Israel, A.; Kirby, M. J. L.
[1] A characterization of equilibrium points of bimatrix games, *Atti Accad. Naz. Lincei Rend. Cl. Sci. Fis. Mat. Natur. Ser. VIII*, **46** (1969), 196–201.

Berman, A.; Plemmons, R. J.
[1] Monotonicity and the generalized inverse, *SIAM J. Appl. Math.*, **22** (1972), 155–161.

Beutler, F. J.
[1] The operator theory of the pseudo inverse I. Bounded operators, *J. Math. Anal. Appl.*, **10** (1965), 451–470.
[2] The operator theory of the pseudo-inverse II. Unbounded operators with arbitrary range, *J. Math. Anal. Appl.*, **10** (1965), 471–493.

Bjerhammar, A.
[1] Rectangular reciprocal matrices, with special reference to geodetic calculations, *Bull. Géodésique* (1951), 188–220.
[2] Applications of calculus of matrices to method of least squares; with special reference to geodetic calculations, *Kungl. Tekn. Högsk. Handl. No. 49* (1951), 36 pp.
[3] A generalized matrix algebra, *Kungl. Tekn. Högsk. Handl. No. 124* (1968), 32 pp.

Björk, Å.
[1] Solving linear least squares problems by Gram-Schmidt orthogonalization, *BIT*, **7** (1967), 1–21.
[2] Iterative refinement of linear least squares solutions I, *BIT*, **7** (1967), 257–278.
[3] Iterative refinement of linear least squares solutions II, *BIT*, **8** (1968), 8–30.
Björk, Å.; Golub, G. H.
[1] Iterative refinement of linear least squares solutions by Householder transformation, *BIT*, **7** (1967), 322–337.
Blattner, J. W.
[1] Bordered matrices, *J. Soc. Indust. Appl. Math.*, **10** (1962), 528–536.
Bohnenblust, F.
[1] A characterization of complex Hilbert spaces, *Portugal. Math.*, **3** (1942), 103–109.
Bonnesen, T.; Fenchel, W.
[1] *Theorie der konvexen Korper*, Springer, Berlin, 1934, vii + 164 pp.
deBoor, C.
[1] The method of Projections as applied to the Numerical Solution of Two Point Boundary Value Problems using Cubic Splines, doctoral disseration in mathematics, University of Michigan, 1966.
Bott, R.; Duffin, R. J.
[1] On the algebra of networks, *Trans. Amer. Math. Soc.*, **74** (1953), 99–109.
Boullion, T. L.; Odell, P. L.
[1] *Proceedings of the Symposium on Theory and Applications of Generalized Inverses of Matrices*, held at the Department of Mathematics, Texas Technological College, Lubbock, Texas, March, 1968. Texas Tech. Press, Lubbock, Texas, 1968, iii + 315 pp.
[2] *Generalized Inverse Matrices*, Wiley-Interscience, New York, 1971, x + 103 pp.
Bounitzky, E.
[1] Sur la fonction de Green des éqations differentielles linéaires ordinaires, *J. Math. Pures Appl. (6)*, **5** (1909), 65–125.
Bourbaki, N.
[1] *Eléments de Mathématique. Livre V. Espaces Vectoriels Topologiques*, Hermann & Cie, Paris, 1953.
[2] *Eléments de Mathématique. Livre II. Algèbre*, Hermann & Cie, Paris, 1958.
Bowdler, H. J.; Martin, R. S.; Peters, G.; Wilkinson, J. H.
[1] Solution of real and complex systems of linear equations, *Numer. Math.*, **8** (1966), 217–239. Republished, pp. 93–110 in Wilkinson and Reinsch [1].
Bowman, V. J.; Burdet C.-A.
[1] On the general solution to systems of mixed-integer linear equations, *SIAM J. Appl. Math.*, **26** (1974), 120–125.
Bradley, J. S.
[1] Generalized Green's matrices for compatible differential systems, *Michigan Math. J.*, **13** (1966), 97–108.
[2] Adjoint quasi-differential operators of Euler type, *Pacific J. Math.*, **16** (1966), 213–237.
Brand, L.
[1] The solution of linear algebraic equations, *Math. Gaz.*, **46** (1962), 203–207.
den Broeder, C. G. Jr.; Charnes, A.
[1] Contributions to the theory of generalized inverses for matrices, Purdue University, Lafayette, Ind., 1957. (Reprinted as *ONR Res. Memo. No. 39*, Northwestern University, Evanston, Illinois, 1962.)

Burmeister, W.
[1] Inversionsfreie Verfahren zur Lösung nichtlinearer Operatorgleichungen, *Zeit. angew. Math. Mech.*, **52** (1972), 101–110.

Burns, F.; Carlson, D.; Haynsworth, E.; Markham, T.
[1] Generalized inverse formulas using the Schur complement, *SIAM J. Appl. Math.*, **26** (1974), 254–259.

Businger, P. A.; Golub, G. H.
[1] Linear least squares by Householder transformations, *Numer. Math.*, **7** (1965), 269–276. Republished, pp. 111–118 in Wilkinson and Reinsch [1].
[2] Algorithm 358: Singular value decomposition of a complex matrix, *Comm. ACM*, **12** (1969), 564–565.

Carlson, D.; Haynsworth, E.; Markham, T.
[1] A generalization of the Schur complement by means of the Moore–Penrose inverse, *SIAM J. Appl. Math.*, **26** (1974), 169–175.

Charnes, A.; Cooper, W. W.
[1] Structural sensitivity analysis in linear programming and an exact product form left inverse, *Naval Res. Logist. Quart.*, **15** (1968), 517–522.

Charnes, A.; Cooper, W. W.; Thompson, G. L.
[1] Constrained generalized medians and hypermedians as deterministic equivalents for two-stage linear programs under uncertainty, *Managment Sci.*, **12** (1965), 83–112.

Charnes, A.; Granot, F.
[1] Existence and representation of Diophantine and mixed Diophantine solutions to linear equations and inequalities, Center for Cybernetic Studies, The University of Texas, Austin, Texas, January 1973.

Charnes, A.; Kirby, M.
[1] Modular design, generalized inverses and convex programming, *Operations Res.*, **13** (1965), 836–847.

Cheney, E. W.
[1] *Introduction to Approximation Theory*, McGraw-Hill Book Company, New York, 1966, xii + 259 pp.

Chernoff, H.
[1] Locally optimal designs for estimating parameters, *Ann. Math. Statist.*, **24** (1953), 586–602.

Chipman, J. S.
[1] On least squares with insufficient observations, *J. Amer. Statist. Assoc.*, **54** (1964), 1078–1111.
[2] Specification problems in regression analysis, pp. 114–176 in Boullion and Odell [1].

Chipman, J. S.; Rao, M. M.
[1] Projections, generalized inverses and quadratic forms, *J. Math. Anal. Appl.*, **9** (1964), 1–11.
[2] On the treatment of linear restrictions in regression analysis, *Econometrica*, **32** (1964), 198–209.

Clarkson, J. A.
[1] Uniformly convex spaces, *Trans. Amer. Math. Soc.*, **40** (1936), 396–414.

Cline, R. E.
[1] Representations for the generalized inverse of a partitioned matrix, *J. Soc. Indust. Appl. Math.*, **12** (1964), 588–600.
[2] Representations for the generalized inverse of sums of matrices, *SIAM J. Numer. Anal.*, **2** (1965), 99–114.
[3] Inverses of rank invariant powers of a matrix, *SIAM J. Numer. Anal.*, **5** (1968), 182–197.

Cline, R. E.; Greville, T. N. E.
[1] An extension of the generalized inverse of a matrix, *SIAM J. Appl. Math.*, **19** (1970), 682–688.

Cline, R. E.; Pyle, L. D.
[1] The generalized inverse in linear programming. An intersection projection method and the solution of a class of structured linear programming problems, *SIAM J. Appl. Math.*, **24** (1973), 338–351.

Coddington, E. A.; Levinson, N.
[1] *Theory of Ordinary Differential Equations*, McGraw-Hill Book Co., New York 1955, xii + 429 pp.

Courant, R.; Hilbert, D.
[1] *Methoden der Mathematischen Physik*, Vol I, Springer, Berlin, 1931.

Cudia, D. F.
[1] Rotundity, pp. 73–97 in *Convexity, Proc. Sympos. Pure Math. Vol. VII* (V. Klee, Editor), Amer. Math. Soc., Providence, R.I., 1963, xv + 516 pp.

Davis, D. L.; Robinson, D. W.
[1] Generalized inverses of morphisms, *Linear Algebra and Appl.*, **5** (1972), 329–338.

Decell, H. P. Jr.
[1] An application of the Cayley-Hamilton theorem to generalized matrix inversion, *SIAM Review*, **7** (1965), 526–528.

Dennis, J. B.
[1] *Mathematical Programming and Electrical Networks*, M. I. T. Press, Cambridge, Mass., 1959, 186 pp.

Desoer, C. A.; Whalen, B. H.
[1] A note on pseudoinverses, *J. Soc. Indust. Appl. Math.*, **11** (1963), 442–447.

Deutsch, E.
[1] Semi-inverses, reflexive semi-inverses, and pseudoinverses of an arbitrary linear transformation, *Linear Algebra and Appl.*, **4** (1971), 313–322.

Drazin, M. P.
[1] Pseudo inverses in associative rings and semigroups, *Amer. Math. Monthly*, **65** (1958), 506–514.

Drygas, H.
[1] Consistency of the least squares and Gauss-Markov estimators in regression models, *Z. Wahrscheinlichkeitstheorie und verw. Gebiete*, **17** (1971), 309–326.
[2] *The Coordinate-Free Approach to Gauss-Markov Estimation.* Springer-Verlag, Berlin, 1970, viii + 113 pp.

Duffin, R. J.
[1] Network models, pp. 65–91 in *Mathematical Aspects of Electrical Network Analysis SIAM-AMS Proc. Vol. III* (Wilf, H. S.; Harary, F., Editors), Amer. Math. Soc., Providence, R.I., 1971, 206 pp.

Dunford, N.; Schwartz, J. T.
[1] *Linear Operators Part I*, Interscience, New York, 1957, xiv + 858 pp.

Duris, C. S.
[1] Optimal quadrature formulas using generalized inverses. Part I. General theory and minimum variance formulas, *Math. Comp.*, **25** (1971), 495–504.
[2] Optimal quadrature formulas using generalized inverses. Part II. Sard "best" formulas, Math. Rpt. No. 70-17, Dept. of Math., Drexel Univ., 1970.

Eckart, C.; Young, G.
[1] The approximation of one matrix by another of lower rank, *Psychometrika*, **1** (1936), 211–218.
[2] A principal axis transformation for non-Hermitian matrices, *Bull. Amer. Math. Soc.*, **45** (1939), 118–121.

Elliott, W. W.
[1] Generalized Green's functions for compatible differential systems, *Amer. J. Math.*, **50** (1928), 243–258.
[2] Green's functions for differential systems containing a parameter, *Amer. J. Math.*, **51** (1929), 397–416.

Englefield, M. J.
[1] The commuting inverses of a square matrix, *Proc. Cambridge Philos. Soc.*, **62** (1966), 667–671.

Erdelsky, P. J.
[1] *Projections in a Normed Linear Space and a Generalization of the Pseudo-Inverse*, doctoral dissertation in mathematics, California Institute of Technology, Pasadena, 1969.

Erdelyi, I.
[1] On partial isometries in finite dimensional Euclidean spaces, *SIAM J. Appl. Math.*, **14** (1966), 453–467.
[2] On the "reverse order law" related to the generalized inverse of matrix products, *J. Assoc. Comput. Mach.*, **13** (1966), 439–443.
[3] The quasi-commuting inverses for a square matrix, *Atti. Accad. Naz. Lincei Rend. Cl. Sci. Fis. Mat. Natur. Ser. VIII*, **42** (1967), 626–633.
[4] On the matrix equation $Ax = \lambda Bx$, *J. Math. Anal. Appl.*, **17** (1967), 119–132.
[5] Partial isometries closed under multiplication on Hilbert spaces, *J. Math. Anal. Appl.*, **22** (1968), 546–551.
[6] Normal partial isometries closed under multiplication on unitary spaces, *Atti. Accad. Naz. Lincei Rend. Cl. Sci. Fix. Mat. Natur. Ser. VIII*, **43** (1968), 186–190.
[7] Partial isometries defined by a spectral property on unitary spaces, *Atti. Accad. Naz. Lincei Rend. Cl. Sci. Fix. Mat. Natur. Ser. VIII*, **44** (1968), 741–747.
[8] Partial isometries and generalized inverses, pp. 203–217 in Boullion and Odell [1].
[9] A generalized inverse for arbitrary operators between Hilbert spaces, *Proc. Cambridge Philos. Soc.*, **71** (1972), 43–50.

Erdelyi, I.; Ben-Israel, A.
[1] Extremal solutions of linear equations and generalized inversion between Hilbert spaces, *J. Math. Anal. Appl.*, **39** (1972), 298–313.

Erdelyi, I.; Miller, F. R.
[1] Decomposition theorems for partial isometries, *J. Math. Anal. Appl.*, **30** (1970), 665–679.

Faddeev, D. K.; Kublanovskaja, V. N.; Faddeeva, V. N.
[1] Linear algebraic systems with rectangular matrices (Russian), *Modern Numerical Methods, No. 1, Computational Methods of Linear Algebra* (Proceedings of the International Summer School on Numerical Methods, Kiev, 1966) (*Math. Rev.*, **39** (1970), #6887).

Fan, Ky
[1] On a theorem of Weyl concerning eigenvalues of linear transformations. I, *Proc. Nat. Acad. Sci. USA*, **35** (1949), 652–655.
[2] On a theorem of Weyl concerning eigenvalues of linear transformations, II, *Proc. Nat. Acad. Sci. USA*, **36** (1950), 31–35.

Fan, Ky; Hoffman, A.
[1] Some metric inequalities in the space of matrices, *Proc. Amer. Math. Soc.*, **6** (1955), 111–116.

Fisher, A. G.
[1] On construction and properties of the generalized inverse, *SIAM J. Appl. Math.*, **15** (1967), 269–272.

Fletcher, R.
[1] Generalized inverse methods for the best least-squares solution of systems of nonlinear equations, *Computer J.*, **10** (1968), 392–399.
[2] A technique for orthogonalization, *J. Inst. Math. Appl.*, **5** (1969), 162–166.
[3] Generalized inverses for nonlinear equations and optimization, pp. 75–85 in Rabinowitz [1].

Forsythe, G. E.
[1] The maximum and minimum of a positive definite quadratic polynomial on a sphere are convex functions of the radius, *SIAM J. Appl. Math.*, **19** (1970), 551–554.

Forsythe, G. E.; Golub, G. H.
[1] On the stationary values of a second-degree polynomial on the unit sphere, *SIAM J. Appl. Math.*, **13** (1965), 1050–1068.

Foulis, D. J.
[1] Relative inverses in Baer*-semigroups, *Michigan Math. J.*, **10** (1963), 65–84.

Frame, J. S.
[1] Matrix functions and applications. I. Matrix operations and generalized inverses, *IEEE Spectrum*, **1** (1964), 209–220.

Franck, P.
[1] Sur la distance minimale d'une matrice régulière donnée au lieu des matrices singulières, pp. 55–60 in *Deux. Congr. Assoc. Franc. Calcul. et Trait. Inform. Paris 1961*, Gauthier-Villars, Paris, 1962 (*Math. Rev.*, **29** (1965), #2953).

Fredholm, I.
[1] Sur une classe d'équations fonctionnelles, *Acta Math.*, **27** (1903), 365–390.

Gabriel, R.
[1] Extension of generalized algebraic complement to arbitrary matrices (Romanian), *Stud. Cerc. Mat.*, **17** (1965), 1567–1581 (*Math. Rev.*, **35** (1968), #6703).
[2] Das verallgemeinerte Inverse einer Matrix deren Elemente einem beliebigen Körper angehören, *J. Reine Angew. Math.*, **234** (1969), 107–122 (*Math. Rev.*, **41** (1971), #1753).
[3] Das verallgemeinerte Inverse einer Matrix über einem beliebigen Körperanalytisch betrachtet, *J. Reine Angew. Math.*, **244** (1970), 83–93.

Gaches, J.; Rigal, J-L.; Rousset de Pina, X.
[1] Distance euclidienne d'une application linéaire σ au lieu des applications de rang r donné. Détermination d'une meilleure approximation de rang r, *C. R. Acad. Sci. Paris*, **260** (1965), 5672–5674.

Gantmacher, F. R.
[1] *The Theory of Matrices*, Vols. I and II, Chelsea, New York, 1959.

Garnett, J. M. III; Ben-Israel, A.; Yau, S. S.
[1] A hyperpower iterative method for computing matrix products involving the generalized inverse, *SIAM J. Numer. Anal.*, **8** (1971), 104–109.

Germain-Bonne, B.
[1] Calcul de pseudo-inverses, *Rev. Française Informat. Recherche Opérationelle*, **3** (1969), 3–14.

Glazman, I. M.; Ljubich, Ju. I.
[1] *Finite Dimensional Linear Analysis* (Russian), Nauka, Moscow, 1969. English translation to be published by M. I. T. Press.

Goldberg, S.
[1] *Unbounded Linear Operators*, McGraw-Hill Book Co., New York, 1966, viii + 199 pp.

Goldman, A. J.; Zelen, M.
[1] Weak generalized inverses and minimum variance linear unbiased estimation, *J. Res. Nat. Bur. Standards Sect. B*, **68B** (1964), 151–172.

Goldstein, A. A.
[1] *Constructive Real Analysis*, Harper and Row, New York, 1967, xii + 178 pp.

Golub, G. H.
[1] Numerical methods for solving linear least squares problems, *Numer. Math.*, **7** (1965), 206–216.
[2] Least squares singular values and matrix approximations, *Aplikace Mathematiky*, **13** (1968), 44–51.
[3] Matrix decompositions and statistical calculations, Tech. Report No. CS 124 Computer Science Dept., Stanford University, March 10, 1969.

Golub, G. H.; Kahan, W.
[1] Calculating the singular values and pseudo-inverse of a matrix, *SIAM J. Numer. Anal.*, **2** (1965), 205–224.

Golub, G. H.; Pereyra, V.
[1] The differentiation of pseudoinverses and nonlinear least squares problems whose variables separate, *SIAM J. Numer. Anal.* **10** (1973), 413–432.

Golub, G. H.; Reinsch, C.
[1] Singular value decomposition and least squares solutions, *Numer. Math.*, **14** (1970), 403–420. Republished, pp. 134–151 in Wilkinson and Reinsch [1].

Golub, G. H.; Styan, G. P. H.
[1] Numerical computations for univariate linear models, STAN-CS-236-71, Computer Science Dept., Stanford University, September, 1971.

Golub, G. H.; Wilkinson, J. H.
[1] Note on the iterative refinement of least squares solution, *Numer. Math.*, **9** (1966), 139–148.

Good, I. J.
[1] Some applications of the singular decomposition of a matrix, *Technometrics*, **11** (1969), 823–831.

Graybill, F. A.; Marsaglia, G.
[1] Idempotent matrices and quadratic forms in the general linear hypothesis, *Ann. Math. Statist.*, **28** (1957), 678–686.

Graybill, F. A.; Meyer, C. D. Jr.; Painter, R. J.
[1] Note on the computation of the generalized inverse of a matrix, *SIAM Rev.*, **8** (1966), 522–524.

Green, B.
[1] The orthogonal approximation of an oblique structure in factor analysis, *Psychometrika*, **17** (1952), 429–440.

Greub, W.; Rheinboldt, W. C.
[1] Non self-adjoint boundary value problems in ordinary differential equations, *J. Res. Nat. Bur. Standards Sect. B.*, **64B** (1960), 83–90.

Greville, T. N. E.

[1] On smoothing a finite table: A matrix approach, *J. Soc. Indust. Appl. Math.*, **5** (1957), 137–154.

[2] The pseudoinverse of a rectangular or singular matrix and its application to the solution of systems of linear equations, *SIAM Review*, **1** (1959), 38–43.

[3] Some applications of the pseudoinverse of a matrix, *SIAM Review*, **2** (1960), 15–22.

[4] Note on fitting functions of several independent variables, *J. Soc. Indust. Appl. Math.*, **9** (1961), 109–115. Erratum, *ibid.* **9** (1961), 317.

[5] Note on the generalized inverse of a matrix product, *SIAM Review*, **8** (1966), 518–521.

[6] Spectral generalized inverses of square matrices, MRC Technical Summary Report #823, Mathematics Research Center, University of Wisconsin, Madison, Wisc., October, 1967.

[7] Some new generalized inverses with spectral properties, pp. 26–46 in Boullion and Odell [1].

[8] Solutions of the matrix equations $XAX = X$ and relations between oblique and orthogonal projectors, *SIAM J. Appl. Math.*, **26** (1974).

Guillemin, E. A.

[1] *Theory of linear physical systems*, Wiley, New York (1963), xvii + 586 pp.

Halmos, P. R.

[1] *Finite-dimensional vector spaces*, (2nd edition), D. Van-Nostrand Co., Princeton, N. J., 1958, vii + 195 pp.

[2] *A Hilbert Space Problem Book*, D. Van-Nostrand Co., Princeton, N. J., 1967, xvii + 356 pp.

Halmos, P. R.; McLaughlin, J. E.

[1] Partial isometries, *Pacific J. Math*, **13** (1963), 585–596.

Halmos, P. R.; Wallen, L. J.

[1] Powers of partial isometries, *J. Math. Mech.*, **19** (1970), 657–663.

Halperin, I.

[1] Closures and adjoints of linear differential operators, *Ann. of Math.*, **38** (1937), 880–919.

Hamburger, H.

[1] Non-symmetric operators in Hilbert space, pp. 67–112 in *Proceedings Symposium on Spectral Theory and Differential Problems*, Oklahoma A & M College, Stillwater, Oklahoma, 1951.

Hansen, G. W.; Robinson, D. W.

[1] On the existence of generalized inverses, *Linear Algebra and Appl.*, **8** (1974), 95–104.

Hartwig, R. E.

[1] 12 inverses and the invariance of $BA^\dagger C$, to appear in *Linear Algebra and Appl.*

[2] Singular values and *g*-inverses of bordered matrices (to appear).

[3] Rank factorization and *g*-inversion (to appear).

[4] Block generalized inverses to appear, in *Arch. Rational Mech. Anal.*

Harwood, W. R.; Lovass-Nagy, V.; Powers, D. L.

[1] A note on the generalized inverses of some partitioned matrices, *SIAM J. Appl. Math.*, **19** (1970), 555–559.

Hawkins, J. B.; Ben-Israel, A.

[1] On generalized matrix functions, *Linear and Multilinear Algebra*, **1** (1973), 163–171.

Hearon, J.Z.

[1] On the singularity of a certain bordered matrix, *SIAM J. Appl. Math.*, **15** (1967), 1413–1421.

[2] Polar factorization of a matrix, *J. Res. Nat. Bur. Standards Sect. B.*, **71B** (1967), 65–67.

[3] Partially isometric matrices, *J. Res. Nat. Bur. Standards Sect. B.*, **71B** (1967), 225–228.

[4] Generalized inverses and solutions of linear systems, *J. Res. Nat. Bur. Standards Sect. B.*, **72B** (1968), 303–308.

Hearon, J. Z.; Evans, J. W.

[1] On spaces and maps of generalized inverses, *J. Res. Nat. Bur. Standards Sect. B.*, **72B** (1968), 103–107.

[2] Differentiable generalized inverses, *J. Res. Nat. Bur. Standards Sect. B.*, **72B** (1968), 109–113.

Hestenes, M. R.

[1] Quadratic forms in Hilbert space, *Pacific J. Math.*, **1** (1951), 525–581.

[2] Inversion of matrices by biorthogonalization and related results, *J. Soc. Indust. Appl. Math.*, **6** (1958), 51–90.

[3] Relative Hermitian matrices, *Pacific J. Math.*, **11** (1961), 225–245.

[4] Relative self-adjoint operators in Hilbert space, *Pacific J. Math.*, **11** (1961), 1315–1357.

[5] A ternary algebra with applications to matrices and linear transformations, *Arch. Rational Mech. Anal.*, **11** (1962), 138–194.

Hilbert, D.

[1] *Grundzüge einer allgemeinen Theorie der linearen Integralgleichungen*, B. G. Teubner, Leipzig and Berlin, 1912, xxvi + 282 pp. Reprint of six articles which appeared originally in the *Göttingen Nachrichten* (1904), 49–51; (1904), 213–259; (1905), 307–338; (1906), 157–227; (1906), 439–480; (1910), 355–417).

Ho, B. L.; Kalman, R. E.

[1] Effective construction of linear state-variables models from input/output functions, *Regelungstechnik*, **14** (1966), 545–548.

Holmes, R. B.

[1] *A Course on Optimization and Best Approximation*, Springer-Verlag, Berlin, 1972, viii + 233 pp.

Householder, A. S.

[1] *The Theory of Matrices in Numerical Analysis*, Blaisdell, New York, 1964, xi + 257 pp.

Householder, A. S.; Young, G.

[1] Matrix approximation and latent roots, *Amer. Math. Monthly*, **45** (1938), 165–171.

Hurt, M. F.; Waid, C.

[1] A generalized inverse which gives all the integral solutions to a system of linear equations, *SIAM J. Appl. Math.*, **19** (1970), 547–550.

Hurwitz, W. A.

[1] On the pseudo-resolvent to the kernel of an integral equation, *Trans. Amer. Math. Soc.*, **13** (1912), 405–418.

Ijiri, Y.

[1] On the generalized inverse of an incidence matrix, *J. Soc. Indust. Appl. Math.*, **13** (1965), 941–945.

John, J. A.

[1] Use of generalized inverse matrices in MANOVA, *J. Roy. Statist. Soc. Ser. B.*, **32** (1970), 137–143.

John, P. W. M.

[1] Pseudo-inverses in the analysis of variance, *Ann. Math. Statist.*, **35** (1964), 895–896.

Jones, J. Jr.

[1] On the Lyapunov stability criteria, *J. Soc. Indust. Appl. Math.*, **13** (1965), 941–945.

[2] Solution of certain matrix equations, *Proc. Amer. Math. Soc.*, **31** (1972), 333–339.

Kakutani, S.
[1] Some characterizations of Euclidean spaces, *Japan J. Math.*, **16** (1939), 93–97.

Kallina, C.
[1] A Green's function approach to perturbations of periodic solutions, *Pacific J. Math.*, **29** (1969), 325–334.

Kalman, R. E.
[1] Contributions to the theory of optimal control, *Bol. Soc. Mat. Mexicana* (2), **5** (1960), 102–119.
[2] A new approach to linear filtering and prediction problems, *Trans. ASME Ser. D. J. Basic Eng.*, **82** (1960), 35–45.
[3] New results in linear filtering and prediction theory, *Trans. ASME Ser. D. J. Basic Eng.*, **83** (1961), 95–107.
[4] Mathematical description of linear dynamical systems, *SIAM J. Control*, **1** (1963), 152–192.

Kalman, R. E.; Ho, Y. C.; Narenda, K. S.
[1] Controllability of linear dynamic systems, in *Contributions to Differential Equations*, Vol. I, 189–213. Interscience, New York, 1963.

Kammerer, W. J.; Nashed, M. Z.
[1] Steepest descent for singular linear operators with nonclosed range, *Applicable Anal.*, **1** (1971), 143–159.
[2] A generalization of a matrix iterative method of G. Cimmino to best approximate solution of linear integral equations of the first kind, *Atti. Accad. Naz. Lincei Rend. Cl. Sci. Fis. Mat. Natur. Ser. VIII*, **51** (1971), 20–25.
[3] Iterative methods for best approximate solutions of linear integral equations of the first and second kinds, *J. Math. Anal. Appl.*, **40** (1972), 547–573.
[4] On the convergence of the conjugate gradient method for singular linear operator equations, *SIAM J. Numer. Anal.*, **9** (1972), 165–181.

Kantorovich, L. V.; Akilov, G. P.
[1] *Functional Analysis in Normed Spaces* (translated from Russian by D. E. Brown), Pergamon Press, Oxford, England, 1964, xiii + 773 pp.

Kantorovich, L. V.; Krylov, V. I.
[1] *Approximate Methods of Higher Analysis*, Interscience, New York, 1958, xii + 681 pp.

Kato, T.
[1] *Perturbation Theory for Linear Operators*, Springer-Verlag, New York, 1966, xix + 592 pp.

Katz, I. J.
[1] Wiegmann type theorems for EP_r matrices, *Duke Math. J.*, **32** (1965), 423–427.
[2] Remarks on a paper of Ben-Israel, *SIAM J. Appl. Math.*, **18** (1970), 511–513.

Katz, I. J.; Pearl, M. H.
[1] On EP_r and normal EP_r matrices, *J. Res. Nat. Bur. Standards, Sect. B.*, **70B** (1966), 47–77.
[2] Solutions of the matrix equations $A = XA = AX$, *J. London Math. Soc.*, **41** (1966), 443–452.

Kirby, M. J. L.
[1] *Generalized Inverses and Chance Constrained Programming*, doctoral dissertation in applied mathematics, Northwestern University, Evanston, Ill., June, 1965.

Kishi, F. H.
[1] On line computer control techniques and their application to re-entry aerospace vehicle control, pp. 245–257 in *Advances in Control Systems Theory and Applications* (C. T. Leondes, Editor), Academic Press, 1964.

Klinger, A.
[1] Approximate pseudoinverse solutions to ill-conditioned linear systems, *J. Optimization Theory Appl.*, **2** (1968), 117–128.

Korganoff, A.; Pavel-Parvu, M.
[1] Interprétation a l'aide des pseudo-inverses de la solution d'équations matricielles linéaires provenant de la discretisation d'operateurs differentielles et integraux, *83ᵉ Congrès de l'Association Française pour l'avancement des Sciences*, Lille, July, 1964.
[2] *Méthodes de calcul numérique-2. Eléments de théorie des matrices carrées et rectangles en analyse numérique*, Dunod, Paris 1967, xx + 441 pp.

Kublanovskaya, V. N.
[1] On the calculation of generalized inverses and projections (Russian), *Z. Vycisl. Mat. i Mat. Fiz.*, **6** (1966), 326–332.

Kuo, M. C. Y.; Kazda, L. F.
[1] Minimum energy problems in Hilbert function space, *J. Franklin Inst.*, **283** (1967), 38–54.

Kurepa, S.
[1] Generalized inverse of an operator with a closed range, *Glasnik Mat.*, **3** (23) (1968), 207–214.

Lancaster, P.
[1] Explicit solutions of linear matrix equations, *SIAM Rev.*, **12** (1970), 544–566.

Lanczos, C.
[1] Linear systems in self-adjoint form, *Amer. Math. Monthly*, **65** (1958), 665–679.
[2] *Linear Differential Operators*, D. Van Nostrand Co., Princeton, N.J., 1961, xvi + 564 pp.

Landesman, E. M.
[1] Hilbert-space methods in elliptic partial differential equations, *Pacific J. Math.*, **21** (1967), 113–131.

Langenhop, C. E.
[1] On generalized inverses of matrices, *SIAM J. Appl. Math.*, **15** (1967), 1239–1246.

Leach, E. B.
[1] A note on inverse function theorems, *Proc. Amer. Math. Soc.*, **12** (1961), 694–697.

Lent, A. H.
[1] *Wiener-Hopf Operators and Factorizations*, doctoral dissertation in applied mathematics, Northwestern University, Evanston, Ill., June, 1971, xii + 112 pp.

Leringe, Ö.; Wedin, P.-Å.
[1] A comparison between different methods to compute a vector x which minimizes $\|Ax - b\|_2$ when $Gx = h$. Department of Computer Sciences, Lund University, Lund, March, 1970.

Lewis, T. O.; Boullion, T. L.; Odell, P. L.
[1] A bibliography on generalized matrix inverses, pp. 283–315 in Boullion and Odell [1].

Locker, J.
[1] An existence analysis for nonlinear equations in Hilbert space, *Trans. Amer. Math. Soc.*, **128** (1967), 403–413.
[2] An existence analysis for nonlinear boundary value problems, *SIAM J. Appl. Math.*, **19** (1970), 199–207.

Lonseth, A. T.
[1] Approximate solutions of Fedholm-type integral equations, *Bull. Amer. Math. Soc.*, **60** (1954), 415–430.

Loud, W. S.
[1] Generalized inverses and generalized Green's functions, *SIAM J. Appl. Math.*, **14** (1966), 342–369.
[2] Some examples of generalized Green's functions and generalized Green's matrices, *SIAM Rev.*, **12** (1970), 194–210.

MacDuffee, C. C.
[1] *The Theory of Matrices*, Chelsea, New York, N.Y., 1956, v + 110 pp.

Marcus, M.; Minc, H.
[1] *A Survey of Matrix Theory and Matrix Inequalities*, Allyn & Bacon, Boston, Mass. 1964, xvi + 180 pp.

Martin, R. S.; Peters, G.; Wilkinson, J. H.
[1] Iterative refinement of the solution of a positive definite system of equations, *Numer. Math.*, **8** (1966), 203–216. Republished, pp. 31–44 in Wilkinson and Reinsch [1].

Maxwell, J. C.
[1] *Treatise of Electricity and Magnetism*, 3rd ed., Vol I, Oxford University Press, Oxford, England, 1892.

McCoy, N. H.
[1] Generalized regular rings, *Bull. Amer. Math. Soc.*, **45** (1939), 175–178.

Meicler, M.
[1] Chebyshev solution of an inconsistent system of $n+1$ linear equations in n unknowns in terms of its least squares solution, *SIAM Rev.*, **10** (1968), 373–375.

Meyer, C. D. Jr.
[1] Representations for (1)- and (1,2)-inverses for partitioned matrides, *Linear Algebra and Appl.*, **4** (1971), 221–232.
[2] Generalized inversion of modified matrices, *SIAM J. Appl. Math.*, **24** (1973), 315–323.

Meyer, C. D.; Painter, R. J.
[1] Note on a least squares inverse for a matrix, *J. Assoc. Comput. Mach.*, **17** (1970), 110–112.

Mihalyffy, L.
[1] An alternative representation of the generalized inverse of partitioned matrices, *Linear Algebra and Appl.*, **4** (1971), 95–100.

Milne, R. D.
[1] An oblique matrix pseudoinverse, *SIAM J. Appl. Math.*, **16** (1968), 931–944.

Minamide, N.; Nakamura, K.
[1] Minimum error control problem in Banach space, *Research Reports of Automatic Control Lab., Faculty of Engineering, Nagoya University*, Vol. 16, April 1969, 51–58.
[2] A restricted pseudoinverse and its applications to constrained minima, *SIAM J. Appl. Math.*, **19** (1970), 167–177.

Mirsky, L.
[1] Symmetric gauge functions and unitarily invariant norms, *Quart. J. Math. Oxford (2)*, **11** (1960), 50–59.

Mitra, S. K.
[1] On a generalized inverse of a matrix and applications, *Sankhyā Ser. A.*, **30** (1968), 107–114.
[2] A new class of g-inverse of square matrices, *Sankhyā Ser. A.*, **30** (1968). 323–330.

Mitra, S. K.; Rao, C. R.
[1] Some results in estimation and tests of linear hypotheses under the Gauss-Markoff model, *Sankhyā Ser. A.*, **30** (1968), 281–290.
[2] Conditions for optimality and validity of simple least squares theory, *Ann. Math. Statist.*, **40** (1969), 1617–1624.

Moler, C. B.
[1] Iterative refinement in floating point, *J. Assoc. Comput. Mach.*, **14** (1967), 316–321.

Mond, B.
[1] Generalized inverse extensions of matrix inequalities, *Linear Algebra and Appl.*, **2** (1969), 393–399.

Moore, E. H.
[1] On the reciprocal of the general algebraic matrix (Abstract), *Bull. Amer. Math. Soc.*, **26** (1920), 394–395.
[2] General Analysis, *Memoirs Amer. Philos. Soc.*, **1** (1935), esp. pp. 147–209.

Moore, R. H.; Nashed, M. Z.
[1] A general approximation theory for generalized inverses of linear operators in Banach spaces, *SIAM J. Appl. Math.* **27** (1974).

Morris, G. L.; Odell, P. L.
[1] Common solutions for *n* matrix equations with applications, *J. Assoc. Comput. Mach.*, **15** (1968), 272–274.

Munn, W. D.
[1] Pseudoinverses in semigroups, *Proc. Cambridge Phil. Soc.*, **57** (1961), 247–250.

Munn, W. D.; Penrose, R.
[1] A note on inverse semigroups, *Proc. Cambridge Phil. Soc.*, **51** (1955), 396–399.

Murray, F. J.; von Neumann, J.
[1] On rings of operators *I*, *Ann. of Math.*, **37** (1936), 116–229.

Nanda, V. C.
[1] A generalization of Cayley's theorem, *Math. Z.*, **101** (1967), 331–334.

Nashed, M. Z.
[1] Steepest descent for singular operator equations, *SIAM J. Numer. Anal.*, **7** (1970), 358–362.
[2] Generalized inverses, normal solvability and iteration for singular operator equations, pp. 311–359 in *Nonlinear Functional Analysis and Applications* (L. B. Rall, Editor), Academic Press, 1971.
[3] *Generalized Inverses and Applications*, to appear Academic Press, 1974.

Nashed, M. Z.; Votruba, G. F.
[1] A unified approach to generalized inverses of linear operators: algebraic, topological and proximal properties, to appear in Nashed [3].

Nelson, D. L.; Lewis, T. O.; Boullion, T. L.
[1] A quadratic programming technique using matrix pseudoinverses, *Indust. Math.*, **21** (1971), 1–21.

von Neumann, J.
[1] Über adjungierte Funktionaloperatoren, *Ann. of Math.*, **33** (1932), 294–310.
[2] On regular rings, *Proc. Nat. Acad. Sci. U.S.A.*, **22** (1936), 707–713.
[3] Some matrix-inequalities and metrization of matric-space, *Tomsk Univ. Rev.*, **1**, (1937), 286–300. Republished in *John von Neumann Collected Works*, MacMillan, New York, 1962, Vol. IV, pp. 205–219.
[4] *Functional Operators. Vol. II: The Geometry of Orthogonal Spaces*, Annals of Math. Studies No. 29, Princeton University Press, Princeton, N.J. 1950.
[5] *Continuous Geometry*, Princeton University Press, Princeton, N.J., 1960, 299 pp.

Newman, T. G.; Odell, P. L.
[1] On the concept of a *p-q* generalized inverse of a matrix, *SIAM J. Appl. Math.*, **17** (1969), 520–525.

Noble, B.
[1] A method for computing the generalized inverse of a matrix, *SIAM J. Numer. Anal.*, 3 (1966), 582–584.
[2] *Applied Linear Algebra*, Prentice-Hall Inc., Englewood Cliffs, N.J., 1969, xvi + 523 pp.
[3] Computational methods for generalized inverses of matrices and related topics in numerical linear algebra, to appear in Nashed [3].

Olivares, J. E. Jr.
[1] Generalized inverse of the tie-set and cut-set matrix for networks with complete graphs, pp. 583–586 in *Proceedings of the 1968 Hawaii International Conference on Systems Science*.

Ortega, J. M.; Rheinboldt, W. C.
[1] *Iterative Solution of Nonlinear Equations in Several Variables*, Academic Press, New York, 1970, xx + 572 pp.

Osborne, E. E.
[1] On least squares solutions of linear equations, *J. Assoc. Comput. Mach.*, 8 (1961), 628–636.
[2] Smallest least square solutions of linear equations, *SIAM J. Numer. Anal.*, 2 (1965), 300–307.

Parlett, B. N.
[1] The LU and QR algorithms, pp. 116–130 in *Mathematical Methods for Digital Computers*, Vol. 2, A. Ralston and H. S. Wilf (editors), J. Wiley, New York, 1967.

Pearl, M. H.
[1] On Cayley's parameterization, *Canad. J. Math.*, 9 (1957), 553–562.
[2] A further extension of Cayley's parameterization, *Canad. J. Math.*, 11 (1959), 48–50.
[3] On normal and EP_r matrices, *Michigan Math. J.*, 6 (1959), 1–5.
[4] On normal and EP_r matrices, *Michigan Math. J.*, 6 (1959), 89–94.
[5] On normal EP_r matrices, *Michigan Math. J.*, 8 (1961), 33–37.
[6] On generalized inverses of matrices, *Proc. Cambridge Phil. Soc.*, 62 (1966), 673–677.
[7] A decomposition theorem for matrices, *Canad. J. Math.*, 19 (1967), 344–349.
[8] Generalized inverses of matrices with entries taken from an arbitrary field, *Linear Algebra and Appl.*, 1 (1968), 571–587.

Pennington, R. H.
[1] *Introductory Computer Methods and Numerical Analysis* (2nd edition), MacMillan Co., New York, 1970, xi + 452 pp.

Penrose, R.
[1] A generalized inverse for matrices, *Proc. Cambridge Philos. Soc.*, 51 (1955), 406–413.
[2] On best approximate solution of linear matrix equations, *Proc. Cambridge Philos. Soc.*, 52 (1956), 17–19.

Pereyra, V.
[1] Stability of general systems of linear equations, *Aequationes Mathematicae*, 2 (1969), 194–206.

Pereyra, V.; Rosen, J. B.
[1] Computation of the pseudoinverse of a matrix of unknown rank, Tech. Rep. CS 13, Computer Sciences Dept., Stanford University, Sept. 1964. (*Comp. Rev.*, 6 (1965), 259 #7948).

Peters, G.; Wilkinson, J. H.
[1] The least squares problem and pseudo-inverses, *Computer J.*, 13 (1970), 309–316.

Petryshyn, W. V.
[1] On generalized inverses and on the uniform convergence of $(I - \beta K)^n$ with application to iterative methods, *J. Math. Anal. Appl.*, 18 (1967), 417–439.

Pietsch, A.
[1] Zur Theorie der σ-Transormationen in lokalkonvexen Vektorräumen, *Math. Nach.*, **21** (1960), 347–369.

Plemmons, R. J.
[1] Generalized inverses of Boolean relation matrices, *SIAM J. Appl. Math.*, **20** (1971), 426–433.

Plemmons, R. J.; Cline, R. E.
[1] The generalized inverse of a nonnegative matrix, *Proc. Amer. Math. Soc.*, **31** (1972), 46–50.

Poole, G. D.; Boullion, T. L.
[1] Weak spectral inverses which are partial isometries, *SIAM J. Appl. Math.*, **23** (1972), 171–172.

Porter, W. A.
[1] *Modern Foundations of System Engineering*, MacMillan, New York, 1966, xii + 493 pp.
[2] A basic optimization problem in linear systems, *Math. Systems Theory*, **5** (1971), 20–44.

Porter, W. A.; Williams, J. P.
[1] A note on the minimum effort control problem, *J. Math. Anal. Appl.*, **13** (1966), 251–264.
[2] Extensions of the minimum effort control problem, *J. Math. Anal. Appl.*, **13** (1966), 536–549.

Price, C. M.
[1] The matrix pseudoinverse and minimal variance estimates, *SIAM Rev.*, **6** (1964), 115–120.

Przeworska-Rolewicz, D.; Rolewicz, S.
[1] *Equations in Linear Spaces*, Polska Akad. Nauk Monog. Mat. 47, PWN Polish Scientific Publishers, Warsaw, 1968, 380 pp.

Pyle, L. D.
[1] Generalized inverse computations using the gradient projection method, *J. Assoc. Comput. Mach.*, **11** (1964), 422–428.
[2] A generalized inverse ε algorithm for constructing intersection projection matrices, *Numer. Math.*, **10** (1967), 86–102.
[3] The generalized inverse in linear programming. Basic structure. *SIAM J. Appl. Math.*, **22** (1972), 335–355.

P. Rabinowitz (Editor)
[1] *Numerical Methods for Nonlinear Algebraic Equations*, Gordon and Breach, London, 1970, xi + 199 pp.

Rabson, G.
[1] The generalized inverse in set theory and matrix theory, Dept. of Mathematics, Clarkson College of Technology, Potsdam, N. Y., 1969.

Rado, R.
[1] Note on generalized inverses of matrices, *Proc. Cambridge Philos. Soc.*, **52** (1956), 600–601.

Rall, L. B.
[1] *Computational Solution of Nonlinear Operator Equations*, Wiley, New York, 1969, viii + 225 pp.

Rao, C. R.
[1] A note on a generalized inverse of a matrix with applications to problems in mathematical statistics, *J. Roy. Statist. Soc. Ser. B.*, **24** (1962), 152–158.
[2] *Linear Statistical Inference and its Applications*, J. Wiley and Sons, New York, 1965, xviii + 522 pp.

[3] Generalized inverse for matrices and its applications in mathematical statistics, *Research Papers in Statistics (Festschrift J. Neyman)*, pp. 263–279, Wiley, London, 1966.

[4] Calculus of generalized inverses of matrices, part I: General theory, *Sankhyā Ser. A.*, **29** (1967), 317–342.

Rao, C. R.; Mitra, S. K.

[1] *Generalized Inverse of Matrices and its Applications*, Wiley and Sons, Inc. New York, 1971, xiv + 240 pp.

[2] Theory and application of constrained inverse of matrices, *SIAM J. Appl. Math.*, **24** (1973), 473–488.

Rayner, A. A.; Pringle, R. M.

[1] A note on generalized inverses in the linear hypothesis not of full rank, *Ann. Math. Statist.*, **38** (1967), 271–273.

Reid, W. T.

[1] Generalized Green's matrices for compatible systems of differential equations, *Amer. J. Math.*, **53** (1931), 443–459.

[2] Principal solutions of non-oscillatory linear differential systems, *J. Math. Anal. Appl.*, **9** (1964), 397–423.

[3] Generalized Green's matrices for two-point boundary problems, *SIAM J. Appl. Math.*, **15** (1967), 856–870.

[4] Generalized inverses of differential and integral operators, pp: 1–25 in Boullion and Odell [1].

[5] *Ordinary Differential Equations*, Wiley-Interscience, New York, 1970, xv + 551 pp.

Rheinboldt, W. C.

[1] A unified convergence theory for a class of iterative processes, *SIAM J. Numer. Anal.*, **5** (1968), 42–63.

Rice, J.

[1] Experiments on Gram–Schmidt orthogonalization, *Math. Comp.*, **20** (1966), 325–328.

Rinehart, R. F.

[1] The equivalence of definitions of a matric function, *Amer. Math. Monthly*, **62** (1955), 395–414.

Robers, P. D.; Ben-Israel, A.

[1] An interval programming algorithm for discrete linear L_1 approximation problems, *J. Approximation Theory*, **2** (1969), 323–336.

[2] A suboptimization method for interval linear programming: A new method for linear programming, *Linear Algebra and Appl.*, **3** (1970), 383–405.

Robert, P.

[1] On the group-inverse of a linear transformation, *J. Math. Anal. Appl.*, **22** (1968), 658–669.

Robinson, D. W.

[1] On the generalized inverse of an arbitrary linear transformation, *Amer. Math. Monthly*, **69** (1962), 412–416.

Rockafellar, R. T.

[1] *Convex Analysis*, Princeton University Press, Princeton, N. J., 1970, xviii + 451 pp.

Rohde, C. A.

[1] *Contributions to the Theory, Computation and Application of Generalized Inverses*, doctoral dissertation, University of North Carolina at Raleigh, May 1964.

[2] Generalized inverses of partitioned matrices, *J. Soc. Indust. Appl. Math.*, **13** (1965), 1033–1035.

[3] Some results on generalized inverses, *SIAM Rev.*, **8** (1966), 201–205.

[4] Special applications of the theory of generalized matrix inversion to statistics, pp. 239–266 in Boullion and Odell [1].

Rohde, C. A.; Harvey, J. R.
[1] Unified least squares analysis, *J. Amer. Statist. Assoc.*, **60** (1965), 523–527.

Rosen, J. B.
[1] The gradient projection method for nonlinear programming. Part I: Linear constraints, *J. Soc. Indust. Appl. Math.*, **8** (1960), 181–217.
[2] The gradient projection method for nonlinear programming. Part II: Nonlinear constraints, *J. Soc. Indust. Appl. Math.*, **9** (1961), 514–532.
[3] Minimum and basic solutions to singular linear systems, *J. Soc. Indust. Appl. Math.*, **12** (1964), 156–162.
[4] Chebyshev solution of large linear systems, *J. Comput. Syst. Sci.*, **1** (1967), 29–43.

Rosenberg, M.
[1] Range decomposition and generalized inverse of nonnegative Hermitian matrices, *SIAM Rev.*, **11** (1969), 568–571.

Rosenbloom, P. C.
[1] The method of steepest descent, pp. 127–176 in *Numerical Analysis. Proceedings of the sixth Symposium in Applied Mathematics*, McGraw-Hill Book Co., New York, 1956.

Rust, B. W.; Burrus, W. R.; Schneeberger, C.
[1] A simple algorithm for computing the generalized inverse of a matrix, *Comm. ACM*, **9** (1966), 381–386.

Schmidt, E.
[1] Zur Theorie der linearen und nichtlinearen Integralgleichungen, I. Entwicklung willkürlicher Funktionen nach Systemen vargeschriebener, *Math. Ann.*, **63** (1907), 433–476.
[2] Zur Theorie der linearen und nichtlinearen Integralgleichungen, II. Auflösung der allgemeinen linearen Integralgleichung, *Math. Ann.*, **64** (1907), 161–174.

Schönemann, P. H.
[1] A generalized solution of the orthogonal Procrustes problem, *Psychometrika*, **31** (1966), 1–10.

Schulz, G.
[1] Iterative Berechnung der reziproken Matrix, *Z. Angew. Math. Mech.*, **13** (1933), 57–59.

Schwartz, J.
[1] Perturbations of spectral operators, and applications, *Pacific J. Math.*, **4** (1954), 415–458.

Schwerdtfeger, H.
[1] *Introduction to Linear Algebra and the Theory of Matrices*, P. Noordhoff, Groningen, 1950, 288 pp.

Scroggs, J. E.; Odell, P. L.
[1] An alternate definition of a pseudoinverse of a matrix, *J. SIAM Appl. Math.*, **14** (1966) 796–810.

Sharpe, G. E.; Styan, G. P. H.
[1] Circuit duality and the general network inverse, *IEEE Trans. Circuit Th.*, **12** (1965), 22–27.
[2] A note on the general network inverse, *IEEE Trans. Circuit Th.*, **12** (1965), 632–633.
[3] A note on equicofactor matrices, *Proc. IEEE*, **55** (1967), 1226–1227.

Sheffield, R. D.
[1] *On pseudo-inverses of linear transformations in Banach spaces*, Oak Ridge National Laboratory Report 2133, 1956.
[2] A general theory for linear systems, *Amer. Math. Monthly*, **65** (1958), 109–111.

Shinozaki, N.; Sibuya, M.; Tanabe, K.
[1] Numerical algorithms for the Moore–Penrose inverse of a matrix: Direct methods, *Ann. Inst. Statist. Math.*, **24** (1972), 193–203.

Showalter, D. W.
[1] Representation and computation of the pseudoinverse, *Proc. Amer. Math. Soc.*, **18** (1967), 584–586.

Showalter, D. W.; Ben-Israel, A.
[1] Representation and computation of the generalized inverse of a bounded linear operator between Hilbert spaces, *Atti Accad. Naz. Lincei Rend. Cl. Sci. Fis. Mat. Natur. Ser. VIII*, **48** (1970), 120–130.

Sibuya, M.
[1] Subclasses of generalized inverses of matrices, *Ann. Inst. Statist. Math.*, **22** (1970), 543–556.

Siegel, C. L.
[1] Über die analytische Theorie der quadratischen Formen III, *Ann. Math.*, **38** (1937), 212–291, in particular pp. 217–229.
[2] Equivalence of quadratic forms, *Amer. J. Math.*, **63** (1941), 658–680.

Singer, I.
[1] *Best Approximation in Normed Linear Spaces by Elements of Linear Subspaces*, Springer-Verlag, Berlin, 1970, 415 pp.

Smithies, F.
[1] The eigen-values and singular values of integral equations, *Proc. London Math. Soc.*, **43** (1937), 255–279.
[2] *Integral equations*, Cambridge University Press, Cambridge, England, 1958, x + 172 pp.

Söderström, T.; Stewart, G. W.
[1] On the numerical properties of an iterative method for computing the Moore–Penrose generalized inverse, *SIAM J. Numer. Anal.*, **11** (1974), 61–74.

Stakgold, I.
[1] Branching of solutions of nonlinear equations, *SIAM Rev.*, **13** (1971), 289–332. Erratum, *ibid.* **14** (1972), 492.

Stallings, W. T.; Boullion, T. L.
[1] Computation of pseudoinverse matrices using residue arithmethc, *SIAM Rev.*, **14** (1972), 152–163.

Stein, P.
[1] Some general theorems on iterants, *J. Res. Nat. Bur. Standards*, **48** (1952), 82–83.

Stern, T. E.
[1] Extremum relations in nonlinear networks and their applications to mathematical programming, *Journées d'Études sur le Contrôle Optimum et Les Systèmes Nonlinéaires*, Saclay, France, Institut National des Sciences et Techniques Nucléaires, pp. 135–156.
[2] *Theory of Nonlinear Networks and Systems*, Addison-Wesley, Reading, Mass., 1965, xiv + 594 pp.

Stewart, G. W.
[1] On the continuity of the generalized inverse, *SIAM J. Appl. Math.*, **17** (1969), 33–45.
[2] Projectors and generalized inverses, Report TNN-97, University of Texas at Austin Computation Center, October 1969.

Tan, W. Y.
[1] Note on an extension of the Gauss–Markov theorems to multivariate linear regression models, *SIAM J. Appl. Math.*, **20** (1971), 24–29.

Taussky, O.
[1] Note on the condition of matrices, *Math. Tables Aids Comput.*, **4** (1950), 111–112.
[2] Matrices C with $C^n \to 0$, *J. Algebra*, **1** (1964), 5–10.

Taylor, A. E.
[1] *Introduction to Functional Analysis*, J. Wiley & Sons, New York, 1958, 423 pp.

Taylor, A. E.; Halberg, C. J. A. Jr.
[1] General theorems about a bounded linear operator and its conjugate, *J. Reine Angew. Math.*, **198** (1957), 93–111.

Tewarson, R. P.
[1] A direct method for generalized matrix inversion, *SIAM J. Numer. Anal.*, **4** (1967), 499–507.
[2] On some representations of generalized inverses, *SIAM Rev.*, **11** (1969), 272–276.
[3] On two direct methods for computing generalized inverses, *Computing*, **7** (1971), 236–239.

Thompson, R. C.
[1] Principal submatrices IX: Interlacing inequalities for singular values, *Linear Algebra and Appl.*, **5** (1972), 1–12.

Thomson, W. (Lord Kelvin)
[1] *Cambridge and Dublin Math. J.*, (1848), 84–87.

Tseng, Y. Y.
[1] *The Characteristic Value Problem of Hermitian Functional Operators in a Non-Hilbert space*, doctoral dissertation in Mathematics, University of Chicago, 1933 (published by the *University of Chicago Libraries*, Chicago, 1936).
[2] Sur les solutions des équations operatrices fonctionnelles entre les espaces unitaires. Solutions extremales. Solutions virtuelles., *C. R. Acad. Sci. Paris*, **228** (1949), 640–641.
[3] Generalized inverses of unbounded operators between two unitary spaces, *Dokl. Akad. Nauk SSSR (N.S.)*, **67** (1949), 431–434 (reviewed in *Math. Rev.*, **11** (1950), p. 115).
[4] Properties and classification of generalized inverses of closed operators, *Dokl. Akad. Nauk. SSSR (N.S.)*, **67** (1949), 607–610 (reviewed in *Math. Rev.*, **11** (1950), p. 115).
[5] Virtual solutions and general inversions, *Uspehi. Mat. Nauk. (N.S.)*, **11** (1956), 213–215 (reviewed in *Math. Rev.*, **18** (1957), p. 749).

Tucker, D. H.
[1] Boundary value problems for linear differentilal systems, *SIAM J. Appl. Math.*, **17** (1969), 769–783.

Urquhart, N. S.
[1] Computation of generalized inverse matrices which satisfy specified conditions, *SIAM Rev.*, **10** (1968), 216–218.
[2] The nature of the lack of uniqueness of generalized inverse matrices, *SIAM Rev.*, **11** (1969), 268–271.

Vogt, A.
[1] On the linearity of form isometries, *SIAM J. Appl. Math.*, **22** (1972), 553–560.

Votruba, G.
[1] *Generalized Inverses and Singular Equations in Functional Analysis*, doctoral dissertation in Mathematics, The University of Michigan, Ann Arbor, Michigan, 1963.

Wahba, G.; Nashed, M. Z.
[1] The approximate solution of a class of constrained control problems, *Proceedings of the Sixth Hawaii International Conference on System Sciences*, 1973.

Ward, J. F.; Boullion, T. L.; Lewis, T. O.
[1] A note on the oblique matrix pseudoinverse, *SIAM J. Appl. Math.*, **20** (1971), 173–175.

[2] Weak spectral inverses, *SIAM J. Appl. Math.*, **22** (1972), 514–518.

Wedderburn, J. H. M.

[1] *Lectures on Matrices*, Amer. Math. Soc. Colloq. Publ. Vol. XVII, American Mathematical Society, Providence, R.I., 1934.

Wedin, P.-Å.

[1] Perturbation bounds in connection with singular value decomposition, *BIT*, **12** (1972), 99-111.

[2] Perturbation theory for pseudo-inverses, *BIT*, **13** (1973), 217–232.

[3] The non-linear least squares problem from a numerical point of view. I. Geometrical properties, Dept. of Computer Sciences, Lund University, Lund, August, 1972.

Wee, W. G.

[1] Generalized inverse approach to adaptive multiclass pattern classification, *IEEE Trans. Electron. Comput.*, **C17** (1968), 1157–1164.

[2] A generalized inverse approach to clustering pattern selection and classification, *IEEE Trans. Inform. Th.*, **17** (1971).

Weyl, H.

[1] Repartición de corriente en una red conductora, *Revista Matemática Hispano-Americana*, **5** (1923), 153–164.

[2] Inequalities between the two kinds of eigenvalues of a linear transformation, *Proc. Nat. Acad. Sci. USA*, **36** (1950), 49–51.

Whitney, T. M.; Meany, R. K.

[1] Two algorithms related to the method of steepest descent, *SIAM J. Numer. Anal.*, **4** (1967), 109–120.

Wilkinson, J. H.

[1] *The Algebraic Eigenvalue Problem*, Oxford University Press, London, 1965, xviii + 662 pp.

[2] The solution of ill-conditioned linear equations, pp. 65–93 in *Mathematical Methods for Digital Computers, Vol. II*, A. Ralston and H. Wilf (Editors), Wiley, New York, 1967.

Wilkinson, J. H.; Reinsch, C. (Editors)

[1] *Handbook for Automatic Computation. Vol. II: Linear Algebra*, Springer-Verlag, Berlin, 1971, viii + 439 pp.

Williamson, J.

[1] A polar representation of singular matrices, *Bull. Amer. Math. Soc.*, **41** (1935), 118–123.

[2] Note on a principal axis transformation for non-Hermitian matrices, *Bull. Amer. Math. Soc.*, **45** (1939), 920–922.

Willner, L. B.

[1] An elimination method for computing the generalized inverse, *Math. Comp.*, **21** (1967), 227–229.

Wimmer, H.; Ziebur, A. D.

[1] Solving the matrix equation $\sum_{\rho=1}^{r} f_\rho(A) X g_\rho(B) = C$, *SIAM Rev.*, **14** (1972), 318–323.

Wyler, O.

[1] Green's operators, *Ann. Mat. Pura Appl.*, **66** (1964), 251–264.

[2] On two-point boundary problems, *Ann. Mat. Pura Appl.*, **67** (1965), 127–142.

Wynn, P.

[1] Upon the generalized inverse of a formal power series with vector valued coefficients, *Compositio Math.*, **23** (1971), 453–460.

Yosida, K.

[1] *Functional Analysis* (2nd edition), Springer-Verlag, Berlin–New York, 1958, xi + 458 pp.

Zacks, S.
[1] Generalized least squares estimators for randomized fractional replication designs, *Ann. Math. Statist.*, **35** (1964), 696–704.

Zadeh, L. A.; Desoer, C. A.
[1] *Linear System Theory*, McGraw-Hill Book Co., New York, 1963, xxi + 628 pp.

Zarantonello, E. H.
[1] Differentioids, *Advances in Math.*, **2** (1968), 187–306.

Zlobec, S.
[1] On computing the generalized inverse of a linear operator, *Glasnik Mat.*, **2** (22) (1967), 265–271.
[2] An explicit form of the Moore-Penrose inverse of an arbitrary complex matrix, *SIAM Rev.*, **12** (1970), 132–134.
[3] *Contributions to Mathematical Programming and Generalized Inversion*, doctoral dissertation, Northwestern University, 1970.
[4] On computing the best least squares solutions in Hilbert spaces, Department of Mathematics, McGill University, Montreal, September, 1972.

Zlobec, S.; Ben-Israel, A.
[1] On explicit solutions of interval linear programs, *Israel J. Math.*, **8** (1970), 12–22.
[2] Explicit solutions of interval linear programs, *Operations Res.*, **21** (1973), 390–393.

Zyskind, G.
[1] On canonical forms, nonnegative covariance matrices and best and simple least squares linear estimators in linear models, *Ann. Math. Statist.*, **38** (1967), 1092–1109.

Zyskind, G.; Martin, F. B.
[1] On best linear estimation and a general Gauss–Markov theorem in linear models with arbitrary nonnegative covariance structure, *SIAM J. Appl. Math.*, **17** (1969), 1190–1202.

GLOSSARY OF SYMBOLS

383

AUTHOR INDEX

SUBJECT INDEX